INTERNATIONAL SERIES OF MONOGRAPHS IN
ANALYTICAL CHEMISTRY

GENERAL EDITORS: R. BELCHER AND H. FREISER

VOLUME 36

CHEMICAL METHODS
OF ROCK ANALYSIS

CHEMICAL METHODS
OF ROCK ANALYSIS

SECOND EDITION

BY

P. G. JEFFERY

PERGAMON PRESS

Oxford · New York · Toronto · Sydney · Braunschweig

84195

Pergamon Press Ltd., Headington Hill Hall, Oxford

Pergamon Press Inc., Maxwell House, Fairview Park, Elmsford, New York 10523

Pergamon of Canada Ltd., 207 Queen's Quay West, Toronto 1

Pergamon Press (Aust.) Pty. Ltd., 19a Boundary Street,
Rushcutters Bay, N.S.W. 2011, Australia

Pergamon Press GmbH, Burgplatz 1, Braunschweig 3300, West Germany

First edition 1970

Second edition 1975

Library of Congress Cataloging in Publication Data

Jeffery, Paul Geoffrey.
Chemical methods of rock analysis.

(International series of monographs in
analytical chemistry, v. 36)
Includes bibliographies.
1. Rocks—Analysis. I. Title.
QE438.J44 1975   552'.06   74–16500
ISBN 0 08 018076 0

Printed in Great Britain by Biddles Ltd., Guildford, Surrey.

CONTENTS

PREFACE TO THE SECOND EDITION

SOME of the older techniques of rock analysis that featured in the first edition of this book—such as the J. Lawrence Smith method for determining the alkali metals—are now so seldom used that they no longer can be justified in a volume of this kind. New material has been added to most of the chapters, this has almost all been based upon recent work using well-established techniques. These include atomic absorption spectroscopy which the author continues to regard as essentially a "chemical method". Purely instrumental techniques, of which electron probe microanalysis is an example, have been subject to extensive development since the first edition, but remain outside the scope of this book.

The author gratefully records the advice and assistance that he has received in preparing this edition. The improvements that have been made to the text arise very largely from the constructive criticism that he has received since the appearance of the first edition.

Stevenage, 1973

ACKNOWLEDGEMENTS

THIS book was first planned in 1956. The long gestation period is due solely to the author's continual, if somewhat ineffective, attempts to cope with the flood of original work that has appeared in this field since that date, and to the difficulties in making an adequate selection of material from it. That it has appeared at all is due to the help and encouragement that the author has received from his co-workers in a number of laboratories. In particular the author is grateful to J. B. Pollock of the Geological Survey of Uganda, G. A. Sergeant and A. D. Wilson of The Laboratory of The Government Chemist, D. J. Ellis, D. R. Hewett and Miss J. Richardson of Warren Spring Laboratory, and especially to G. R. E. C. Gregory, also of Warren Spring Laboratory, who read and criticised the completed manuscript.

The author's thanks are extended to The Director, Warren Spring Laboratory and to the Ministry of Technology for permission to publish this book, and to the various authors and editors for permission to reproduce material published elsewhere.

Stevenage, 1968

CHAPTER 1

THE COMPOSITION
OF ROCK MATERIAL

Introduction

Since early times man has speculated upon the origin and composition of the earth and the great variety of rocks and minerals of which it is composed. For many of the eminent chemists of the eighteenth and nineteenth centuries, the uncharacterised minerals provided the challenge that led to the identification and subsequent isolation of the elements missing from the periodic table. By the end of the nineteenth century, Berzelius, Lothar Meyer, Lawrence Smith and others had laid the foundations of the classical scheme of silicate rock analysis as we know it today and, by the end of the century methods for the determination of all elements present in major amounts had been proposed and evaluated. By 1920, when Washington had issued the third edition of his book *Manual of the Chemical Analysis of Rocks*[1] and Hillebrand his *The Analysis of Silicate and Carbonate Rocks*[2] (itself a revised and enlarged version of earlier texts), interest in silicate rock analysis had spread to those elements present in only minor amounts. Barium, zirconium, sulphur and chlorine—elements that could all be determined gravimetrically by well-established methods— were soon added to the list of major components required for a "complete analysis". Elements such as titanium, vanadium and chromium were recognised as essential components of certain silicates, and new procedures were devised for their determination.

This interest in the minor components of silicate rocks has continued almost without a break to the present day, extending to elements at lower and lower concentration as more and more sensitive techniques have become available.

As with other well-defined applications of analytical chemistry, the

1

ability to undertake a good analysis depended upon the skill of the analyst in making his separations and in completing his determinations gravimetrically or titrimetrically, although for manganese a visual comparison of colours provided an early example of the use of a colorimetric method. The general sensitivity of photometric methods, coupled with the improvements in the design of instruments available from about 1950 onwards has resulted in a considerable extension of the use of such methods. At first this extension was limited to the minor and trace components such as titanium, phosphorus and fluorine, but this was later extended also to those elements present in major amounts—silicon, iron and aluminium.

Some considerable effort has been devoted by a number of analysts to devising new schemes of rock analysis based largely upon spectrophotometric methods, but including also the determination of calcium and magnesium by complexometric titration. These schemes are considered in some detail in Chapter 5. Most of the proposed schemes suffer from some disadvantage—some of the procedures are analytically unsound, some require the services of a skilled analyst, and most if not all are too inflexible to be applied to a wide range of rocks without modification.

Although many chemists regard the schemes for "complete analysis" of silicate rocks by spectrophotometry with suspicion, the prospect of obtaining large numbers of such analyses cheaply and rapidly has been welcomed by many geologists. Unfortunately this enthusiasm has not always been accompanied by an understanding of the chemistry (and the errors!) of the processes involved, or of the difficulties in making precise spectrophotometric measurement. The ease with which agreement between duplicate results can be obtained is often taken as an indication of the accuracy of the determination. What is often forgotten is that the "rapid" (sometime approximate) analyses, valuable in a series of similar analyses for comparative studies, may later be used by other workers, and then given equal weight with analyses obtained by more rigorous methods.

The extensive introduction of spectrophotometric methods to silicate rock analysis was followed by the use of other instrumental methods. Emission spectrography, known also as optical spectrography and previously used extensively for the qualitative examination of minerals, became a valuable additional technique in many rock analysis laboratories. In some of these it became the practice to make a qualitative examination of all silicate rocks prior to chemical analysis. This served to identify elements of interest that might subsequently warrant determination by other means. It also gave the analyst a guide to the approximate values that he could expect to find. Emission spectrography has provided the geologist with his dream of large numbers of rapid, cheap analyses—at least for the minor and trace components of silicates. Attempts to use it for obtaining "complete analyses"[3] have not been widely followed.

One of the most tedious of the determinations in the classical scheme for the complete analysis of silicate rocks is that of the alkali metals, involving a difficult decomposition procedure and a number of subsequent separation stages. It is therefore easy to see why the use of flame photometry was widely adopted, even before the difficulties associated with its use were properly understood and defined. Gravimetric methods for the separation of the alkali metals soon became unnecessary. The determination of the rarer alkali metals, previously seldom attempted and even more rarely successfully achieved, was now possible on a routine basis. Calcium, strontium and barium, elements with characteristic flame emission, were also determined by this technique, although rather less readily than sodium and potassium, and possibly also with less enthusiasm on the part of the rock analyst with other techniques available.

Schemes of rapid rock analysis usually included titrimetric procedures for calcium and magnesium, although difficulties were sometimes encountered in the presence of much manganese. In recent years atomic absorption spectroscopy has provided an acceptable alternative technique for both calcium and magnesium, as well as for manganese, and iron. A small number of elements for which this technique is particularly sensitive, such as zinc and copper which are present in trace amounts in most silicate rocks, have also beeen determined in this way.

The difficulties inherent in collecting and determining all the silica by the classical method can be avoided by using a combined gravimetric and photometric method.[4] The major part of the silica is recovered following a single dehydration with hydrochloric acid, and is then determined by volatilisation with hydrofluoric acid in the usual way. The minor fraction that escapes collection is determined in the filtrate by a photometric molybdenum-blue method.

At this stage in the development of rock analysis, the greatest remaining problem is that of the determination of aluminium. In the classical scheme of analysis aluminium was determined by difference. As a consequence, errors in the determination of several other constituents were reflected in the values for aluminium. Direct gravimetric and titrimetric methods for aluminium after a separation of interfering elements (iron, titanium, manganese, chromium, vanadium, zirconium and possibly phosphorus) are frequently tedious, and have been adopted only for want of something better. Spectrophotometric methods for aluminium are non-selective, making some form of separation essential, whilst for atomic absorption spectroscopy a high temperature burner must be used.

This introduction would not be complete without a reference to the "Co-operative investigation of the precision and accuracy of chemical, spectrochemical and modal analysis of silicate rocks" reported in the United States *Geological Survey Bulletin* No. 980[5] and in a number of later papers. This exercise involved the distribution of two ground rock samples, a granite, G-1, and diabase, W-1, to a number of laboratories regularly making rock analyses. A detailed examination was then made of the large number of determinations subsequently reported. One of the most important points to emerge from this investigation was that the agreement between analysts was not of the order that could be expected from individual estimates of the accuracy of the procedures involved. There was more than one reason for this lack of agreement, and the precision was better for some elements than for others. It is to be hoped that the overall standard of rock analysis has improved as a result of this investigation.

International Standards

It is difficult to compare the results of one laboratory with those of another unless an adequate series of standards are available covering the range of determinations currently being performed in the laboratories concerned.[6] The granite G-1 and diabase W-1 distributed by the United States Geological Survey have been widely referred to as "standard samples", although at the time of the initial distribution the United States Geological Survey refrained from referring to them as such.

These were not the first samples of geological material to be prepared for distribution. Both the National Bureau of Standards (U.S.A.) and the Bureau of Analysed Standards (U.K.) had prepared a number of materials of prime interest in the ceramic industry, which were also of interest to the rock analyst. Until recently none of these samples had achieved a world-wide distribution and lacked the authority required of international geochemical standards.

The pressing need for such standard materials and the long interval between the distribution of the earlier samples G-1 and W-1 and the later standards, has resulted in what may be termed a "rash" of new standards. Not all of these have been prepared with the care and attention to detail required of international standards. It does not appear to have been realised that the preparation, including crushing, grinding and sampling of a large bulk of material, in a state of homogeneity and free from contamination, is a task of some considerable magnitude.

The suspicion that contamination may have occurred, although not disasterous for the rock analyst concerned with an inter-laboratory comparison, can be upsetting to the geologist and geochemist. For the benefit of the subsequent users of the material, the procedures used for crushing, grinding and sampling should be described, and if any steps have been taken to limit or even to measure the extent and nature of the contamination—these should be reported in detail. Only then will it be possible to consider the petrological significance of any unusually high values, such as 5 ppm molybdenum in G-1 or 50 ppm tin in T-1.

The great variety of rock, mineral and allied material standards that are now available can be seen from Table 1, adapted from that given by Flanagan and Gwym,[7] but with addition of new materials and deletion of those known to be now no longer available. A larger list of over 100 "standard" materials has been given by Flanagan,[8] describing materials from twenty-three sources in eleven different countries. As pointed out by Abbey,[9] not only has there been a lack of consistency in identification symbols, but a tendency for each group to choose its own scheme of sample selection and of methods of preparing them. Thus we now have an overabundance of granite samples, characterised by relatively small differences in composition, whilst the few ultrabasic samples are similarly bunched close together in chemical composition. Some of these "standard samples" are concerned with the economic applications of geological materials and have been available for some years. Others are relatively new, with little compositional data available so far.

It is difficult to see how samples produced in varying amounts, under differing conditions, in laboratories often isolated from each other, can result in anything but an inadequate selection of material. The criticisms made in the first edition—neglect of sedimentary rocks and lack of rock-forming minerals—remain, although there has been some improvement in this direction.

The value of these materials tends to fall as the stocks become depleted, and any measure that extends the life of individual standards is therefore to be recommended. Flanagan,[10] for example, suggests that any recipient of these standards should obtain about 10 kg of two or three rocks from his area, process these as "in-house standards", and calibrate them against the international standards.

Elements Determined

In his examination of silicate rocks the petrologist is primarily concerned with the mineralogical composition, and his interest in the chemical analysis is largely directed towards the major components of the rock forming minerals, that is towards those elements present in major proportion. There is a small group of elements which, calculated as oxides, account for 99 per cent or more by weight of a large number of silicate rocks. All analyses of igneous rocks that claim to be complete must include values for these thirteen constituents:

> silicon
> aluminium
> iron (ferrous and ferric)
> magnesium
> calcium
> manganese
> titanium
> phosphorus
> sodium
> potassium
> water (water evolved above and below 105°)

TABLE 1. GEOCHEMICAL STANDARDS AND THEIR SOURCES

Type of sample	Sample no.	Source	Remarks
Anhydrite	AN	6	
Andesite	AGV-1	12	
Basalt	BR	5	
	BM	6	
	4978	11	certified for radium content only
	BCR-1	12	
	JB-1	13	
Bauxite	69a	11	
	BX-N	5	
Biotite	1-B	8	
	Mica-Fe	5	
Brick, fire	269	3	plus sample 315
Brick, silica	267	3	plus sample 314
	102	11	
Calcsilicate	M-3	4	
Cement	1011	11	plus samples 1013 to 1016
Chrome ore	308	3	
Clay shale	TB	6	
Disthene	DT-N	5	
Diabase (dolerite)	I-3	4	
	4984	11	certified for radium content only
	W-1	12	
Diorite	DR-N˙	5	
Dolomite	400	10	plus limestone and magnetite blends
Dunite	DTS-1	12	
	D	14	
Gabbro-diorite	4982	11	certified for radium content only
Glass	89	11	plus samples 91, 92 and 93
	VS-N	5	
Granite	G-B	2	for Carpatho-Balkan Geological Association
	I-1	4	
	G	14	
	GA	5	
	GH	5	
	GM	6	
	G-2	12	replacement for G-I
Granodiorite	GSP-1	12	
	JG-1	13	
Haematite	453	10	plus dolomite and magnetite blends
Hornblende	1H	8	
Iron ore	175/1	3	plus samples 301 and 302
	27e	11	
	28a	11	
Kyanite	DT-N	5	

TABLE 1.—*continued.*

Type of sample	Sample no.	Source	Remarks
Limestone	KH	6	
	401	10	plus dolomite blends
	1a	11	
Lujarrite	L	14	
Manganese ore	176/1	3	
	25c	11	
Magnetite	450	10	plus dolomite and haematite blends
Magnesite, burned	104	11	
Muscovite	—	8	
	P-207	12	for K-Ar and Rb-Sr dating
Nepheline syenite	—	7	
Norite	N	14	
Peridotite	PCC-1	12	
Petalite	182	11	
Phlogopite	Mica-Mg		
Phosphate	120a	11	
Potash felspar	70a	11	
	376	3	
Pyroxenite	P	14	
Refractories	76	11	plus samples 77, 78, 103a, 198 and 199
Sand, glass	165	11	
Schist	M-2	4	
Serpentine	UB-N	5	
Shale	TS	6	
Sillimanite	309	3	
Slag, basic	174/1	3	plus samples 248/1 and 249/1
Soda felspar	99a	11	
	375	3	
Spodumene	181	11	
Sulphide ore	SU-1	9	
Syenite	S	14	
Tin ore	138	11	
Tonalite	T-1	1	
Zinc ore	113	11	

Sources:
1. Commissioner, Geological Survey, P.O. Box 903, Dodoma, Tanzania.
2. E. Aleksiev and R. Boyadjieva, Geological Institute, Bulgarian Academy of Sciences, Sofia, Bulgaria.
3. Bureau of Analysed Standards, Newham Hall, Middlesbrough, England.
4. A. B. Poole, Department of Geology, Queen Mary College, Mile End Road, London E.1, England.
5. H. de la Roche, or K. Govindaraju, Centre de Recherches Pétrographiques et Géochimiques, B.P. 682, Nancy, France.
6. K. Schmidt, Zentrales Geologisches Institut, Invalidenstrasse 44, 104 Berlin, Deutsche Demokratische Republik.

7. A. A. Koukharenko, Department of Mineralogy, Leningrad State University, Leningrad V-164, U.S.S.R.
8. T. Hugi, Mineralogisch-petrographisches Institut, Universität Bern, Bern, Switzerland.
9. A. H. Gillieson, Canadian Standard Reference Materials Project, Mineral Sciences Division, Mines Branch, 555, Booth Street, Ottawa, Canada.
10. G. Frederick Smith Chemical Company, P.O. Box 23344, Columbus, Ohio 43223, U.S.A.
11. Office of Standard Reference Materials, National Bureau of Standards, Washington D.C., 20234, U.S.A.
12. F. J. Flanagan, U.S. Geological Survey, Washington D.C., 20242, U.S.A. For muscovite P-207, M. A. Lanphere or G. B. Dalrymple, U.S. Geological Survey, Menlo Park, California, 94025, U.S.A.
13. A. Ando, Geochemical Research Section, Geological Survey of Japan, 135, Hisamoto-cho, Takatsu, Kawasaki, Japan.
14. T. W. Steele, National Institute for Metallurgy, Private Bag 7, Auckland Park, South Africa.

For many rock analysts, a complete analysis will include not only these thirteen components, but also a number of other elements that are occasionally present in rock specimens in amounts of up to several per cent. Those frequently reported include:

> sulphur (sulphide and sulphate)
> carbon (carbonate and non-carbonate)
> chlorine
> fluorine
> chromium
> vanadium
> barium

This list can be extended considerably by the inclusion of certain other elements less frequently reported, zirconium, strontium, lithium for example, that are also occasionally present in more than trace amounts (in this context usually considered to be more than $0 \cdot 02$ per cent) and which have been largely neglected in earlier analyses. Emission spectrography is often employed to establish the presence of these and other even rarer constituents.

For many rocks the likelihood of finding unusually high amounts of certain elements can sometimes be indicated by the petrologist, following an examination of a thin section. Another useful guide can be obtained by studying earlier analyses of similar rock types. It is for this reason that where the figures have been available, average values for the occurrence of individual elements in particular rock types have been included in each chapter of this book. These values should be used with caution as many of them have been calculated from a large number of individual values that covered a very wide range.

The elements reported in sedimentary rocks are essentially the same as those noted above for igneous silicate rocks. In sandstones and quartzites, silica is the dominant and sometimes only major component, all other elements being present in minor or trace proportion only. Shales, muds and slates resemble the igneous silicates, with the same group of major elements present in somewhat similar proportions, although carbon dioxide, organic matter and pyritic sulphur are likely to be present in increased amounts. Some limestones are little more than calcium carbonate, but others contain major amounts of magnesium and iron. Those limestones with an arenaceous fraction may contain appreciable amounts of silica, aluminium, iron and other elements.

Some of the most difficult rocks to analyse are the carbonatites. These igneous carbonates vary considerably in mineralogical composition, but often include appreciable amounts of certain silicates particularly pyroxenes and micas, oxide minerals such as magnetite, phosphates such as apatite and monazite, and sulphides. Many of the carbonatite occurrences are of economic importance as sources of niobium (pyrochlore), iron ore (magnetite), phosphate (apatite), copper (sulphide minerals) or vermiculite. The difference between the known deposits is so great that it is not possible to draw up a list of elements that should be determined in a "complete analysis".

With the introduction of more sensitive methods of analysis, it is now clear that the list of elements that can be reported from igneous silicate rocks could, if methods were available, be extended to include all the naturally occurring elements of the periodic table.

Reporting an Analysis

Published chemical analyses of igneous rocks are often put to uses that were not considered by the analyst. The potential value of an analysis is therefore greater than the sum of the determinations, and this should be increased by including with the analysis full details of the origin of the specimen and notes on the petrographic examination.

The name of the analyst, address of the laboratory where the analysis was made and the date of the analysis should also be included. These notes will give the means of recovering more detailed information, such as the procedures used, if these are wanted at a later date. Hamilton[11] making a plea for more information of this sort gives the following example from Shaw[12] of the type of petrographic information that should accompany the analysis:

L 62 Staurolite schist

A silvery crumpled schist. Porphyroblasts of staurolite, with quartz-graphite inclusions, and of occasional garnets, in a matrix of quartz, muscovite and biotite, with crumpled schistosity planes, Staurolite crystals have been fragmented by shearing along micaceous folia. Principal opaque mineral is graphite, but iron oxides are also present. Minor felspar and sphene. Grade: staurolite zone.

In many cases it will not be possible to give such a detailed description of the specimen, but what information is available should be recorded in such a form as to leave no doubt as to what was analysed.

The conventional way of reporting the detailed analysis of a silicate rock is to express each element in the form of its oxide, and to give the results to the second decimal place. This can lead to certain difficulties, as for example in reporting the ferrous iron content of rocks containing much pyrite or carbonaceous matter. Attempts have been made to depart from these traditions by giving results to only the first decimal place and by expressing the constituents as elements in place of oxides. Neither of these suggestions has so far been widely adopted. The summation of the oxides is widely regarded as a test of the skill of the analyst, and for this reason alone is unlikely to be discarded. The analyst should, however, be aware of the possibility of compensating errors occurring in his analysis, giving a fortuitously good total. Chalmers and Page[13] have suggested that where the results are made the basis for comparisons, each complete chemical analysis should include an estimate of the precision and accuracy of the results. It will usually be found that not more than three significant figures can be justified. Claims to accuracies comparable with, or better than, those of the accepted atomic weights of the elements should be resisted.

Constituents that are present in only trace amount are usually reported as parts per million of the element, rather than of the oxide. At lower concentrations parts per billion (1 in 10^9) are sometimes preferred.

References

1. WASHINGTON H. S., *Manual of the Chemical Analysis of Rocks*, Wiley, New York, 3rd edition, 1918.
2. HILLEBRAND W. F., The Analysis of Silicate and Carbonate Rocks, *U.S. Geol. Surv. Bull.* 700, Washington, 1919.
3. AHRENS L. H., *Quantitative Spectrochemical Analysis of Silicates*, Pergamon, Oxford, 1954.
4. JEFFERY P. G. and WILSON A. D., *Analyst* (1960) **85,** 478.
5. FAIRBAIRN H. W. and others, *U.S. Geol. Surv. Bull.* 980, 1951.
6. INGERSON E., *Geochim. Cosmochim. Acta* (1958) **14,** 188.
7. FLANAGAN F. J. and GWYM M. E., *Geochim. Cosmochim. Acta* (1967) **31,** 1211.
8. FLANAGAN F. J., *Geochim. Cosmochin. Acta* (1970) **34,** 121.
9. ABBEY S., *Geol. Surv. Canada*, Paper 72–30, 1970.
10. FLANAGAN F. J., *Geochim. Cosmochim. Acta* (1973) **37,** 1189.
11. HAMILTON W. B., *Geochim. Cosmochim. Acta* (1958) **14,** 253.
12. SHAW D. M., *Bull. Geol. Soc. Amer.* (1956) **126,** 493.
13. CHALMERS R. A. and PAGE E. S., *Geochim. Cosmochim. Acta* (1957) **11,** 247.

CHAPTER 2

SAMPLE PREPARATION

Selection of Material

Problems involved in the selection of the material are largely the concern of the field geologist, but the geochemist and rock analyst should appreciate the difficulties of ensuring that material which arrives in the laboratory is representative of the exposure from which it has been taken. Great care must be taken in the choice of material, in the collection of a suitable bulk and in the proper labelling and storage of the material before despatch to the laboratory. The importance of keeping full detailed field notes concerning the rock exposure and in the proper indexing of all specimens cannot be over-emphasised.

In general, where there is no shortage of material, it is easier to collect too much at the first visit, than to return later to collect more. It will be required for petrographic and possible mineralogical studies as well as for chemical and spectrographic analysis. Reserve specimens should also be retained for further study and also for future reference.

At this stage it is important to understand fully what the sample is intended to represent. An outcrop of granite has sometimes been represented by a single specimen. Neither this nor a series of chips taken over the exposed surface is likely to be typical of the granite in depth—each specimen collected represents the granite only at the place from which it was taken. The practice of combining chips from as much as possible of the exposed area to give a composite sample has only one merit—it reduces the analytical effort required. The results for composite samples tend to reflect the way in which the composing was done and may suggest an overall composition that occurs nowhere in the outcrop. Wherever possible such outcrops should be sampled over the whole of the exposed area, but the specimens taken should be kept separate and if possible analysed separately. The results for constituents of interest are of greater value if they indicate both a mean value and the extent of variation from it.

13

It will be appreciated that the size of the sample necessary to give a representative specimen will vary with the mineral grain size. A far smaller quantity will be required of a fine-grained rock with no pheno-crysts, such as dolerite, than of a coarse-grained or porphyritic rock. For this reason it is difficult to lay down any rule as to the weight of rock material that should be taken. In general, it is only for the very coarsely crystalline rocks, such as pegmatites, that a sample size of greater than 20 kg is necessary. For dolerites and other fine grained rocks a minimum of $2\frac{1}{2}$ kg should be collected. Table 2, adapted from a suggestion of Dr. C. G. B. DuBois, was compiled to indicate to geologists the approximate size of sample that should be collected if chemical analysis was envisaged.

Chemical analysis can of course be made on a very much smaller sample weight—amounting to no more than a gram or two if necessary. But in such instances the task of the analyst is rendered more difficult by limiting his choice of methods and by leaving no margin for repeat determinations. It is highly undesirable to use all the material leaving none for reference or future work.

TABLE 2. THE WEIGHT OF ROCK MATERIAL
THAT SHOULD BE COLLECTED

Maximum grain size (mm)	Weight suggested (kg)
1	$2\frac{1}{2}$
5	5
10	10
20	20

Due care should be taken to see that the rock is as fresh as possible and that no skin of altered material is included. Likewise, fragments with obvious mineral veins and inclusions should be excluded (or preferably be collected and analysed separately). Paint should not be used to label a specimen, as this can give rise to contamination of one or more trace constituents. If it is necessary to use paint—as, for example, in tropical areas where paper labels or containers are likely to be eaten by ants—it should be removed in the laboratory before the specimen is crushed. Fresh bags should be used to contain the rock material—previous use of sample containers is a frequent source of contamination. It is not unknown for a petrologist to spend time examining small white crystals adhering to a specimen, only to identify them as sugar from an earlier use of the container.

Crushing and Grinding

The first step in preparing the sample is to examine the total bulk of material and to select the portion required for chemical and spectrographic analysis. This may more conveniently be done by removing the portions for the petrographic examination and for the reserve collection. In the case of coarse-grained and porphyritic rocks, a sample of not less than 10 kg should then be available for the preparation of the material for chemical analysis, and proportionately less of the fine-, even-grained rocks. At this and all subsequent stages in the sampling, crushing and grinding, an intelligent approach is necessary to ensure that the introduction of extraneous matter is kept to a minimum, and that any elements introduced do not appreciably alter the composition of the material. Only then can the results of the chemical analysis of the prepared sample be taken to represent the chemical composition of the material collected.

The following notes are based upon the procedure used by the author to prepare igneous silicate and carbonate rocks for analysis. A simplified procedure is used for friable rocks such as unconsolidated sediments which can usually be fed directly to a mechanical agate mortar and pestle. All other samples are fed first to a small jaw-crusher. The product is screened and any oversize material returned to the jaw-crusher, now set with the jaws giving a slightly smaller gap. The whole of the sample can be reduced in this way to pass a No. 5 mesh sieve. This rock material is then riffled to give about 500 g which is sieved on a No. 10 mesh sieve, the oversized material then being cram-fed to the jaw-crusher on its finest setting. The product is riffled once more to give 75 to 100 g of material, all of which is subsequently ground to give the sample for analysis. The grinding is done by feeding small quantities at a time to a mechanical agate mortar and pestle. The grinding is stopped from time to time to remove the 100-mesh material by sieving through bolting cloth.

Once the grinding is complete, the sample material is transferred to a large bottle—an 8-oz bottle is a convenient size—and is thoroughly mixed by shaking and rolling. After this the material is transferred to a smaller bottle and labelled with details of the specimen, locality from which it was taken, serial or catalogue number and the notebook reference. The details recorded in the notebook should include notes on the sampling, crushing and grinding procedures, sieving operations, weight of the prepared material and the date. The sample is then ready for analysis.

The practice of coning and quartering is not recommended for the sampling of the small amounts of material collected for rock analysis. A series of riffles of varying sizes of the type shown in Fig. 1, can be kept for this purpose, or alternatively a rotary sampling machine can be used.

Samples produced by grinding in a mechanical agate mortar and pestle are similar to those ground by hand in that they contain a great deal of fine material. This can be seen from Table 3 which gives screen analyses of a number of silicate rocks that had been prepared for complete analysis by grinding to pass a No. 72 mesh sieve.

TABLE 3. SCREEN ANALYSIS OF SILICATES PREPARED FOR ANALYSIS

Sieve mesh (British Standard sizes)	% retained		
	Quartzite	Granite	Diorite
85	0·06	1·92	0·03
100	20·64	27·27	21·53
120	12·60	10·95	11·02
150	12·00	11·65	10·69
170	6·16	6·99	7·16
200	4·22	6·26	13·15
240	8·68	10·17	17·91
300	3·53	4·23	5·22
350	2·64	2·57	1·23
bottom pan	29·47	17·99	11·86

The total number of particles present in these and other similar ground specimens exceeds 10^6 per gram, and the surface area as determined by nitrogen absorption is in the range $0·5–1·5$ m^2 per gram.

Contamination

Agate mortars and pestles are a frequent source of contamination, but introduction of extraneous material can occur at all stages in the preparation of the sample. Contamination from painted labels has already been noted. If such labels have been used, they should be removed by chipping or grinding before the sample is crushed. Jaw-crushers must be cleaned particularly carefully and thoroughly if cross contamination is to be avoided. If the crusher is fitted with jaws of

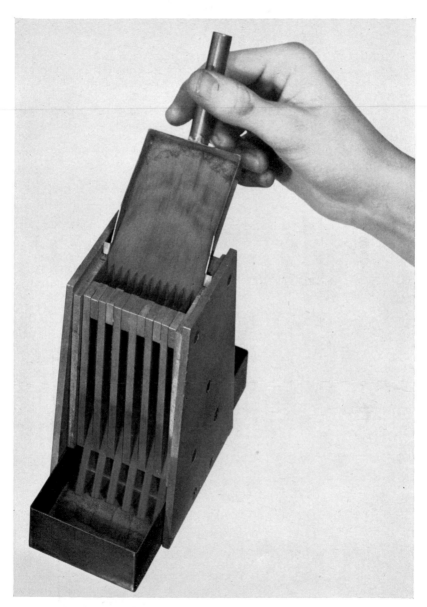

FIG. 1. Laboratory riffle (Crown copyright)

mild steel, small fragments may be shorn from the faces giving appreciable errors in the determination of ferrous iron. The amounts of tramp iron introduced in this way can be considerably reduced by fitting jaws of hardened manganese steel, but this may in turn introduce small amounts of other elements such as chromium into the sample.

The practice of using iron jaws to crush the sample, followed by the removal of the introduced iron fragments with a magnet, is not to be recommended, as any magnetite present in the rock will be similarly removed together with smaller quantities of other iron minerals such as pyrrhotite and ilmenite. This would materially affect the composition of some samples—carbonatites for example.

The use of nylon bolting cloth supported in a ring of plastic material cut from an acrylic pipe, can eliminate metallic contamination at the sieving stages, but care must be taken that loss by dusting is kept to the minimum.

If many silicate analyses are to be made it is preferable to reserve a special agate mortar for grinding these samples. If ore minerals are ground in it, traces of these minerals can usually be found in subsequent samples no matter how carefully the cleaning is done. This has been shown by grinding 5 g of ferberite in a mechanical agate mortar until the whole of the material passed a 150-mesh sieve. This material was then discarded and the mortar cleaned and dried. 5 g of crushed quartz was then ground for 5 minutes, and the mortar again cleaned and dried. Ten further portions of quartz were similarly ground with intervening cleaning and drying. These ground samples were examined for tungsten, when the results given in Table 4 were obtained. It is virtually impossible to be sure that trace elements will not be introduced into the sample if the same mortar is used for both rocks and ores.

The introduction of material is not the only change occurring in rock samples during grinding. Water may be lost, or in some cases gained, whilst both ferrous iron and sulphur may undergo partial oxidation. These effects are enhanced by excessive grinding. For this reason Hillebrand[1] recommends that silicate rocks should be reduced in size only to pass a 70-mesh sieve. This may reduce oxidation changes, but at the same time increases the difficulty of decomposing the rock. This was clearly shown by French and Adams,[2] who reported that a sample of diabase (dolerite), I_3 containing approximately 10 per cent FeO, gave a progressively lower FeO content as the grinding period was prolonged. After 20 minutes the FeO content had decreased by more than 0·5 per cent and this rate of oxidation was maintained throughout the grinding period (see Fig. 2). Grinding of the rock material in the same agate mortar for 10 minutes but with continuous moistening with acetone produced a sufficiently fine material with no detectable oxidation.

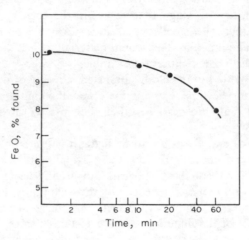

FIG. 2

TABLE 4. CONTAMINATION OF QUARTZ
ON GRINDING

Material	W, ppm
Quartz, unground	0·7
1st ground portion quartz	1300
2nd ,, ,, ,,	61
3rd ,, ,, ,,	31
4th ,, ,, ,,	22·6
5th ,, ,, ,,	16·1
6th ,, ,, ,,	7·4
7th ,, ,, ,,	3·8
8th ,, ,, ,,	2·9
9th ,, ,, ,,	3·1
10th ,, ,, ,,	2·5
11th ,, ,, ,,	3·1

Fine Grinding of Micas

Anyone who has attempted to reduce specimens of mica minerals or rock samples containing large amounts of mica, will have experienced the difficulty of reducing platy minerals to an impalpable powder. Where mechanical mortars are used, the harder, more brittle minerals are preferentially ground, leaving the mica minerals to enrich the latter fractions. Care must be taken to ensure that none of the mica fraction is lost or discarded, and that the powdered sample is thoroughly mixed before portions are taken from it.

Abbey and Maxwell[3] have reported that pre-ignition of mica samples makes the grinding stage easier, but that the ignited product slowly gains weight, making accurate weighing virtually impossible. These authors recommend the use of a "blender" with blades rotating at 15,000 rpm for size reduction of mica as follows:

> Place the bulk sample of up to 15 g of mica in the blender jar and add sufficient water to cover the blades. Operate the blender for 3 to 5 minutes, stopping several times to rinse down coarse particles spattered on the walls and lid. Evaporate the resultant slurry to dryness in a silica dish on a steam bath and dry the sample at 100°. Cool to room temperature in the oven, spread the residue on a clean paper, break up any lumps with a spatula, and mix thoroughly by rolling.

Sample Material for Trace-element Determinations

A method of fragmenting silicate rocks that avoids much of the contamination from crushing and grinding machinery was described by Hawley and MacDonald.[4] This involves heating the silicate rock previously broken into lumps about 2–3 cm in size to a temperature of about 600° in an electric furnace and quenching them in cold water. Granitic rocks are readily pulverised in this way, whilst other rocks such as syenites are embrittled, allowing easier grinding in conventional agate mortars. More than one treatment may be necessary for some silicates.

Rock material prepared in this way should not be used for the determination of water, carbon dioxide, sulphur or ferrous iron, and preferably also not for any of the major constituents.

References

1. HILLEBRAND W. F., The Analysis of Silicate and Carbonate Rocks, *U.S. Geol. Surv. Bull.* 700, Washington, 1919.
2. FRENCH W. J. and ADAMS S. J., *Analyst* (1972) **97**, 828.
3. ABBEY S. and MAXWELL J. A., *Chem. in Canada* (1960) **12**, 37.
4. HAWLEY J. E. and MACDONALD G., *Geochim. Cosmochim. Acta* (1956) **10**, 197.

CHAPTER 3

SAMPLE DECOMPOSITION

ROCK material is characterised by a certain degree of order in the crystal forms of its constituent minerals. The process of sample decomposition —the first step of all analyses—consists of the destruction of some or all of the original minerals as part of or prior to the dissolution of the constituent of interest. The processes of decomposition vary considerably—from simple extraction with water, organic solvents or mineral acids to the more elaborate techniques of sintering or fusion. Few of these techniques will decompose completely all types of rock material, nor is this always desirable. Many of the decomposition procedures serve to remove the major part of the constituent minerals but leave a minor fraction as a residue that can be removed from the solution by filtration. Whether or not this residue will require separate decomposition will depend upon the amount of residue and more particularly whether it can be expected to contain the elements of present interest.

Thus, for example, in the course of an analysis of a basic rock, the ground material was evaporated with hydrofluoric and sulphuric acids and after all soluble sulphates had been removed, a small residue amounting to 0·88 per cent of the rock material was recovered. This residue was found to consist entirely of chromite. It was ignored in the determination of the alkali metals, and was only of marginal importance in the determination of total iron. Although this residue contained most of the chromium present in the sample taken, it was not of direct interest, as the chromium content was determined in a separate sample portion.

In a further example, a carbonatite rock was decomposed by digestion with hot hydrochloric acid. The residue contained biotite, magnetite, perovskite, pyrochlore and barite. A determination of calcium, magnesium and strontium in the acid extract gave the alkaline earth carbonate content of the rock material, whilst the barium, titanium, tantalum and niobium values were almost entirely contained in the residue.

Extraction Procedures

EXTRACTION WITH WATER

Extraction with water is of very limited application to silicate rocks and minerals, and then only where the material contains evaporite minerals such as sodium chloride or magnesium sulphate. A small number of minerals such as copperas, $FeSO_4.7H_2O$ will dissolve in water, but the extracts tend to be cloudy and to deposit hydrated iron oxide unless a few drops of sulphuric acid are added.

EXTRACTION WITH ORGANIC SOLVENTS

Although there is a wide range of organic solvents available only a few minerals can be extracted with them. In silicate rock analysis the greatest use appears to be in the recovery of carbon, nitrogen and sulphur compounds from sedimentary rocks. Elemental sulphur is not a common constituent of rocks, but when present it can usually be removed by extraction with carbon disulphide or more safely with pyridine.

Decomposition with Mineral Acids

DECOMPOSITION WITH HYDROCHLORIC ACID

With the exception of the minerals of the scapolite group, all carbonate minerals are decomposed, either in the cold or on digestion at an elevated temperature with dilute hydrochloric acid. This method of decomposition is therefore of particular use for carbonate and carbonatite rocks, where it serves to separate the carbonate fraction from the silicate and oxide fractions. However, certain silicate minerals, particularly calc-silicates such as wollastonite occurring in metamorphic limestones, can be wholly or partially decomposed by prolonged heating with hydrochloric acid. Certain sulphide minerals are also attacked by hydrochloric acid.

DECOMPOSITION WITH NITRIC ACID

Concentrated nitric acid serves to decompose not only carbonate minerals, but also any sulphide minerals present. This is probably its most important application in rock analysis, leading to one method for the determination of sulphide sulphur. In this procedure nitric acid is usually added with hydrochloric acid to assist the decomposition of any carbonate minerals present. Metals such as lead and zinc, occurring as sulphide minerals in a silicate matrix, can then be determined by polarography or atomic absorption spectroscopy.

DECOMPOSITION WITH HYDROFLUORIC ACID

Hydrofluoric acid has long been used for the decomposition of silicate rocks, usually in combination with nitric, perchloric or sulphuric acid. This combination enables substantially all the fluorine as well as all the silica to be removed in the evaporation, leaving a residue that can be dissolved in water or dilute acid and used for the determination of the alkali metals, alkaline earths, iron, aluminium, titanium, manganese, and phosphorus. With many rocks a small residue consisting of acid-resistant minerals such as zircon, topaz, corundum, sillimanite, tourmaline and rutile may remain, together with barium sulphate, particularly if the sample material contains. much barium and sulphuric acid is used for the decomposition.

The use of hydrofluoric acid without the addition of other mineral acid has recently been recommended by Langmyhr and Sveen[1] and by May and Rowe.[2] High temperatures and pressures are necessary if a reasonably complete decomposition is to be obtained; Langmyhr and Sveen, for example, used temperatures of up to 250° in a PTFE-lined aluminium bomb, whilst May and Rowe used up to 525° and an estimated pressure of 30,000 psi in a platinum-lined bomb. Advantages claimed for the high temperature–high pressure decomposition with hydrofluoric acid are that the procedure is much more effective than evaporation with a mixture of sulphuric and hydrofluoric acids in decomposing refractory minerals, and that because silicon is not volatilised in the closed system, it can be determined photometrically at a later stage. Antweiler[3] has used hydrofluoric acid to decompose large fragments of silicate rock (up to 50 g) by digestion at a temperature of 85° for 24 hours in a polythene or other suitable container with a close-fitting lid.

Most authors have, however, preferred to use hydrofluoric acid in the presence of some other mineral acid. This serves to moderate the initial reaction between hydrofluoric acid and finely powdered silicate material, although it is recommended that all powdered rock material should be moistened with water prior to adding hydrofluoric acid for this reason. Failure to do this can result in overheating and consequent loss of material by spitting. Nitric acid is often added to decompose any traces of carbonate minerals, to oxidize sulphides and organic matter and to convert iron and other elements into their higher valency states.

Evaporation with perchloric–hydrofluoric acid mixtures has frequently been recommended for the decomposition of silicates. This evaporation is a great deal more easy to carry out than the similar evaporation with sulphuric acid, there being less tendency for the solution to spit, as the perchlorate salts crystallise more cleanly than the corresponding sulphates. The perchlorate residue, unlike the sulphate residue, is readily soluble in water—aluminium and ferric sulphates in particular, once dehydrated, can only be dissolved with difficulty. In addition the perchlorate ion, unlike the sulphate ion, does not have a depressant effect upon the flame emission of the alkali metals, and must for this reason alone be preferred for this particular determination.

The evaporation with a mineral acid additional to hydrofluoric serves also to remove the fluorine ion which otherwise interferes with the determination of aluminium, titanium and certain other elements. The order of effectiveness in removing residual amounts of fluorine increases in the order nitric—perchloric—sulphuric acid. Langmyhr[4] has shown that a double evaporation with perchloric acid at a temperature of 180° reduces the fluorine level to a value that can be reached in a single fuming with sulphuric acid at a temperature of 250°, and that only microgram amounts can then be recovered from the residue.

Work in the author's laboratory has broadly confirmed these observations, except that larger amounts of fluorine were recovered in each case, and that the only really effective way of removing these traces of fluorine was to add potassium pyrosulphate to the residue obtained by evaporating the excess sulphuric acid and to convert the evaporation into a fusion. This further stage has an additional advantage in that the pyrosulphate melt is readily soluble in hot dilute hydrochloric acid, in contrast to the sulphate residue, which is soluble only with difficulty.

Certain authors have recommended that the silicate rock material should be allowed to stand overnight with hydrofluoric acid, either at room temperature or at the temperature of a steam bath. The addition of perchloric or other mineral acid and subsequent evaporation is then undertaken on the following day. This procedure is particularly effective for decomposing those rocks that are rich in magnesium.

Fusion Procedures

FUSION WITH ALKALI FLUORIDE

Fusions with ammonium fluoride have been recommended for the decomposition of beryl and other silicate minerals.[5] Not all silicates are attacked and attempts to decompose sillimanite, kyanite and zircon were ineffective.[6] In most instances where alkali fluoride is used the fluoride melt is converted to a pyrosulphate melt by heating with sulphuric acid. This serves to decompose complex fluorides, to convert all metal fluorides to sulphates and to remove fluorine more or less completely from the melt.

FUSION WITH POTASSIUM PYROSULPHATE

Potassium bisulphate has been advocated for certain purposes; it is converted to pyrosulphate in the earlier stages of the fusion and its use is not recommended as considerable spitting can occur in the conversion stage. Moreover, very little attack of oxide minerals can occur until this removal of water has taken place. Silicate minerals are not decomposed by direct fusion with potassium pyrosulphate, which should be retained for the decomposition of the residue remaining after an evaporation with hydrofluoric acid. This can be done immediately after the evaporation as described above, or after the major part of the metallic constituents present as perchlorates or sulphates have been removed by dissolution in dilute acid. The residue then obtained is often quite small, but it frequently contains a variety of minerals, some silicate (zircon, tourmaline, andalusite, etc.), some oxide (rutile, ilmenite, cassiterite, chromite, etc.). With most silicate rocks this assemblage is best decomposed by fusion with anhydrous sodium carbonate, but if certain minerals preponderate (ilmenite or rutile, for example), then potassium pyrosulphate can be used. Chromite, cassiterite and zircon, some of the commonest accessory minerals, are not appreciably attacked in a pyrosulphate fusion.

Platinum crucibles and dishes, although used and recommended for pyrosulphate fusions, are not the first choice of vessels for this purpose. Sulphur trioxide is readily lost from the melt leaving potassium sulphate which is no longer effective in the decomposition of oxide minerals. Moreover, the platinum is appreciably attacked in the course of the fusion, introducing platinum into the rock solution. This can interfere with subsequent determinations, as for example that of vanadium, p. 489. For this determination it is suggested that the rock material be decomposed by evaporation with hydrofluoric acid in a PTFE (polytetrafluoroethylene) vessel and the residue transferred to a silica crucible for the pyrosulphate fusion.

FUSION WITH SODIUM CARBONATE

All silicate rocks are decomposed more or less completely by prolonged fusion with anhydrous sodium carbonate, which is usually conducted in a platinum crucible. Crucibles of a platinum–iridium alloy, which has much higher mechanical strength and larger resistance to deformation, have been used. Palau crucibles (a gold–palladium alloy) can also be used, being not only more rigid than pure platinum, but also much cheaper. The amounts of platinum or other noble metal introduced into the melt are very small and can usually be ignored.

Platinum crucibles usually become iron-stained after a few fusions of rock material with sodium carbonate. This indicates that some reduction of iron to the metallic state has occurred, which has then become alloyed with the platinum. Some of this iron can often be removed by heating with 6 N hydrochloric acid, some more by fusing potassium pyrosulphate in the crucible, but the last traces of iron are very difficult, if not impossible to remove. In the analyses of acidic and intermediate rocks, this small amount of alloying can be ignored, but with basic rocks a small amount of potassium nitrate or chlorate can be added to maintain the melt in an oxidised condition. This addition increases slightly the extent to which platinum in attacked and removed from the crucible. Reducing melts can be obtained from rock samples containing much sulphide or carbonaceous matter; these elements should be removed by roasting prior to adding the sodium carbonate, although small amounts of sulphide minerals or organic matter can be tolerated as these will be oxidised by the added potassium nitrate or chlorate.

Complete fusion of 1 g of silicate rock is obtained by using 5 g of sodium carbonate. Larger quantities are not justified, even for basic rocks, whilst as little as 3 g will give a fluid melt with some acidic rock materials. After fusion for 1 hour at a temperature of about 1000°, the silicate rock matrix and most of the accessory minerals will be completely decomposed, although further heating at 1200° for an additional period of about 10 minutes is recommended for the decomposition of the small amounts of zircon, rutile and chromite that are sometimes present. Details of this procedure are given in Chapter 4, p. 35.

Although fusions with sodium carbonate are usually recommended, certain authors have noted that sintering will often suffice. Finn and Klekotka,[7] for example, sintered 0·5 g of silicate rock material with 0·6 g of anhydrous sodium carbonate. This method of decomposition has the advantage of reducing the volumes of acids and other reagents that have to be added in subsequent stages of the analysis, of reducing considerably the amount of sodium salts that has to be washed from later precipitates, of reducing the contamination from introduced platinum and any impurities present in the sodium carbonate (perhaps no longer as important as it may have been!), and more particularly of reducing the time necessary for the complete analysis.

Hoffman[8] used 0·5 g of sodium carbonate with 0·5 g of rock material and sintered in a 75-ml platinum dish at a temperature of 1200°. The addition of hydrochloric acid to the sinter gave an insoluble silica residue that could be dehydrated in the same 75-ml dish, in place of the clear solution usually obtained by treating the fused melt, and which requires evaporation and dehydration in a much larger basin.

This 1:1 ratio of sample weight to weight of anhydrous sodium carbonate is not recommended for the decomposition of kyanite, sillimanite, andalusite or silicate rock containing large amounts of aluminosilicates. These minerals tend to fuse and form glassy-melts with a well-ordered structure not readily broken down by the addition of hydrochloric acid. This difficulty does not arise if larger amounts of sodium carbonate are used, and 4 g of flux should be used for a 1 g portion of these silicates.

FUSION WITH ALKALI HYDROXIDE

Sodium and potassium hydroxides are extremely efficient fluxes for the decomposition of silicate minerals. This decomposition occurs rapidly at temperatures very much less than those required for fusions with sodium carbonate. The ease with which silicate minerals dissolve in molten alkali is deceptive in that the accessory mineral fraction is likely to remain unattacked unless the fusion is prolonged. Although 5 minutes fusion is more than sufficient for felspars and other silicate minerals, a full hour is recommended for silicate rocks.

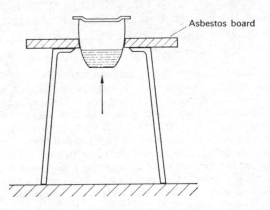

Asbestos board

FIG. 3. Apparatus for fusion with alkali hydroxide.

As molten alkalis are particularly corrosive, this fusion should be carried out at as low a temperature as possible, with the bottom of the crucible at only a faint red heat. Earlier workers recommended a spirit lamp for this decomposition, and positioned the crucible in a hole cut in asbestos board (Fig. 3). This served to keep the upper parts of the crucible cold, preventing "alkali-creep" over the edge of the crucible.

Platinum crucibles are subject to considerable attack from molten alkalis and should not be used. Silver and gold crucibles have been suggested, as the attack by molten alkali is very much less. However, some attack of metal does occur and the silver or gold introduced into the analysis in this way should be removed from the solution at a later stage. It must also be remembered that silver and gold have somewhat lower melting points (960° and 1063° respectively) than platinum, and crucibles can easily be damaged by overheating.

For many purposes iron or nickel crucibles can be used for these fusions. Although there is an appreciable attack of the metal, most crucibles will stand up to at least a dozen fusions before becoming porous. They cannot be used for determinations where the introduced iron or nickel would interfere with the subsequent analysis, but have long been used for the determinations of such elements as chromium and vanadium that form anions in their higher valency states. These crucibles have also been used for the determination of silica by a photometric method, where a rapid, effective decomposition of the silicate fraction is adequate. Both sodium and potassium hydroxides may contain traces of absorbed water and should in the first place be fused in the crucible without sample material.

Procedure. Clean a nickel or iron crucible with hot dilute hydrochloric acid, rinse with water and dry on a hot plate. Using a rough balance, weigh into the crucible the amount of sodium or potassium hydroxide required for the fusion, cover with a lid and heat gently over a small flame for a few minutes. Allow to cool in a desiccator. Weigh the sample of rock material directly on to the melt and transfer the crucible to a hot plate set at full heat. The temperature of most hot plates at full heat is sufficient just to fuse the melt, when the reaction with the sample can occur. Once the rock powder is completely wetted by the alkali, transfer the crucible to a hole cut in a piece of asbestos board, Fig. 3, and continue the fusion over a low flame for the prescribed period.

The procedure avoids difficulties that can occur by "dusting", that is, loss of sample material as a fine fume. This usually occurs when the rock sample is fused directly with pellets of alkali hydroxide taken straight from the reagent bottle.

Procedure. Place the warm, but not hot crucible on its side in a 250-ml borosilicate glass beaker (or stainless steel if silica is to be determined), and cover the melt with water. If the temperature of the melt is sufficient, a rapid steady decomposition of the melt will take place. If no action is apparent, transfer the beaker to a steam bath and heat until the decomposition is complete, leaving a crucible free from adhering melt.

Nickel crucibles have been preferred for the determination of silica, but they should not be used if the determination of iron is required, as some loss of iron occurs when silicates are fused with sodium hydroxide.[9]

SINTER OR FUSION WITH SODIUM PEROXIDE

Sodium peroxide is particularly useful in mineral analysis, as it is the only flux that can easily and readily be used for the complete decomposition of cassiterite or chromite. Earlier authors have tended to avoid its more general use, partly because of the uncertain quality of the reagent then available and partly because of the corrosive action of sodium peroxide on the materials used for crucibles—that is platinum, gold, silver, nickel and iron. Where the obvious advantages of using sodium peroxide could not be overlooked, as for example in the analysis of silicates containing appreciable amounts of chromite, then iron or nickel crucibles were used and discarded after a few determinations. In more recent years these difficulties have been largely overcome and the use of sodium peroxide is now more generally possible. Certain batches of reagent have been found to contain calcium, and these should be avoided if complete analyses are to be made.

One method of avoiding excessive attack of platinum is to line the crucible with a thick layer of fused anhydrous sodium carbonate before adding and mixing the sodium peroxide flux with the sample material. This technique is successful only if the subsequent fusion is not unduly prolonged. Nickel crucibles can be protected from excessive corrosion by a similar lining of the base of the crucible with fused sodium hydroxide.

Zirconium crucibles have been shown[10] to have superior resistance to molten sodium peroxide, although old crucibles may contribute appreciable amounts of zirconium to the melt, particularly if fusions have been conducted at temperatures in excess of 700°. This temperature is not necessary. Rafter and Seelye[11] have shown that most minerals occurring in silicate rocks are rapidly decomposed by sintering with sodium peroxide at a temperature of $480° \pm 20°$. This operation can be conducted at temperatures of up to 540° in platinum crucibles without introducing platinum into the rock sinter or solution. Rafter[12] has recommended that samples for decomposition in this way should be ground to pass a 240-mesh sieve, but this fine grinding is not necessary for most silicate rocks which are readily attacked at 100-mesh size by sintering at the recommended temperature.

Sodium peroxide melts or sinters are readily disintegrated by reaction with water, giving a highly alkaline solution containing much of the silica and aluminium, and a residue containing iron, titanium and other metals as hydroxides. If silica is to be determined then, as with alkali hydroxide fusions, the use of glass beakers must be avoided. Beakers of stainless steel or polypropylene should be used. The reaction of sodium peroxide with water can be violent and on no account must water be added directly to the melt in the crucible, as this may give rise to local overheating and the spitting of caustic alkali.

Procedure. Accurately weigh approximately 0·5 g of the sample material into a clean platinum crucible, add 1 g of sodium peroxide and mix with a platinum rod. Brush any particles adhering to the rod back into the crucible and cover the charge with a thin layer of sodium peroxide. Cover the crucible with a platinum lid and transfer to an electric furnace set at a temperature of 500°.

After 10 minutes remove the crucible, allow to cool and place on its side in a 250-ml polypropylene or stainless-steel beaker. Add the lid and then pour in sufficient water to cover the sinter, cautiously but not too slowly. Once the reaction is complete, rinse and remove the crucible and lid, cover the beaker with a polypropylene clock glass and boil gently for about 10 minutes to decompose peroxides.

FUSION WITH BORIC OXIDE AND ALKALI BORATES

Boric oxide and boric acid, although apparently attractive fluxes for the decomposition of silicate rocks, have never been widely used. This may be due in part to the extremely viscous nature of the melts which makes them difficult to use, and in part to the necessity of removing boron at a later stage of the analysis.

The use of borax (sodium tetraborate), or combinations of boric oxide, boric acid or borax with sodium carbonate has achieved some prominence in the analysis of materials rich in alumina, and has been recommended[13] for the decomposition of refractory minerals such as corundum, and chromium- and zirconium-bearing materials. It can be used with advantage for the analysis of kyanite, sillimanite and other aluminosilicates.

Procedure. Accurately weigh approximately 1 g of the finely powdered aluminosilicate material into a platinum crucible and mix intimately with 3 g of anhydrous sodium carbonate and 0·5 g of borax glass. Cover the crucible with a platinum lid and heat, either over a Bunsen burner or in a muffle furnace at a temperature of about 650° for 30 minutes to allow carbon dioxide to escape and the rock material to sinter with the flux. Increase the temperature gradually until fusion occurs and finally heat at about 1000° for a period of 10 minutes to complete the decomposition. Allow the melt to cool around the walls of the crucible.

Borate-carbonate melts disintegrate readily in dilute hydrochloric acid giving solutions that can be evaporated for the determination of silica. Methyl alcohol is added to the solution before commencing the evaporation, in order to remove boron as the volatile methyl borate. Failure to remove the boron at this stage will give high values for silicon, as some boron will be trapped with the silica on dehydration and subsequently be lost in the evaporation of the weighed silica with hydrofluoric and sulphuric acids. This evaporation with methyl alcohol is not necessary if silica is to be determined photometrically, as boron does not interfere with either the silicomolybdate or the molybdenum blue methods.

Biskupsky[6] has suggested using a flux composed of boric acid and lithium fluoride for the decomposition of silicate rocks and minerals. Lithium tetraborate is formed in the fusion, whilst silica is removed as the volatile tetrafluoride. Both boron and excess fluoride are removed by heating the melt with concentrated sulphuric acid. Advantages claimed are that only 12 to 13 minutes fusion time is required and that zircon, sillimanite, topaz, spinel, corundum, rutile, kyanite and other refractory minerals are decomposed without difficulty.

Procedure. Mix 2 g of boric acid with 3 g of lithium fluoride in a platinum crucible, and accurately weight 0·5 g of the sample material into a small hollow in the mixture. Cover the sample portion with the flux by gently tapping the sides of the crucible. Cover with a platinum lid and heat gently for 2 to 3 minutes over a small flame. Gradually increase the temperature and finally ignite over the full flame of a Meker burner for 10 minutes. Allow to cool. Add 10 ml of concentrated sulphuric acid to the crucible and heat gently until the evolution of all gas bubbles has ceased. Increase the heating and bring to copious fumes of sulphuric acid for 2 to 3 minutes. Allow to cool, transfer the crucible and lid to a 250-ml beaker containing 150 ml of water, and boil with 5 ml of concentrated hydrochloric or nitric acid for 10 to 15 minutes until complete solution is obtained.

Lithium metaborate ($LiBO_2$) has been suggested by Ingamells[14, 15] as a suitable flux for the decomposition of silicate rocks preparatory to determining silicon, phosphorus, iron, titanium, manganese, nickel and chromium by spectrophotometry. Sodium and potassium can be determined by flame photometry[16] and other elements by an emission spectrographic solution technique giving an essentially complete analysis (less FeO, CO_2, H_2O and certain minor components) from one sample portion (see Chapter 5, p. 56).

Miscellaneous Decomposition Procedures

FIRE ASSAY FOR GOLD AND SILVER

Fire assay is the term used for the determination of gold and silver in rocks, ores, concentrates and bullion by pyrometallurgical processes. A number of unit operations are involved including "pot fusion", "scorification" and "cupellation". Pot fusion is the initial stage in which the material is fluxed with sodium carbonate or a mixture of sodium carbonate and borax, together with an oxide of lead, usually red lead Pb_3O_4, and a reducing agent such as charcoal, flour or argol. Special provision must be made for pyritic samples or ores containing arsenic or tellurium. In the fusion process, red lead is reduced to lead metal which forms a button at the base of the crucible in which the values of gold and silver are concentrated.

This lead button is reduced in size by the process of scorification, which is an oxidising fusion of the lead button used to remove impurities present. In cupellation the lead button, now reduced to 15–20 g in weight, is placed on a heated cupel of bone ash or magnesia which absorbs molten litharge produced in a further oxidising fusion, until finally a small bead or "prill" of gold–silver alloy is left. This can be weighed and the gold and silver separated and determined.

These procedures have been revised and extended to include also the determination of the platinum metals, but in essence they remain what they always have been—exercises in simple manipulation combined with fine judgement based upon long practice. Although still used extensively for ore analysis, they have been displaced to some extent by atomic absorption spectrophotometry and by neutron activation analysis.

MISCELLANEOUS FLUXES

A number of special fluxes and sintering procedures have been used for the determination of particular elements. These include a mixture of calcium carbonate and ammonium chloride for the determination of alkali metals (Lawrence Smith procedure), a procedure given in detail in the first edition of this book, but now considered to be obsolete, and ammonium iodide for the recovery of tin present as cassiterite, given on p. 454.

Other fluxes with particular uses include alkali cyanides for the decomposition of cassiterite:

$$SnO_2 + 2NaCN = 2NaCNO + Sn$$

and a mixture of elemental sulphur and potassium carbonate for the conversion of arsenic, antimony and tin to their thiosalts. Neither procedure has any direct application to silicate rock analysis.

References

1. LANGMYHR F. J. and SVEEN S., *Anal. Chim. Acta* (1965) **32**, 1.
2. MAY I. and ROWE J. J., *Anal. Chim. Acta* (1965) **33**, 648.
3. ANTWEILER J. C., *U.S. Geol. Surv. Prof. Paper* 424-B (1961), p. 322.
4. LANGMYHR F. J., *Anal. Chim. Acta* (1967) **39**, 516.
5. CHEAD A. C. and SMITH G. F., *J. Amer. Chem. Soc.* (1931) **53**, 483.
6. BISKUPSKY V. S., *Anal. Chim. Acta* (1965) **33**, 333.
7. FINN A. N. and KLEKOTKA J. F., *Bur. Std. J. Res.* (1930) **4**, 813.
8. HOFFMAN J. I., *J. Res. Nat. Bur. Std.* (1940) **25**, 379.
9. BENNETT H., EARDLEY R. P. and THWAITES I., *Analyst* (1961) **86**, 135.
10. BELCHER C. B., *Talanta* (1963) **10**, 75.
11. RAFTER T. A. and SEELYE F. T., *Nature* (1950) **165**, 317.
12. RAFTER T. A., *Analyst* (1950) **75**, 485.
13. BENNETT H. and HAWLEY W. G., *Methods of Silicate Analysis*, Academic Press, 1965 (2nd ed.), p. 41.
14. INGAMELLS C. O., *Talanta* (1964) **11**, 665.
15. INGAMELLS C. O., *Analyt. Chem.* (1966) **38**, 1228.
16. SUHR N. H. and INGAMELLS C. O., *Analyt. Chem.* (1966) **38**, 730.

CHAPTER 4

CLASSICAL SCHEME FOR THE ANALYSIS OF SILICATE ROCKS

Introduction

In the classical scheme for the analysis of silicate rocks, provision was always made for the determination of a total of the thirteen most commonly occurring constituents. Of these, the alkali metals were determined in a separate portion of rock material, as were moisture, total water and ferrous iron. Most rock analysts preferred also to determine manganese, titanium, phosphorus and total iron in separate portions, leaving only silica, "mixed oxides", calcium and magnesium to be determined in what was known as the "main portion". Where the silicate rock sample was available in only small amounts, the sample portion used for the determination of moisture was used also for the elements in the main portion, as well as for total iron and sometimes also for titanium. Strontium, when present in more than trace amount, was precipitated with calcium as oxalate, and then separated and determined gravimetrically.

One of the most serious criticisms of the classical scheme is that any error in the determination of some of the constituents—iron, titanium or phosphorus, for example—was reflected in a similar error in the aluminium content, which was always obtained by difference. For this reason alone, where the classical separations are still used, the determination of iron, titanium and phosphorus should be made in separate portions of rock material, rather than in separated fractions.

This chapter is concerned primarily with the analysis of the main portion, that is with the determination of silica, the total of elements precipitated with ammonia and known collectively as the "mixed oxides", "ammonia group" or by certain analysts as the "R_2O_3 precipitate", together with calcium and magnesium. In the original classical scheme of analysis manganese appeared partly with magnesium in the phosphate precipitate and partly with iron and other elements in the ammonia precipitate.[1] Procedures have been devised to collect all the manganese in one fraction, but these are not entirely successful. Chromium, vanadium, zirconium and other elements are also precipitated with ammonia and, when present in more than trace amounts, can introduce errors into the reported aluminium content.

The scheme of analysis is given in outline form in Fig. 4. It is based upon the use of a 1 g portion of ground silicate rock material. The broad outline of the scheme and some of the methods used were devised in the nineteenth century, but continuous development has occurred since then, largely by way of refining procedures in the light of subsequent knowledge and experience.

Decomposition of the Sample

Procedure. Accurately weigh approximately 1 g of the finely powdered silicate rock material into a platinum crucible of about 25-ml capacity, add 3–5 g of anhydrous sodium carbonate (see p. 26) and mix with a platinum or glass rod. Brush any particles of rock material or flux from the rod back into the crucible, cover with a platinum lid and heat over a Bunsen burner or in an electric furnace to a dull red heat (furnace set at about 700°), and maintain at this temperature for about 30 minutes. Slowly raise the temperature to about 1000° and maintain at this temperature for a further 30 minutes, finally transfer the crucible to a Meker burner (not a blast Meker) or to an electric furnace set at 1200°, and heat for a further 10 minutes. Allow to cool, rotating the crucible held in a pair of platinum-tipped tongs, so as to allow the melt to solidify in a layer around the walls of the crucible.

Fill the crucible almost to the brim with water, add 2 or 3 drops of ethanol and allow to stand overnight. On the following day rinse the solution and residue from the crucible into a large (6-inch) platinum dish, wash the crucible with water and set it aside. In the presence of much manganese the melt is tinged green with alkali manganate, but this is reduced by the ethanol on standing overnight.

This procedure serves to decompose completely all the minerals present in silicate rocks. As noted in Chapter 3, p. 26, the quantity of sodium carbonate used is now regarded as excessive, and can be reduced, as for example by using the sintering technique of Hoffman,[2] as follows:

Procedure. Accurately weigh approximately 0·5 g of the finely powdered silicate rock material and 0·5 g of anhydrous sodium carbonate into a 75-ml platinum dish and mix together with a glass rod. If the rock material contains

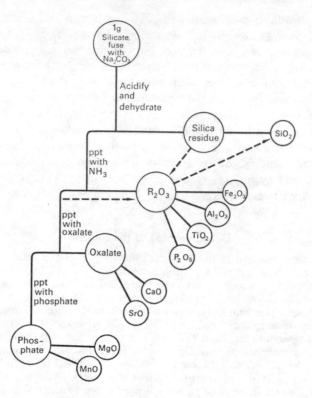

FIG. 4. Classical scheme for the analysis of the main portion.

much ferrous iron add also 0·05 g of potassium nitrate. Brush the mixture into the centre of the dish and then spread out in the form of a disc of about 3 cm diameter. Cover the mixture as evenly as possible with a further 0·5 g of anhydrous sodium carbonate. Transfer the dish to an electric muffle furnace and heat, slowly at first then more strongly until a temperature of 1200° is reached. Maintain the dish at this temperature for 15 minutes, then allow to cool with a cover over the dish to prevent loss of material by spitting in the cooling stage.

The advantages of this method of decomposition have been listed on p. 26. It has found application in the field of glass technology, where it has been strongly recommended by Chirnside.[3] A sodium peroxide sinter in a platinum crucible can also be used for the decomposition of the main portion. Care must be taken to obtain a good grade of sodium peroxide, free from calcium. The procedure for the decomposition is given in detail on p. 28.

The Determination of Silica

SEPARATION AND COLLECTION OF SILICA

The rock material, decomposed as described above, is acidified with hydrochloric acid, and the chloride solution evaporated to dryness. Most silicate rock analysts prefer to use platinum apparatus for this evaporation, but porcelain dishes can also be used. The major part of the silica present in the solution is recovered by dehydration and filtration, leaving aluminium, iron, alkali and alkaline earth elements together with the minor part of the silica in the filtrate. In the classical scheme of analysis, the filtrate is returned to the platinum basin for a second evaporation and dehydration to recover an additional silica fraction. Only a few milligrams of silica remain in solution after this second evaporation and these cannot be recovered by a third evaporation. These traces will be precipitated together with iron, aluminium, titanium and other elements by adding ammonia.

In the procedure described by Hoffman,[2] the evaporation and dehydration of silica are conducted in the 75-ml platinum dish used for the sample decomposition, and the silica fractions recovered by filtration are returned to this dish for ignition and subsequent treatment.

Procedure. Cover the platinum dish with a large clock glass, and by displacing it slightly, carefully add 15 ml of concentrated hydrochloric acid. Replace the cover and allow to stand for a few minutes until all vigorous action has ceased.

Add 5 ml of concentrated hydrochloric acid to the platinum crucible used for the decomposition of the sample, cover with a small clock glass and transfer to a steam bath for 10 minutes. Allow to cool and then rinse the contents into the large platinum basin with a jet of water. Carefully wipe the crucible out with damp filter paper to remove all traces of silica adhering to it, add these pieces of paper to the solution in the dish. If the same crucible is to be used for the ignition of the silica precipitates, ignite the crucible over a Meker burner, allow to cool, weigh after exactly 25 minutes, and then set it aside until required. (*Note:* platinum–iridium crucibles although suitable for sodium carbonate fusions should not be used for ignitions as they tend to lose weight at high temperatures.)

Remove the cover, rinse down with water and then replace it over the dish. Transfer the dish to a steam bath and heat until no further effervescence is apparent, then rinse down and remove the clock glass, and evaporate the solution to dryness on the steam bath. As the last traces of water and acid are removed, the deep yellow colour of the residue is replaced by a paler tint. When this stage is complete, test for complete expulsion of hydrogen chloride with the stopper of an ammonia bottle. Leave the dish on the steam bath for 30 minutes after fumes of ammonium chloride can no longer be detected.

Remove the dish from the steam bath, allow to cool and add 10 ml of concentrated hydrochloric acid, tilting the dish to ensure that all the residue is wetted with acid. Rinse down the clock glass and the sides of the dish, adding sufficient water to give a total volume of about 100 ml. Stir the solution with a stout glass rod and warm on a steam bath until all soluble salts have dissolved, leaving only a gelatinous residue of silica.

Collect the silica residue on a small medium-textured filter paper and wash at least six times with cold water and twice with hot water, to remove all soluble chlorides from the residue. Rinse the filtrate and washings back into the large platinum dish, transfer to the steam bath and again evaporate to dryness. As the evaporation proceeds, break up all crystals of sodium chloride with the flattened end of a glass rod. When all moisture and hydrochloric acid have been removed, transfer the dish to an electric oven and dry at a temperature of 105° to 110° for 1 hour.

Moisten the residue with 10 ml of concentrated hydrochloric acid and dissolve the chloride salts in about 100 ml of water as before. Collect the small residue, consisting largely of silica on a small close-textured filter paper and wash first with cold, then hot water, as described above. Carefully wipe the large platinum dish with wet filter paper to collect any silica adhering to the dish, and add this paper to the residue in the filter funnel before washing. Reserve the combined filtrate and washings for the subsequent analyses.

IGNITION AND VOLATILISATION OF SILICA

The silica residues contain small amounts of iron, aluminium and even smaller amounts of other elements of the ammonia group—titanium, zirconium and phosphorus. Calcium, magnesium, strontium and barium are not likely to be present, and if the washing has been correctly and adequately performed, the alkali elements are also unlikely to be present.

The total weight of the silica residue is determined after ignition in platinum. Silica is then removed by evaporation with hydrofluoric and sulphuric acids:

$$SiO_2 + 4HF = SiF_4 + 2H_2O$$

Iron, aluminium and other elements present in small amounts are converted to sulphates, but on strong ignition these are again converted to oxides. The difference in weight corresponds to the silica lost in the evaporation with hydrofluoric acid. A small correction arising from the presence of a trace of silica in the filter papers, and from involatile residue in the hydrofluoric acid, should be determined. With present grades of hydrofluoric acid, this correction should be very small, amounting to no more than about 1 mg. Before calculating the silica content of the sample material, the "silica traces" must be added. These traces are recovered from the ammonia precipitate at a later stage of the analysis.

Procedure. Clean, ignite over a Meker burner and weigh a platinum crucible of about 25–30-ml capacity. Transfer to it the moist filter papers containing the silica residues and dry carefully over a small flame or in an electric oven. Allow to cool, moisten with 4 or 5 drops of 20 N sulphuric acid, and continue the heating over a low flame, burning the paper away and giving a white residue. Transfer the crucible to a Meker burner, cover with a platinum lid—slightly displaced—and heat strongly for 20 minutes. Allow to cool in a desiccator and weigh after exactly 25 minutes. Repeat the ignition for a period of 10 minutes, cooling and weighing as before; repeat the ignition as necessary to obtain constant weight.

Moisten the residue with 1 ml of water and add 5 drops of 20 N sulphuric acid and 10 ml of concentrated hydrofluoric acid. Transfer the crucible to a hot plate and evaporate the silica and excess hydrofluoric acid. Raise the temperature towards the end of the evaporation to remove free sulphuric acid. Allow to cool. Transfer the crucible to a silica triangle and heat over a Bunsen burner to decompose sulphates, and then over a Meker burner until constant weight is obtained. The loss in weight is the uncorrected main fraction of the silica. Set the crucible aside for the ignition of the ammonia precipitate.

To determine the correction, transfer filter papers equal in number to those used in the silica determination, to a clean, weighed platinum crucible, burn off the carbon and ignite over a Meker burner. Allow the crucible to cool and then weigh the residue obtained. This gives the weight of filter paper ash. Now add 5 drops of 20 N sulphuric acid and 10 ml of concentrated hydrofluoric acid, transfer the crucible to a hot plate and evaporate the hydrofluoric and sulphuric acids as with the silica evaporation. Finally ignite over a Meker burner, cool and weigh. There is usually a small increase in weight after the ignition, corresponding to the arithmetic total of a small loss by volatilisation of silica from the filter paper ash, combined with a gain in weight from the non-volatile residue in the hydrofluoric acid. The overall increase in weight must be *added* to the silica value previously obtained.

The Determination of "Mixed Oxides"

PRECIPITATION OF THE "MIXED OXIDES"

The mixed oxides are precipitated in the filtrate from the silica determination by adding ammonia to the hot solution until it is just alkaline to methyl red or bromocresol purple indicator, i.e. at a pH of about 7. Iron, aluminium, phosphorus, zirconium, vanadium and chromium are precipitated together with a number of other elements present in only minor or trace amounts including beryllium, gallium, indium, thorium, scandium and the rare earths. The very small amounts of nickel, cobalt and zinc present in most silicate rocks are not precipitated. but accompany calcium, strontium and magnesium into the filtrate. If nickel is present in more than trace amounts, some will be caught in the ammonia precipitate.[4] Some small amount of calcium and magnesium will be entrained in the ammonia precipitate, but these amounts are recovered by dissolving the precipitate in dilute hydrochloric acid and reprecipitating with ammonia.

Although the bulk of the aluminium is precipitated with ammonia, some small amount is found in the filtrates ("aluminium traces"), from which it can be recovered and added to the ammonia precipitate.

Small amounts of manganese are not usually precipitated with ammonia, but pass into the alkaline filtrate and are subsequently precipitated as phosphate with the magnesium. Larger amounts of manganese are divided between the two fractions, the major part with the magnesium. It has been noted that manganese will be precipitated with the elements of the ammonia group if oxidising agents are added. Bromine water has been used for this by Holt and Harwood[5] and ammonium persulphate by Hillebrand.[6] Neither technique gives complete precipitation of manganese.[1]

The use of oxidising agents such as persulphate converts chromium into a higher valency state which is then not precipitated with ammonia. Moreover, the addition of oxidising agents increases the difficulty of adjusting the pH of the solution. In the author's opinion, the simplest and best procedure is not to attempt to precipitate manganese with the ammonia group, but to collect most of it with the magnesium and then determine it photometrically in the phosphate residue. This, together with a determination of total manganese in a separate portion, enables the minor fraction in the ammonia precipitate to be calculated. The amount of manganese collected in the oxalate precipitate can generally be ignored. There is little point in removing manganese from the filtrate after the precipitation of the ammonia group by treatment with bromine water, or by co-precipitation with zirconium, as described by Peck and Smith[7] as the small amount of manganese incorporated in the ammonia precipitate would still remain to be determined.

Procedure. Add 5 ml of concentrated hydrochloric acid to the combined filtrate and washings from the removal of silica (for rocks rich in magnesium, add 10 ml), and concentrated ammonia solution until the formation of a precipitate that only just dissolves on stirring. Heat the solution just to boiling, add a few drops of bromocresol purple indicator solution and continue adding ammonia until complete precipitation is obtained, and the supernatant liquid is purple in colour. Avoid adding an excess of ammonia.

With rocks containing much iron, the end point will be obscured, and the addition of ammonia is best continued until the precipitation is essentially complete before adding single drops of indicator solution. The colour of the indicator can then be observed as it mixes with the solution. Bring the solution to the boil again and allow the precipitate to settle somewhat. If the supernatant liquid is not purple in colour, add more ammonia, drop by drop, until the purple colour is restored. Stir in a macerated filter paper and allow to stand for 1 minute.

Collect the residue on a large open-textured filter paper and wash six times with a wash solution containing 20 g of ammonium nitrate per litre and made just alkaline with ammonia to bromocresol purple indicator. Reserve the filtrate and washings. Transfer the filter paper and residue back to the beaker in which the precipitation was made, add 50 ml of water and 15 ml of concentrated hydrochloric acid. Cover the beaker with a clock glass and digest on a steam bath until complete dissolution is obtained, then dilute to 250–300 ml with water. Again precipitate with ammonia as described above, collect the ammonia precipitate on a large open-textured paper and wash as before. Combine the filtrate and washings with those obtained from the first precipitation with ammonia, and reserve for the subsequent stages of the analysis. Transfer the filter to the platinum crucible previously used for the volatilisation of silica.

RECOVERY OF THE "ALUMINIUM TRACES"

The small amount of aluminium present in the filtrate after the precipitation with ammonia can be recovered by evaporating the solution to small volume and re-precipitating with ammonia. If ammonium persulphate has been added, chromium will also have passed into the filtrate, and will be collected with the aluminium traces.

Procedure. Add concentrated hydrochloric acid drop by drop to the combined filtrates and washings until the solution is just acid, then transfer the beaker to a steam bath or hot plate and evaporate to a volume of about 200 ml. Now add concentrated ammonia drop by drop until the solution is just alkaline to bromocresol purple, cover the beaker with a clock glass and digest on the steam bath for 10 minutes. The traces of aluminium form a small gelatinous precipitate. Collect this precipitate on a small open-textured filter paper, wash four or five times with the ammonium nitrate wash solution, dissolve in dilute hydrochloric acid and re-precipitate with ammonia. Collect the precipitate as before, wash well and add to the ammonia precipitate in the platinum crucible. Reserve the combined filtrates and washings for the determination of calcium and magnesium.

IGNITION OF THE "MIXED OXIDES"

The ammonia precipitate is ignited in the crucible used for the volatilisation of silica, and still containing the small amounts of iron, aluminium and other elements co-precipitated with the silica. Opinions differ as to the best temperature for the ignition. The conversion of ferric oxide to magnetite at temperatures in excess of 1100°, have led some authors to suggest that 1100° is the maximum that should be used. However it is known that alumina is not completely dehydrated at this temperature, and for this reason other authors have recommended 1200°. For ammonia precipitates consisting largely of alumina, this temperature of 1200° can safely be used, but for precipitates rich in iron, this temperature should be used only if an oxidising atmosphere can be ensured.

Procedure. Dry the filters in the platinum crucible, and using a Bunsen burner with a low flame, burn off the carbon at a low temperature. Increase the flame and ignite in the uncovered crucible over the full flame of the burner for 30 minutes. Transfer the crucible to an oxidising muffle furnace and ignite to constant weight at a temperature of 1200°. If an oxidising muffle is not available, use a blast Meker burner, but as far as possible keeping the flame away from the upper parts of the partly covered crucible.

RECOVERY OF "SILICA TRACES"

The ignited mixed oxides are particularly refractory after ignition at a temperature of 1200°. They can however be brought into solution following a fusion with potassium pyrosulphate. A small amount of silica known as the "silica traces", is usually recovered from the mixed oxides and can be filtered off and determined as before. If required, iron, titanium, vanadium and phosphorus can be determined photometrically in the sulphate solution although, as noted earlier, these determinations are better made on separate portions of the sample material.

Procedure. After weighing the mixed oxides, add 6–7 g of potassium pyrosulphate, cover the crucible with a close-fitting platinum lid and fuse gently over a small Bunsen flame for a protracted period of at least an hour. Too high a flame should not be used, as this results in the rapid loss of sulphur trioxide from the melt. Finally heat the crucible over a full Bunsen flame for 5–10 minutes and allow to cool. Place the crucible on its side in a 250-ml beaker and add 100 ml of 4 N sulphuric acid. Cover the beaker with a clock glass, transfer to a steam bath and digest until the solid melt has completely disintegrated. Rinse and remove the crucible and lid and also the clock glass.

Transfer the beaker to a hot plate, evaporate to fumes of sulphuric acid and allow to fume copiously for 10 minutes. Allow to cool. Cautiously dilute with about 100 ml of water and digest on a steam bath until all soluble material has passed into solution. At this stage a clear solution should be obtained, in which a few milligrams of silica are visible as a slight residue. Collect this residue on a small medium-textured filter paper, wash with cold water and determine the silica by volatilisation with hydrofluoric and sulphuric acids as described previously. Combine the filtrate and washings and dilute to volume in a 200-ml volumetric flask for the determination of total iron, titanium, etc., if these are required.

DETERMINATION OF ALUMINIUM BY DIFFERENCE

The aluminium content of the rock material, as determined by the difference method, is obtained by subtracting from the total of mixed oxides expressed as a percentage of the sample taken, the combined total of other elements present, each determined separately. These elements include total iron, calculated as Fe_2O_3, titanium as TiO_2, phosphorus as P_2O_5, vanadium as V_2O_3, chromium as Cr_2O_3, zirconium as ZrO_2, that part of the manganese present in the residue, calculated as Mn_3O_4 and "silica traces" as SiO_2. Other elements in the ammonia precipitate are seldom present in amounts sufficient to be totalled in this way.

The Determination of Calcium

Calcium is separated from the filtrates remaining after the collection of the "aluminium traces", by precipitation as calcium oxalate. Much of the strontium present in the rock material will also be precipitated.

Although the classical scheme for silicate rock analysis includes provision for the separation and separate determination of strontium, this method is no longer considered adequate, and determination by atomic absorption spectroscopy in a separate portion of the rock material is recommended, see p. 427. The amount of strontium present in most silicate rocks does not introduce significant errors into the calcium determination by its precipitation as oxalate, and for most rocks, the determination of co-precipitated strontium is not necessary.

Procedure. The combined filtrates and washings from the removal of the aluminium traces should have a volume of about 300 ml. Make this solution just alkaline to bromocresol purple indicator and heat to boiling. Add a solution of 5 g of ammonium oxalate in 100 ml of hot water, bring to the boil again, digest on a steam bath for 10 minutes and then allow to stand overnight.

Collect the precipitated calcium oxalate on a close-textured filter paper and wash five times with a cold solution containing 1 g of ammonium oxalate per litre. Reserve the filtrate and washings. Rinse the precipitate back into the beaker used for the precipitation and dissolve by warming with 5 ml of concentrated hydrochloric acid and 100 ml of water. Heat the solution to boiling and filter back through the paper, collecting the filtrate in a clean 400-ml beaker and washing the paper seven or eight times with warm water.

Add 2 g of ammonium oxalate dissolved in about 50 ml of warm water followed by concentrated ammonia solution until a slight precipitate forms that does not redissolve on stirring. Clear this precipitate with 2 drops of concentrated hydrochloric acid, heat the solution to boiling and precipitate calcium oxalate by adding 4 N ammonia until the solution is just alkaline to methyl red or bromocresol purple indicator. Transfer the beaker to a steam bath for 30 minutes, then allow to cool and stand for 3–4 hours or overnight. Collect the precipitate on a close-textured filter paper and wash six times with the ammonium oxalate wash solution. Combine the filtrate and washings with those obtained from the first precipitation of calcium oxalate, and reserve for the determination of magnesium.

Transfer the filter paper and contents to a clean, ignited and weighed platinum crucible and dry in an electric oven set at 105°, or over a very low flame. Burn off the filter paper by heating over a Bunsen burner at dull red heat and then ignite over the full flame of a Meker burner. Cool, and weigh as calcium oxide, CaO. Calcium oxide residues are hygroscopic and if a muffle furnace is available, as an alternative method the crucible can be ignited at a temperature of 500°, and the residue weighed as calcium carbonate.

The Determination of Magnesium

Magnesium is determined by precipitation as magnesium ammonium phosphate hexahydrate, and weighing as magnesium pyrophosphate:

$$2Mg(NH_4)PO_4.6H_2O = 13H_2O + Mg_2P_2O_7 + 2NH_3$$

Any manganese present in the solution will also be precipitated as an ammonium phosphate and be ignited and weighed as pyrophosphate, $Mn_2P_2O_7$. Other contaminants of the magnesium pyrophosphate include barium and traces of strontium and calcium not collected in the oxalate precipitate.

Although some analysts precipitate magnesium in the presence of the large amounts of ammonium salts that have been added to the rock solution in the course of the earlier separations, this cannot be recommended, as the ammonium salts tend to prevent the precipitation of small amounts of magnesium, and to give incomplete precipitation with larger amounts. In the procedure described, these are removed by evaporation with concentrated nitric acid.

The ignition of magnesium pyrophosphate is one of the most difficult operations in the classical scheme of silicate rock analysis. If the burning off is done at too high a temperature, some of the pyrophosphate may be reduced and the precipitate invariably becomes impregnated with carbon, which is then very difficult to burn off. The composition of the precipitate does not always correspond exactly to the composition $Mg(NH_4)PO_4.6H_2O$—some $Mg(NH_4)_4(PO_4)_2$ may be included. On ignition this forms $Mg(PO_3)_2$, which can only be converted to pyrophosphate by ignition at temperatures in the range 1150° to 1200°. At these temperatures the pyrophosphate itself slowly loses P_2O_5. If the magnesium ammonium phosphate precipitate contains traces of calcium or other elements, the ignition may cause fusion of the residue, even at temperatures as low as 1000°.

Procedure. Combine the filtrates and washings from the calcium oxalate precipitations in a large beaker and evaporate on a steam bath to give a solution volume of about 200 ml. Rinse this into a 600-ml beaker, allow to cool and cover with a clock glass. Displace the clock glass slightly and add 100 ml of concentrated nitric acid. Replace the cover glass and return the beaker to the steam bath, heating until all reaction has ceased. Rinse and remove the cover glass and evaporate to dryness.

Dissolve the residue in about 100 ml of water, acidifying with a few drops of hydrochloric acid if necessary. Make the solution just alkaline to methyl red or bromocresol purple indicator, digest on a steam bath for 10 minutes and collect any small residue on an open-textured filter paper. Wash this residue with the ammonium nitrate wash solution, transfer to a platinum crucible, dry, ignite and weigh. The weight of residue should not exceed 2 mg, and is mostly alumina. Dilute the filtrate to about 250 ml with water and make just acid with hydrochloric acid. Add 6 g of diammonium hydrogen phosphate dissolved in about 50 ml of water, followed by 30 ml of concentrated aqueous ammonia, added with continuous stirring. Cover the beaker with a clock glass and set it aside overnight, preferably in a refrigerator. If the sample material contains only very small amounts of magnesium, allow to stand for 48 hours.

Collect the precipitate on a close-textured filter paper and wash six times with N ammonia solution (approximately 50 ml of concentrated ammonia per litre). Reserve the combined filtrate and washings. Rinse the precipitate back into the beaker used for the precipitation, add 1 ml of concentrated hydrochloric acid and 50 ml of water and warm to give a clear solution. Filter this solution back through the filter paper used for the first precipitate, and wash well with water. Collect the filtrate and washings in a 400-ml beaker and dilute to about 200 ml with water. Add 1 g of diammonium hydrogen phosphate dissolved in a little water, add then concentrated aqueous ammonia drop by drop until the precipitation of magnesium appears to be complete (or until the solution is alkaline, if only traces of magnesium are present), and then 10 ml of concentrated ammonia. Allow the solution to stand overnight.

Collect the precipitate on a close-textured filter paper and wash with dilute ammonia solution as before. The filtrate and washings can be combined with those obtained from the first precipitation of magnesium, and used for the determination of nickel, if this is required. Otherwise these filtrates and washings are discarded. Transfer the paper containing the magnesium ammonium phosphate precipitate to a clean, ignited and weighed platinum crucible and heat carefully over a Bunsen burner to dry the precipitate, char the paper and burn off the carbon at as low a temperature as possible. Heat over a full Bunsen flame until a completely white residue is obtained, then over a Meker burner or in an electric furnace set at a temperature of 1050° to obtain constant weight. Weigh as $Mg_2P_2O_7$.

DETERMINATION OF MANGANESE IN THE MAGNESIUM RESIDUE

The ignited magnesium pyrophosphate residue is fully soluble in dilute hydrochloric acid only with some difficulty. It is, however, soluble in concentrated sulphuric acid, to give a solution that can be readily used for the photometric determination of manganese.

Procedure. Moisten the magnesium pyrophosphate residue with 5 ml of water, add 2 ml of concentrated nitric acid and 5 ml of 20 N sulphuric acid. Transfer the crucible to a hot plate and evaporate to fumes of sulphuric acid. Allow to cool, dilute with water and rinse into a small beaker for the photometric determination of manganese by oxidation to permanganate with potassium periodate as described on p. 329.

References

1. JEFFERY P. G. and WILSON A. D., *Analyst* (1959) **84**, 663.
2. HOFFMAN J. I., *J. Res. Nat. Bur. Stds.* (1940) **25**, 379.
3. CHIRNSIDE R. C., *J. Soc. Glass Technol.* (1959) **43**, 5T.
4. HARWOOD H. F. and THEOBALD L. S., *Analyst* (1933) **58**, 673.
5. HOLT E. V. and HARWOOD H. F., *Mineral. Mag.* (1928) **21**, 318.
6. HILLEBRAND W. F. and LUNDELL G. E. F., *Applied Inorganic Analysis*, 2nd ed., Wiley, New York, 1953, p. 870.
7. PECK L. C. and SMITH V. C., *U.S. Geol. Surv. Prof. Paper* 424-D, p. 401, 1961.

CHAPTER 5

THE RAPID
ANALYSIS OF SILICATE ROCKS

Introduction

The complete chemical analysis of silicate rocks using classical methods is very time consuming and requires the services of a skilled analyst. This means that such analyses are expensive. A geologist studying a particular exposure is likely to be restricted by cost to those analyses that are illustrative rather than to those that are informative. Any steps towards cheaper and more rapid silicate analyses are therefore likely to be welcomed by both the field geologist and petrographer. The rock analyst, familiar with the difficulties and inaccuracies of the classical scheme, is also likely to welcome the opportunity of using newer methods of analysis, although he may not entirely agree with some of the ideas, particularly those incorporated in some of the earlier proposed schemes.

Classical rock analysis is based firmly upon the technique of gravimetric analysis. This itself is time consuming, but is rendered all the more so by the use of a particular sequence of lengthy precipitations, many of which have to be repeated to obtain even reasonably quantitative separation. Schemes for the rapid analysis of silicate and other rocks are based upon a replacement of these gravimetric procedures with other techniques that are simpler and more rapid in themselves. Even more important than this, these determinations are frequently made in the presence of other elements, avoiding lengthy and tedious separation stages.

What is surprising in some of the earlier schemes is that the authors, when compelled to use separations, have failed to take advantage of newer methods—ion exchange or solvent extraction, for example—but preferred to retain the imperfect separations used in the classical scheme, and sometimes even use them in a way that no classical rock analyst would contemplate or tolerate. This can be illustrated by a single example quoted by Chirnside.[1] What classical analyst would make a single precipitation of the elements of the ammonia group by adjusting the pH to $4 \cdot 8$–$5 \cdot 1$ by dropwise addition of ammonia, and then hope to determine the whole of the calcium in the filtrate? (Shapiro and Brannock,[2] original scheme for rapid rock analysis.)

The classical scheme for the analysis of silicate rocks as we understand it today was not designed, it is the result of continuous evolution. Schemes for the rapid analysis of silicate rocks are also undergoing this process of evolution, and too much time should not be spent in considering and condemning the earlier attempts, at the expense of examining the latter schemes.

In some of the published reports too much attention has, in the author's view, been paid to speed and to the grade of analyst necessary. The use of less-skilled analysts invariably produces inferior results, and even the more-skilled analysts would have difficulty with some of the published schemes. Speed in rock analysis, whatever scheme is used, is largely a function of the organisation of the work, and detailed instructions to the analyst as to how he should arrange his day are largely wasted efforts. If the analyst can organise his work he will, if he cannot, then detailed time tables are useless.

The points to look for in any new scheme for rock analysis must surely be these:

A. Separation procedures should be avoided wherever possible; the determination of individual components should be made in the presence of the remaining components.
B. The method used for determining an individual component should be specific for that component and give results of acceptable accuracy and precision.
C. The number of sample portions taken for the analysis should be reduced to the minimum.
D. The scheme should be capable of application to a wide range of silicate rocks, and no commonly occurring silicate rock should give inaccurate, misleading, or ambiguous results by the methods as described.

Thus the considerable amount of effort spent in devising new schemes for the analysis of silicate rocks based upon complex ion formation and ion-exchange separation,[3-5] although useful, does not appear to be in the main stream of development in rapid, instrumental methods of rock analysis.

Schemes that have been introduced have a number of features in common. Firstly, although provision is made for the determination of a total of thirteen most abundant components of silicate rocks, "moisture", total water and ferrous iron are determined in separate portions of the rock material. The methods of analysis adopted for these determinations are almost always the same as those used in the classical scheme.

Secondly, instrumental methods are employed wherever possible. These include flame photometric procedures for sodium and potassium and spectrophotometric procedures for silicon, aluminium, total iron, titanium, manganese and phosphorus. A number of different titrimetric procedures have been described for determining calcium and magnesium. Many of these are likely to disappear, being replaced largely by newer methods based upon atomic absorption spectroscopy.

Thirdly, although there are notable exceptions to this, most of the schemes for rapid analysis rely upon the decomposition of two sample portions for the determination of the ten components. Silicon and some-times also aluminium are determined in one fraction, a small sample weight decomposed by fusion with sodium hydroxide, whilst iron, titanium, manganese, phosphorus, calcium and magnesium are deter-mined in a somewhat larger sample portion decomposed by evaporation with hydrofluoric acid.

Finally, the techniques themselves are suitable for "batch operation", that is six or eight rock samples can be processed at one time with very little extra time to that required for processing a single sample. To the beginner in rock analysis, or to the analyst who only occasionally is called upon to undertake this kind of work, the opportunity of includ-ing a previously analysed sample can be most valuable in preventing gross errors, such as those arising from faulty mathematics.

Although there are considerable variations in detail between one scheme for the rapid analysis of silicate rocks and another, in outline at least there is a striking uniformity. Thus the details shown in Fig.5, although taken from a later version of the scheme by Shapiro and Brannock,[6] apply equally well to the scheme of Blanchet and Malaprade[7] and have more than a passing resemblance to those of Corey and Jackson,[8] Riley,[9] Riley and Williams[10] and others.

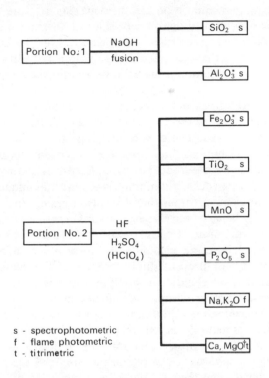

FIG. 5. Scheme for the rapid analysis of silicate rocks. *Also by titrimetry, †also by photometry.

Portion no. 1. This portion, used by Shapiro and Brannock[6] for the
determination of both silica and aluminium, and by Riley[9] for silica
only, is prepared by fusing a small sample weight with sodium hydroxide
in a nickel crucible. This operation is described in detail on p. 28. As
silica is always present in major amounts, and is frequently greater than
all the remaining elements put together, great care is necessary if
accurate results are to be obtained, both in the measuring of volumes
and optical densities and in the preparation of the solution. This is
especially necessary at the fusion stage where loss by "dusting" can
occur. Even when this great care is taken, the accuracy is limited by
the inherent inaccuracies of spectrophotometric methods, although

some improvement can be obtained by making replicate analyses and by taking a series of readings on each sample.

Most authors have described molybdenum blue methods for silica, mostly resembling that described in detail on p. 410, although methods based upon the formation of the yellow silicomolybdate have also been suggested.

There is more variety in the methods used for aluminium, which can be determined in either the first or the second portion of rock material taken for the analysis. These methods range from photometric determination with aluminon,[8] alizarin red-S,[6] or 8-hydroxyquinoline,[9] to titrimetric methods based upon complex formation with EDTA.[11] With many silicate rocks, aluminium is the most abundant constituent after silicon, and great care is therefore also required in making precise measurements. None of the reagents so far suggested is specific or even selective for aluminium. For this reason some authors separate aluminium from interfering elements,[8] whilst others add complexing reagents to limit or prevent this interference. The methods adopted in some schemes of rapid rock analysis for this determination have received considerable criticism, and a great deal more work is necessary before the determination of aluminium can be made as easily, precisely and accurately as many of the others.

Portion no. 2. This portion is used for determining total iron, titanium, manganese, phosphorus, calcium, magnesium and the alkali metals. It is prepared for the analysis by evaporation with hydrofluoric and either sulphuric or perchloric acid. Difficulties arise when, as frequently happens, a residue remains after this treatment. Oxide minerals such as chromite, rutile, or corundum, do not contain appreciable amounts of alkali metals, and no significant error will be introduced into their determination if such residues are discarded. This unattacked portion can, however, contain quite an appreciable amount of the total titanium content of the rock, together with significant amounts of other minor constituents. Iron, titanium and these other minor elements can be recovered following a fusion of the residue with sodium carbonate or potassium pyrosulphate.

Resistant silicate minerals containing alkali metals are more difficult to prepare for analysis and Riley[9] has recommended a procedure based upon heating with hydrofluoric and perchloric acids in a PTFE (polytetrafluoroethylene) bomb at a temperature of 150°. Somewhat similar procedures have been suggested using only hydrofluoric acid in PTFE or platinum-lined bombs by Langmyhr and Graff[11] and by May and Rowe[12] for the decomposition of this portion of the silicate rock.

Photometric methods are commonly used for the determination of titanium, manganese and phosphorus, elements that are present in small or minor amounts, where the accuracy and precision of spectro-photometric methods is adequate. Hydrogen peroxide is the most frequently recommended reagent for titanium, but it is of barely adequate sensitivity. Tiron (catechol-3:5:disulphonic acid) and dianti-pyrylmethane are a great deal more sensitive, and are therefore better for acidic and intermediate rocks which contain only small amounts of titanium. Manganese is determined as permanganate after oxidation with either potassium periodate or ammonium persulphate. The two methods commonly used for phosphorus are based upon the formation of a yellow phosphovanadomolybdate and upon a molybdenum blue given by the reduction of phosphomolybdate respectively. All these methods are described in detail later in this book.

Although photometric methods are frequently recommended for the determination of total iron, the precision obtainable is barely adequate for basic and other rocks rich in ferrous or ferric iron. For these a titrimetric method using potassium dichromate, potassium permanga-nate or ceric sulphate solution provides an acceptable alternative to the photometric methods suggested. For rocks containing only small amounts of iron, photometric methods using 2:2′-dipyridyl or 1:10-phenanthroline are preferred to thioglycollic acid,[13] hydrochloric acid,[2] tiron, salicylic acid or other reagents suggested for this application. A titrimetric method is still preferred for iron when present in major amounts.

In schemes of rapid rock analysis, the alkali metals are usually determined by flame photometry, although atomic absorption methods have since become popular. For calcium and magnesium, the EDTA procedures suggested have required a great deal of experience and skill if correct results are to be obtained. The end point of the calcium determination in particular is very subjective, and far from repeatable in the presence of much iron and manganese. For certain applications, particularly carbonate rocks and silicate rocks rich in calcium and magnesium and poor in iron and manganese, EDTA titrimetric methods are still preferred.

The unsatisfactory nature of some of the determinations in these schemes of rapid rock analysis has long been apparent—photometric methods for aluminium and titrimetric methods for calcium and magnesium being the most deserving of criticism. It is not surprising therefore that the introduction of atomic absorption spectroscopy, with its promise of providing alternative methods for the determination of at least some of the major components of silicate rocks, was seized as the basis for yet further alternative schemes of rock analysis, rapid and otherwise.

The decomposition procedures introduced for the earlier combined spectrophotometric/titrimetric schemes, and gradually improved and refined by later workers, were found to be adaptable to this new technique, which now appears to have displaced entirely some of the earlier rapid techniques in some laboratories—titrimetric calcium and magnesium in particular. As with other schemes, rapid rock analysis based upon atomic absorption must be supplemented with the more traditional methods for "moisture", total water and ferrous iron.

The determination of certain elements present to a minor or trace extent in silicate rocks is particularly sensitive by this technique. This has led to their inclusion in some of the proposed schemes. These include vanadium, chromium and zinc, elements that are not always present in silicate and other rocks in amounts sufficient to be observed and recorded by this technique in its present state of development.

Individual Schemes of Rapid Rock Analysis
SCHEMES OF SHAPIRO AND BRANNOCK[2, 6, 14]

Although not the first of their kind, the schemes of Shapiro and Brannock probably did more to whet the appetite of the geologist for this type of analysis than those of either their predecessors or their imitators. Criticism of certain of the methods has led to a great improvement in both technique and in the methods selected, although most of the procedures are now obsolete.

SCHEME OF RILEY[9]

This scheme, the outline of which is shown in Fig. 6, differs from others proposed at or before that time in making good use of modern separation processes where they could conveniently contribute to the precision and accuracy of the determination. Thus for example before titrating calcium and magnesium with EDTA, the interfering elements are removed by extraction of their complexes with 8-hydroxyquinoline into chloroform. Similarly an anion exchange separation is used to remove iron, aluminium and titanium before the flame photometric determination of the alkali metals. Riley's scheme has been widely followed, possibly because the methods used were capable of somewhat greater accuracy and precision than some of the other methods suggested, but possibly also because a wide range of silicate rocks and minerals could be analysed without difficulty by following the detailed instructions given.

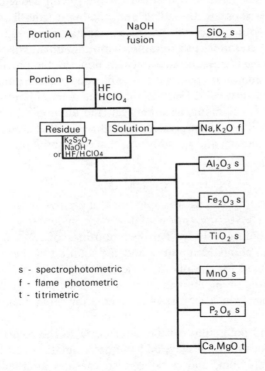

Fig. 6. Scheme for the rapid analysis of silicate rocks (Riley, *Anal. Chim. Acta* (1958) **19**, 413).

SCHEME OF LANGMYHR AND GRAFF[11]

This scheme is more recent than Riley's, and has yet to become as widely known. It is based upon the use of two sample portions for the total of ten constituents, but both portions are decomposed with hydrofluoric acid. The first portion, for silica only, is decomposed in a closed PTFE vessel at an elevated temperature and pressure with hydrofluoric acid alone. After adding aluminium chloride to complex the excess hydrofluoric acid, the silica is determined photometrically as the yellow silicomolybdate. The second portion is decomposed in an open PTFE vessel with a mixture of hydrofluoric and sulphuric acids. The sulphate solution obtained is used directly for the photometric determination of total iron, titanium, manganese and phosphorus, and to provide a reagent blank solution for the silicon determination.

In order to separate the alkali and alkaline earths from iron, aluminium and other elements of the ammonia group, Langmyhr and Graff employ a double precipitation with ammonia. The value of this separation stage is questionable since, as noted in the previous chapter, some of the aluminium passes into the filtrate from the ammonia precipitation, giving low values for the aluminium determination. Moreover, the distribution of manganese between the ammonia precipitate and the filtrate will certainly present difficulties in the subsequent determination of calcium, and possibly also that of aluminium.

SCHEME OF INGAMELLS[15, 16]

A considerable advance in rapid rock analysis was that introduced by Ingamells, involving fusion of the rock material with anhydrous lithium metaborate ($LiBO_2$). The decomposition of a $0 \cdot 1$–$0 \cdot 2$ g portion is complete in about 10 minutes, and the solution of the melt in nitric acid can be used for the photometric determination of silicon, aluminium, total iron, titanium, manganese and phosphorus, as well as nickel and chromium (which feature in a scheme for rapid rock analysis for the first time).

Sodium and potassium can be determined in the solution by flame photometry in the usual way, and the determination can be extended to include rubidium, and possibly also caesium after adding potassium.[17] Shapiro[18] has carried this scheme a stage further and used the acid solution of the melt for the determination of calcium and magnesium (and also sodium and potassium) by atomic absorption spectroscopy. If an emission spectrograph is available, then strontium, barium, chromium, copper, zinc, nickel and zirconium can all be determined in this same solution using a rotating wheel electrode.[17]

Although platinum crucibles were originally suggested for this fusion, the melt adheres to the metal and can only be removed from the crucible with difficulty. Shapiro[18] has recommended using new graphite crucibles, as the melt does not then adhere to the graphite and can easily be removed for dissolution in dilute mineral acid. These crucibles are, however, expensive. In order to prevent the polymerisation of silica in the acid extract, the silica concentration should not exceed about 150 ppm. This means that all other constituents of the rock material will also be present in the solution at high dilution, and particularly sensitive methods of determination are required. For example, the phosphovanadomolybdate method for phosphorus used by Shapiro and Brannock[6] is no longer sufficiently sensitive, and is replaced by a molybdenum blue method. The following detailed instructions for the preparation of the rock solution have been adapted from the work of Shapiro[18] and Ingamells.[16]

Procedure. Accurately weigh approximately 0·1 g of the finely powdered silicate rock material into a new graphite ("vitreous carbon") crucible and add 0·6 g of anhydrous lithium metaborate. Allow to stand before the open door of a muffle furnace set at a temperature of 1000° for a few minutes, and then insert into the furnace on a silica tray. Fuse for 15 minutes, then remove the silica tray containing the crucible and allow to cool. The melt does not wet the crucible, and it can be readily detached from the graphite.

Transfer the melt from the crucible to a 1500-ml polythene or polypropylene beaker containing approximately 950 ml of water and 15 ml of concentrated hydrochloric acid. The melt will dissolve slowly over a period of 2 to 3 hours, but this process may be hastened by mechanical stirring. When dissolution is complete, transfer the solution to a 1-litre volumetric flask and dilute to volume with water. If the determination of silica is not to be made immediately, transfer the solution to a clean, dry, polythene bottle for storage.

This solution can be used for the determination of silica by a molybdenum blue method (p. 410), aluminium by a calcium alizarin red-S method, [6] titanium using tiron (p. 471), total iron with 1:10-phenanthroline (p. 289) and phosphorus by a molybdenum blue method (p. 381). Calcium, magnesium, manganese, sodium and potassium can all be determined in the solution by atomic absorption spectroscopy, with lanthanum solution added as releasing agent.

Schemes based upon Atomic Absorption Spectroscopy

The gradual introduction of atomic absorption techniques, first for the alkaline earth elements calcium and magnesium, then for iron, manganese, sodium and potassium, paved the way for a number of schemes of rock analysis. Some of the more important are considered in greater detail below.

SCHEME OF BELT[19]

This, one of the earliest using atomic absorption spectroscopy, was developed before the widespread use of the high-temperature nitrous oxide-acetylene flame. A mixture of hydrofluoric, nitric and perchloric acids was used to effect a decomposition of the silicate matrix. The residue was taken up in hydrochloric acid, lanthanum added as releasing agent for calcium and magnesium in the presence of aluminium and phosphorus, and the determination of sodium, potassium, manganese, iron, magnesium and calcium made using an air–acetylene flame.

SCHEME OF BERNAS[20]

In this scheme a single portion of the rock material is used for the analysis, and the decomposition is accomplished with hydrofluoric acid in a sealed PTFE vessel held at a temperature of 110° for 30–40 minutes. Boric acid is added to complex excess fluoride ion and to dissolve insoluble fluorides. After dilution to a suitable volume, the solution is aspirated into a nitrous oxide–acetylene flame for the determination of silicon, aluminium, titanium, vanadium, calcium and magnesium, and into an air–acetylene flame for total iron, sodium and potassium. The concentration of all these elements is determined by reference to calibration graphs or by narrow range bracketing.

Although it is known that most silicate rocks and minerals are completely decomposed by this hydrofluoric acid digestion, further work would appear to be desirable on its effectiveness in respect of the wide range of accessory minerals that occur in such rocks.

In view of the extent to which vanadium occurs in many silicate rocks (q.v.), the inclusion of this element in a general scheme by using a much larger sample portion is curious, especially in view of the absence of the much more abundant manganese from the scheme.

No attempt is made to prevent inter-element effects by the addition of releasing agents, nor to suppress ionisation of certain elements by alkali addition. Bernas regarded his fluoborate system as having the "ability to compensate for inter-element effects and thus to eliminate interference phenomena". Whilst this may be true for many commonly occurring silicates, it is doubtful if it holds for the full range of silicate and other rocks that may be encountered.

SCHEME OF LANGMYHR AND PAUS[21]

A variety of decomposition procedures were used by Langmyhr and Paus in their scheme for the analysis of silicate rocks, which includes the determination of silicon, aluminium, titanium, calcium and magnesium with a nitrous oxide–acetylene flame and iron, manganese, sodium and potassium with an air–acetylene flame. The calibration is made by using a bracketing technique. Although the use of lanthanum as a releasing agent and alkali salts as ionisation suppressants are advocated in the paper, their use is not described in any detail.

SCHEMES OF ABBEY[22–25]

In a series of four monographs published by the Geological Survey of Canada, Abbey describes both a hydrofluoric-, nitric-perchloric acid decomposition and a lithium borate fusion for the determination of selected groups of elements present in silicate rocks. These are summarised in Table 5.

TABLE 5. SCHEMES FOR DETERMINING SELECTED GROUPS OF ELEMENTS

Reference	Decomposition procedure	Elements determined
22	$HF/HNO_3/HClO_4$	Lithium, magnesium, zinc, iron
23	$HF/HNO_3/HClO_4$	Iron, magnesium, calcium, sodium, potassium
24	Lithium fluoborate	Silicon, aluminium, iron, magnesium, calcium, sodium, potassium
25	$HF/HNO_3/HClO_4$	Barium, strontium, plus iron, magnesium, calcium, sodium, potassium (and lithium, rubidium and caesium by flame photometry)

In these schemes, strontium is added as releasing agent, particularly for the determination of magnesium in the presence of aluminium, to serve as an ionisation suppressant and to eliminate other less explicable inter-element effects. A final concentration of 3000 ppm Sr being recommended in the solution used to determine silicon, aluminium and small amounts of calcium and potassium, 1500 ppm Sr for the solution used for iron, magnesium, sodium and larger amounts of calcium and potassium. The choice of flame follows that of other workers: silicon, aluminium and small amounts of calcium using a nitrous oxide–acetylene flame, iron, magnesium, sodium, potassium and larger amounts of calcium using an air–acetylene flame.

The possibility of including both manganese and titanium in the scheme was noted, together with chromium, barium, and strontium if present in sufficient quantity. The procedure given below has been adapted from the third of Abbey's monographs, describing the preparation of solutions for the determination of the seven major constituents of silicate rocks referred to above.

Procedure. Accurately weigh approximately 0·2 g of the finely ground rock material into a platinum crucible and ignite at a temperature of 550–600° in a muffle furnace to remove all readily volatile material. Allow to cool and add 1 g of lithium metaborate (note 1). Cover the crucible with a platinum lid and heat in the muffle furnace to a temperature of 950°. Maintain at that temperature for 20 minutes. Remove the crucible and by rotating it allow the melt to solidify around the walls. Replace the crucible in the furnace for a further period of 5 minutes, then again remove and allow to cool whilst rotating the crucible to spread the melt.

Lay the crucible on its side in a 250-ml polyethylene beaker, add 60 ml of diluted (1+9) hydrofluoric acid and warm until all soluble material has passed into solution. Add 100 ml of boric acid solution (50 g per litre) and warm until the precipitated fluorides have passed into solution. Rinse and remove the crucible.

Transfer the solution to a 200-ml volumetric flask, dilute to volume with water, mix well and transfer to a polyethylene bottle. Pipette 50 ml of this sample solution into a 100-ml volumetric flask, add 20 ml of the strontium solution (note 2), dilute to volume with water, mix well and transfer to a polyethylene bottle. Use this solution for the determination of silicon, aluminium and both calcium and potassium if present in only small amounts.

Prepare a more dilute solution of the rock sample by transferring 10 ml of the sample solution to a 100-ml volumetric flask, add 10 ml of the strontium solution (note 2), dilute to volume with water, mix well and transfer to a polyethylene bottle. Use this solution for the determination of total iron, magnesium, sodium and both calcium and potassium if present in more than small amounts.

Notes: 1. Some batches of lithium metaborate have shown an anomalous behaviour on heating. Abbey reports that a sufficiently pure grade of lithium metaborate is now commercially available.

2. Some strontium carbonate supplies were found to contain too much calcium. Strontium nitrate was found to be satisfactory. Dissolve 36 g in water and dilute to 1 litre. Store in a polyethylene bottle.

A somewhat similar scheme to that of Abbey[24] involving decomposition by fusion with lithium metaborate was described by Van Loon and Parissis.[26] Silicon, aluminium, titanium, iron, calcium, magnesium, sodium, potassium and manganese are determined by atomic absorption spectroscopy, leaving only "moisture", total water, ferrous iron and phosphorus of the thirteen commonly occurring major constituents to be determined in the more traditional ways.

References

1. CHIRNSIDE R. C., *J. Soc. Glass Technol.* (1959) **43**, 5T.
2. SHAPIRO L. and BRANNOCK W. W., *U.S. Geol. Surv. Circ.*, 165, 1952.
3. OKI Y., OKI S. and HIDEKATA S., *Bull. Chem. Soc. Japan* (1962) **35**, 273.
4. SHIBATA H., OKI Y. and SAKAKIBARA Y., *Chishitsugaku Zasshi* (1960) **66**, 195.
5. MAYNES A. D., *Anal. Chim. Acta* (1965) **32**, 211.
6. SHAPIRO L. and BRANNOCK W. W., *U.S. Geol. Surv. Bull.* 1144-A, 1962.
7. BLANCHET M. L. and MALAPRADE L., *Chim. Anal.* (1967) **49**, 11.
8. COREY R. B. and JACKSON M. L., *Analyt. Chem.* (1953) **25**, 624.
9. RILEY J. P., *Anal. Chim. Acta* (1958) **19**, 413.
10. RILEY J. P. and WILLIAMS H. P., *Mikrochim. Acta* (1959) (4), 516.
11. LANGMYHR F. J. and GRAFF P. R., A contribution to the analytical chemistry of silicate rocks: A scheme of analysis for eleven main constituents based on decomposition by hydrofluoric acid, *Norges Geol. Undersokelse*, 230, Oslo, 1965.
12. MAY I. and ROWE J. J., *Anal. Chim. Acta* (1965) **33**, 648.
13. MERCY E. P. L., *Geochim. Cosmochim. Acta* (1956) **9**, 161.
14. SHAPIRO L. and BRANNOCK W. W., *U.S. Geol. Surv. Bull.* 1036-C, 1956.
15. INGAMELLS C. O., *Talanta* (1964) **11**, 665.
16. INGAMELLS C. O., *Analyt. Chem.* (1966) **38**, 1228.
17. SUHR N. H. and INGAMELLS C. O., *Analyt. Chem.* (1966) **38**, 730.
18. SHAPIRO L., *U.S. Geol. Surv. Prof. Paper* 575-B, p. 187, 1967.
19. BELT C. B. Jr., *Analyt. Chem.* (1967) **39**, 676.
20. BERNAS B., *Analyt. Chem.* (1968) **40**, 1682.
21. LANGMYHR F. J. and PAUS P. E., *Anal. Chim. Acta* (1968) **43**, 397.
22. ABBEY S., *Geol. Surv. Canada* Paper 67–37 (1967).
23. ABBEY S., *Geol. Surv. Canada* Paper 68–20 (1968).
24. ABBEY S., *Geol. Surv. Canada* Paper 70–23 (1970).
25. ABBEY S., *Geol. Surv. Canada* Paper 71–50 (1972).
26. VAN LOON J. C. and PARISSIS C. M., *Analyst* (1969) **94**, 1057.

CHAPTER 6

SOME STATISTICAL CONSIDERATIONS

AT ONE time the rock analyst was content to determine the constituents of silicate rocks one at a time, and to repeat the individual determinations only if the total for the sample fell outside the range 99·75 to 100·25 per cent. If the total was then less than 99·75 per cent, then and only then did the analyst start looking for chromium, nickel and other constituents that are occasionally present in minor amounts. It is now realised that a good total is not evidence of a good analysis,[1] and that negative errors (in the silica determination, for example) can be balanced by positive errors elsewhere (in the alumina content). Recently introduced procedures in which each component is determined separately in the same sample solution without recourse to extensive separations will do much to remove this balancing of errors. However, errors do arise in the course of making all determinations, and every analyst should know not only how they arise, but how to assess their magnitude and how they can be compared with those of other analysts.

This is much more easily said than done because, as silicate rock analysis is very time consuming, analysts simply do not have the time to carry out the grandiose schemes of replicate rock analyses necessary for rigorous statistical treatment. However, much can be done with the small numbers of repeat analyses that are often available for certain constituents. Fortunately the existence of international rock standards will enable rock analysts to avoid certain gross errors, and to improve their accuracy by processes of simple comparison.

Classification of Errors

An *error* can be defined as the discrepancy between the "true" value of any constituent and the result obtained experimentally, without regard to the magnitude or the reason for the discrepancy.

A *mistake* is an error produced by incorrect reading or using of a method. Examples of mistakes will be well known to most analysts; they include such incidents as omitting to reduce ferric iron before photometric determinations with 1:10-phenanthroline, using a 200-ml volumetric flask in place of a 250-ml flask, and the inability to subtract correctly one weighing from another. Mistakes are usually apparent by the magnitude of the discrepancy, and such results can frequently be discarded before a systematic statistical analysis of the remainder.

Random errors are those that arise whenever a subjective measurement is made. Examples of these include measurement of volume where the position of a meniscus is to be recorded, or of optical density where the position of a line or pointer against a scale is to be noted. Where a large number of determinations are made, the random errors will be both positive and negative and give an average value substantially free from such errors.

However, a random error introduced in the preparation of a standard solution—used, for example, for photometric or atomic absorption standardisation, will be repeated in the subsequent determinations made with that solution, i.e. this now becomes a systematic error.

Systematic errors or *bias* are errors that exist throughout a series of determinations, always in the same direction. They cannot be removed by averaging the results for the series, and the mean value will contain the same error. Systematic errors can be reduced by averaging the mean value with other values obtained by using independent methods.

The errors that arise in the course of a silicate rock analysis by classical methods have been considered by Chalmers and Page,[2] who, from the examples given, find that retention of more than three significant figures can hardly be justified in reporting the results. The retention of the second decimal place (4th significant figure for silica, usually also for alumina and sometimes for other constituents) has, however, been advocated by Chayes,[3] who pointed out that once rounded off, this figure cannot be reconstructed by the user.

Precision and Accuracy

After all that has been said and written about precision and accuracy, it is surprising that confusion still exists. *Accuracy* can be defined as the extent to which an analysis is capable of giving the "true" value for any component, i.e. $(X - \bar{x})$, where X is the "true" value and $\bar{x}$ the mean value $(= \Sigma \, x/n)$ of the results calculated from a series of n determinations. Accuracy, although easy to define, is difficult to measure, as the "true" value, X is seldom known, and frequently can only be inferred from other more rigorous methods of analysis.

Precision is the extent to which the results of a given series of determinations are scattered about the mean value. One way in which this scatter can be reported is to note the maximum and minimum values for the series, together with the calculated mean value. This does not give any idea of the way the results spread out or bunch together about the mean value, and it is preferable to calculate the parameter known as the *standard deviation, s*, given by

$$ s = \sqrt{\left(\frac{\Sigma \, (\bar{x} - x)^2}{\bar{n} - 1} \right)} $$

The standard deviation is an experimentally determined parameter and therefore relates to the results from which it has been calculated and not necessarily to any other set of results, even though they may have been produced by the same method. Any change of method, sample material, analyst or laboratory may affect the value for the standard deviation. The standard deviation is referred to by some authors as the *standard error*, but this term is best avoided as it has been used for the expression $(s/\sqrt{n})$ (standard error of the mean).

The expression given above is frequently used for calculation of the standard deviation of a given set of results, particularly when a new method of analysis is being evaluated or compared with an older or established method. It is not always possible, however, for schemes of replicate analyses to be carried out. In these cases the standard deviation can be calculated from duplicate analyses, where these have been carried out on a large number of samples (at least twenty-five and preferably more), by using the expression

$$ s = \sqrt{\left(\frac{\Sigma d^2}{2n} \right)} $$

where d is the difference between the duplicate results, and n the number of duplicates.

Other parameters used to compare one set of results with another include the *variance* (s^2), and the *coefficient of variation* given by $100 \, s/\bar{x}$, known sometimes as the *relative deviation* and denoted by the symbol C.

The relative deviation is used to compare two sets of results on a percentage basis.

What is sometimes confusing is that a series of results can be precise but inaccurate, or alternatively may be accurate but imprecise. An analyst using a volumetric flask with an error in the graduation will obtain inaccurate results no matter how carefully he works and how precise his results are. Conversely, with properly calibrated apparatus, any errors in diluting to the mark will be averaged out over a series of determinations and give an accurate value from what may be a group of imprecise results. This is illustrated by the results in the following table, giving three series of iron determinations in the granite rock

TABLE 6. THE IRON CONTENT OF GRANITE R 117
Fe₂O₃ per cent

	Analyst A	Analyst B	Analyst C
	1·27	1·28	1·20
	1·28	1·36	1·18
	1·28	1·30	1·18
	1·27	1·20	1·21
	1·29	1·22	1·18
	1·28	1·27	1·20
$\bar{x}$	1·28	1·27	1·19
s	0·008	0·059	0·021

R 117, obtained by three independent analysts all using the same photometric method.

The results of analyst A are good results, being both accurate and precise (in an inter-laboratory study using this material, the average value obtained from twelve independent analyses was 1·27 per cent; Mercy and Saunders[4] obtained values of 1·28–1·30 per cent, mean 1·30 per cent by five different methods). The results of analyst B (an inexperienced analyst) gave an accurate mean value, but are too widely spread for individual acceptance. The results obtained by analyst C have the precision that can be expected of this method, but are inaccurate. This particular inaccuracy was eventually traced to a mistake in the weighing of the standard material.

The terms *reproducibility* and *repeatability* are frequently encountered in references to precision. In current usage[5] *repeatability* is calculated as the standard deviation of a series of results obtained for a particular determination, by a single analyst making all the determinations at or about the same time. In contrast, the *reproducibility* is the standard deviation of a series of results by a number of analysts working in different laboratories, often at different times but certainly with different sets of apparatus. This distinction appears to be a fine one, but is made by certain authors.

These ideas of reproducibility and repeatability can be illustrated by two series of results for the chromium content of W-1, a diabase from Centerville, Va, U.S.A. The first series, Table 7, column A,

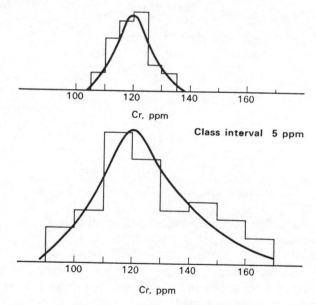

FIG. 7. Histograms of results for chromium in W-1.

consists of fifteen separate determinations made by the author using only one method, that described on p. 204 (photometric measurement of alkali chromate solution after a fusion of the rock material with a

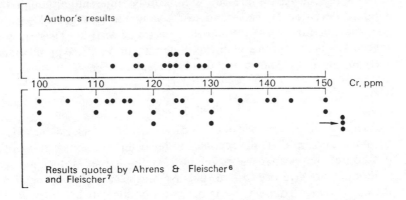

Fig. 8. This distribution of results for chromium in W-1.

TABLE 7. THE CHROMIUM CONTENT OF DIABASE W-1

	A Cr, ppm			B Cr, ppm			
128	117	113	100	110	120	130	140
138	133	124	150	150	100	116	130
123	126	129	120	170	120	105	100
123	126	118	130	124	160	144	115
122	124	117	112	141·5	110	135	125
			125	113	116·3	154	

	A		B	
n	15		29	
$\bar{x}$	124	ppm	126	ppm
range	113–138	ppm	100–170	ppm
s	6·5	ppm	19	ppm
C	5·2	per cent	15	per cent

mixture of anhydrous sodium carbonate and potassium chlorate, followed by an aqueous extraction of the melt). The second series of results, column B, are those reported by Ahrens and Fleischer[6] and by Fleischer,[7] and were obtained by a number of analysts working with a variety of methods in separate laboratories over a number of years.

The first set, column A, demonstrates the *repeatability* of the results obtained by a single method, whereas the second set, column B, shows the *reproducibility* of the results for chromium. The mean values for the two sets of results are not significantly different, although the standard deviation of the second set is, as would be expected, greater than that for the first set, obtained by a single analyst. These results are plotted as histograms with class intervals of 5 and 10 ppm respectively in Fig. 7, and shown as individual results in Fig. 8.

Confidence Limits

As noted earlier, errors in analysis can be either systematic or random. Systematic errors affect the accuracy of the determination, that is of the position of the mean value in relation to the "true" value, whereas random errors give rise to both positive and negative deviations from the mean value, from which the precision is calculated. Errors of this latter kind are usually distributed normally about the mean value. Normal distribution curves are well known in experimental science and are well characterised statistically; they can be obtained from histograms where the total number of results is very large and the class interval very small. In silicate rock analysis there are rarely if ever sufficient determinations available for this treatment, although an approximation to the normal distribution curve can be obtained by fitting a gaussian-type distribution curve to the histogram of those results that are available. This has been done for the two sets of results for chromium in Fig. 7.

From the normal curve, it is possible to calculate what proportion of the total number of results will lie within a certain distance of the mean value. Thus, for example, two-thirds of all the results will be within one standard deviation of the mean value, about 95 per cent of the results within two standard deviations and over 99 per cent within three standard deviations of the mean value. This enables confidence limits to be given for particular determinations. Thus a result expressed as "11·04 per cent Fe_2O_3, ± 0·06 per cent at the 95 per cent confidence level", indicates that a series of determinations have been made with a mean value of 11·04 per cent Fe_2O_3 and a standard deviation of 0·03 per cent. It enables us to predict that 95 per cent of all such determinations will give results in the range 10·98–11·10 per cent and 99 per cent in the range 10·95–11·13 per cent. These figures are approximations appropriate to a normal distribution of results, but they do enable reasonable predictions to be made from the small number of results that are usually all that can be provided, and which are assumed to be part of a normal distribution.

The Significance of Results

It is comparatively easy to compare two sets of results by an inspection of their mean values, standard deviations and coefficients of variation. What is more difficult is to decide whether or not the observed differences are significant. This must to some extent be a subjective assessment, but some help can be obtained by calculating the probability that the observed differences can arise by chance. As an example of this calculation, consider the work of an analyst determining chromium in W-1 by a method similar to that given on p. 204, and obtaining the results 111, 113 and 113 ppm Cr. These results are very close to those obtained by the author reported in Table 7, column A, but the average value of 112 ppm is just outside the range given. Are these new results significantly different? What is the probability that they could have arisen by chance, and that further results by this new analyst would indicate that his results do belong to the original distribution of results?

This is a problem that can be examined by using Student's* t distribution.[8] This distribution defines the limit of variation that can be expected with a given probability for each degree of freedom, and exists in the form of tables in textbooks of statistics and in an abbreviated form in textbooks of analytical chemistry such as Wilson and Wilson.[9] In this example we are concerned with establishing whether the new sample of results with $\bar{X} = 112$, and $n = 3$, is consistent with the set of results with $\bar{x} = 124$, and $s = 6 \cdot 5$.

From this information we can calculate t

$$t = \frac{\mid \bar{X} - \bar{x} \mid \sqrt{(n-1)}}{s}$$

$$= \frac{112 - 124 \times \sqrt{2}}{6 \cdot 5}$$

$$= 2 \cdot 7$$

Entering the table of t values with 2 degrees of freedom $[= (n-1)]$ gives $t = 2 \cdot 7$ between probability $P = 0 \cdot 10$ and $P = 0 \cdot 20$. This means that there is a probability of between 10 and 20 per cent that these

* Pen name of W. S. Gosset.

new results are not different from the earlier set. A probability of between 5 and 10 per cent would have been considered significant and less than 5 per cent highly significant. Thus although we may have our suspicions regarding the new results, and although we may consider further determinations desirable, there is no evidence that the results are significantly different.

Another somewhat similar problem is that of comparing a set of results for one sample material with those obtained from another, of which the following exercise is an example.

In 1951 the results of an inter-laboratory comparison were published. This gave detailed results of about thirty complete analyses of two silicate rocks, a diabase W-1 to which reference has already been made and a granite G-1. A similar inter-laboratory comparison on a much more limited scale was undertaken by the author in the years 1960–2, using a different granite rock, R 117, from Shetland. The sample materials were thus different for the two surveys. The problem now is to decide if there is statistical basis for any claim that the later results for R 117 are an improvement over the earlier ones for G-1.

These two sets of results can be compared by considering the variances (s^2) for each constituent. The ratio of the greater variance to the lesser variance is known as the "F value", and the significance of these F values can be determined from tables in the same way as for t values. Individual results of the determination of total iron in the two sets of complete analyses are given in Table 8, together with the parameters for the two distributions calculated from these individual values. (The results given for G-1 are the original results only, from the paper by Schlecht and Stevens.[10])

The next stage is to set up a *null hypothesis*, i.e. that there is no significant difference between the distribution of iron values for R 117 and the distribution of iron values for G-1, which would occur if the ratio of variances were equal to unity. The problem is now reduced to the calculation of F, and the significance of its departure from unity.

Before doing this, however, it is necessary to consider whether or not the standard deviations of these sets of results are related to the magnitude of the mean value. Experimental evidence has shown that with many (but by no means all) methods of analysis, there is some sort of relationship between these two parameters. As a first approximation this relation can be considered to be linear and a correction can be

TABLE 8. THE TOTAL IRON CONTENT OF G-1 AND R 117

	G-1				R 117	
	1·29	1·91	2·04	2·34	1·26	1·28
	2·47	1·86	1·92	1·94	1·02	1·32
	1·99	2·01	2·10	2·16	1·20	1·28
	2·26	2·99	1·84	1·84	1·29	1·30
	1·86	1·91	1·91	2·13	1·30	1·30
	2·20	2·27	1·88	1·90	1·30	1·36
	1·92	1·94	1·97	2·00		
	1·83	1·83				
n	30				12	
$\bar{x}$	2·02 per cent				1·27 per cent	
s	0·28 per cent				0·086 per cent	
C	14 per cent				6·8 per cent	

applied to enable the two variances to be compared at the same mean value, which in this example is 1·27 per cent Fe_2O_3. The corrected standard deviation for G-1 is therefore given by:

$$S_{2(G-1)} = \frac{S_{1(G-1)} \times \bar{x}_{(R117)}}{\bar{x}_{(G-1)}}$$

$$= 0·18 \text{ per cent}$$

Applying Bessel's correction:

$$S_{2(G-1)}^2 = \frac{(0·18)^2 \times n}{n-1} \qquad \text{where } n = 30$$

$$= 0·0335$$

and

$$S_{(R117)}^2 = \frac{(0·086)^2 \times n}{n-1} \qquad \text{where } n = 12$$

$$= 0·0080$$

$$F = \frac{0·0335}{0·0080} = 42$$

F tables do not list values for 29 degrees of freedom (i.e. 30–1), and the nearest values are for 30 degrees of freedom, for which $F = 2·57$ at the 5 per cent level, and 3·94 at the 1 per cent level. It may therefore be concluded that there is less than 1 per cent chance that the null hypothesis is correct, and from this that there is a highly significant improvement in the later results.

THE x^2 TEST

Tests of significance based upon the t-distribution or F-values are not the only means available for assessing the significance of geochemical data. Many of the sets of numerical values now available can be converted into frequency distributions, which can then be examined by the x^2 test. This is used to evaluate the probability that a given set of experimental results are in agreement with a particular theory, as for example the probability that a series of replicate analyses of one sample material are distributed normally about the mean value, as would be expected if the method used was free of bias. In the following example, a series of tungsten values in a class of silicate rocks is examined using the x^2 test to assess the validity of a lognormal distribution law.

Table 9A contains values for the reported tungsten content of a total of fifty-eight granitic rocks.[11] This is a small number of results upon which to base any conclusion concerning the distribution pattern, but

TABLE 9A. THE TUNGSTEN CONTENT OF SOME GRANITE ROCKS, ppm

1·4	0·5	0·8	0·5	1·9	1·3	1·5	2·2	1·5	1·5
1·5	1·5	1·1	3·1	1·4	3·1	1·9	3·3	3·7	3·5
3·7	1·3	3·1	2·4	1·6	3·7	2·6	9·2	7·7	11·9
12·0	4·5	0·6	0·3	0·5	0·2	0·4	1·1	n.d.	0·1
0·2	1·0	0·7	0·1	1·2	1·1	2·3	1·7	1·7	1·1
0·4	0·2	1·5	0·2	0·5	1·1	1·2	1·9		

n.d. = not detected. The original table gives details of the localities from which the specimens were obtained.

the skewness of the distribution is immediately apparent. This is confirmed by a histogram of the results,[11] and leads to the supposition that they are distributed lognormally. The average of the fifty-eight results in 2·1 ppm, and the standard deviation is 2·5 ppm. The geometric mean, $\bar{x}_g$ (mean of the log values), is 1·2 ppm, and the standard deviation of the log values, s_g, is 0·49 log ppm.

It is now possible to turn the results given in Table 9A into a frequency distribution with class intervals equal to the standard deviation. This is shown in Table 9B as the "observed frequencies".

Before proceeding with the calculation of χ^2, it is necessary to consider what limitations, if any, there are on the calculation. For applications such as these, there are only two limitations that must be considered. The total number of observations must be large, and no single frequency cell should contain small numbers. The bigger the total number of values the better; $n = 58$ is not a very large number, but it is considered sufficient ($n = 50$ can be considered as the lower limit). However, cells numbered 1, 5 and 6 in Table 10 each contain less than five values which is the smallest number that can be accepted in any cell. This difficulty can be avoided by combining cell 1 with 2 and 5 with 6.

TABLE 9B. TUNGSTEN VALUES AS A FREQUENCY DISTRIBUTION

Frequency cell	Observed frequency	Expected frequency*
1 $(\bar{x}_g - 3s_g)$ to $(\bar{x}_g - 2s_g)$ (0·041 ppm to 0·102 ppm)	3 $\Big\}$ 8	1·3 $\Big\}$ 9·2
2 $(\bar{x}_g - 2s_g)$ to $(\bar{x}_g - s_g)$ (0·102 ppm to 0·4 ppm)	5	7·9
3 $(\bar{x}_g - s_g)$ to $(\bar{x}_g)$ (0·4 ppm to 1·2 ppm)	18	19·8
4 $(\bar{x}_g)$ to $(\bar{x}_g + s_g)$ (1·2 ppm to 3·7 ppm)	25·5	19·8
5 $(\bar{x}_g + s_g)$ to $(\bar{x}_g + 2s_g)$ (3·7 ppm to 11·6 ppm)	4·5 $\Big\}$ 6·5	7·9 $\Big\}$ 9·2
6 $(\bar{x}_g + 2s_g)$ to $(\bar{x}_g + 3s_g)$ (11·6 ppm to 35 ppm)	2	1·3

* Calculated assuming lognormality from the known distribution of frequencies in a normal distribution (68 per cent of the results with $\pm s$ of $\bar{x}$, 95 per cent within $2s$ and over 99 per cent within $3s$).

The value of χ^2 can now be calculated as

$$\chi^2 = \sum \left(\frac{\text{observed frequency} - \text{expected frequency}}{\text{expected frequency}} \right)^2,$$

$$= \frac{(1·2)^2}{9·2} + \frac{(1·8)^2}{19·8} + \frac{(5·7)^2}{19·8} + \frac{(2·7)^2}{9·2},$$

$$= 2·75.$$

As with Student's t-distribution and the distribution of F-values, the χ^2-distribution is available in textbooks of statistical analysis. It exists both as a table of values and as a graph giving values of χ^2 for a number of probabilities with different degrees of freedom. In the example there are four frequency cells, but the establishment of three of these automatically establishes the fourth. There are therefore only three degrees of freedom. Entering the χ^2 table of values at three degrees of freedom, $\chi^2 = 2 \cdot 75$ is given between $P = 0 \cdot 30$ and $P = 0 \cdot 50$. In other words, there is between 30 and 50 per cent probability that these results for tungsten in a series of granitic rocks form part of a lognormal distribution. This is considered to be very good agreement between observed and predicted frequencies.

It should perhaps be pointed out that too good an agreement is suspicious. Although possible, it rarely happens that the observed frequencies exactly equal the predicted frequencies ($P = 1 \cdot 0$), and that when it does, the data should be examined for evidence of cooking.

These examples are intended only as an introduction to the use of statistical methods, and for more detailed or more extensive analysis of results reference should be made to standard textbooks such as *Statistical Methods for Research Workers* by R. A. Fisher (Oliver and Boyd, 13th edition, 1963) or *Design and Analysis of Industrial Experiments* by O. L. Davies (Oliver and Boyd, 2nd edition, 1956).

References

1. FAIRBAIRN H. W., *Geochim. Cosmochim. Acta* (1953) **4**, 143.
2. CHALMERS R. A. and PAGE E. S., *Geochim. Cosmochim. Acta* (1957) **11**, 247.
3. CHAYES F., *Amer. Mineral.* (1953) **38**, 784.
4. MERCY E. L. P. and SAUNDERS M. J., *Earth Planet. Sci. Lett.* (1966) **1**, 169.
5. HINCHEN J. D., *J. Gas Chromatog.* (1967) **5**, 641.
6. AHRENS L. H. and FLEISCHER M., *U.S. Geol. Surv. Bull.* 1113, p. 89, 1960.
7. FLEISCHER M., *Geochim. Cosmochim. Acta* (1965) **29**, 1263.
8. GOSSET W. S., *Biometrika* (1908) **6**, 1.
9. WILSON C. L. and WILSON D. W., *Comprehensive Analytical Chemistry*, Elsevier, Vol. 1A, 1959.
10. SCHLECHT W. G. and STEVENS R. E., *U.S. Geol. Surv. Bull.* 980, p. 7, 1951.
11. JEFFERY P. G., *Geochim. Cosmochim. Acta* (1959) **16**, 278.

CHAPTER 7

THE ALKALI METALS

Occurrence

A complete review of the occurrence, geochemistry, and distribution of these metals in silicate and other rocks has been presented by Heier and Adams,[1] who summarised and analysed the data then available for all five elements.

Lithium. Although most silicate rocks contain 20–40 ppm of lithium, many examples are known of rocks containing greater amounts, e.g. samples from the Mourne Mountains granite mass[2] with 0·023 and 0·014 per cent Li_2O. Taylor[3] regards a concentration of greater than 100 ppm as indicative of extreme fractionation, and implying that the rock is a late-stage, high-level product. This fractionation appears in its extreme form in pegmatites with the crystallisation of lithium minerals such as lepidolite, spodumene, zinnwaldite, petalite and others of less frequent occurrence.

Sodium and potassium. All silicate rocks and minerals contain both sodium and potassium in amounts varying from less than 100 ppm in some ultrabasic rocks such as dunite and peridotite, to as much as 10 per cent K_2O or 15 per cent Na_2O in felspar minerals. Rocks containing large amounts of potassium or sodium are rare, and most silicate rock specimens contain both alkalis in somewhat similar amounts in the ranges 1–6 per cent Na_2O and 0·5–6 per cent K_2O. Both elements occur as major constituents of many rock-forming minerals, particularly the alkali felspar group, and are always determined where a complete chemical analysis of a silicate rock or mineral is required.

Rubidium. The close resemblance of rubidium to potassium in chemical behaviour, ionic radius, electronegativity and ionisation potential, results in a very close association of these two metals and a fairly constant K:Rb ratio of about 230:1.[4]

Basic and ultrabasic rocks low in potassium contain little rubidium, values of up to 30 ppm commonly being reported, whilst granitic and other rocks comparatively rich in potassium may contain 100–200 ppm of rubidium. The small difference in ion size results in a slight enrichment of the later rocks relative to potassium, and as a consequence the late-stage granites may contain several hundred ppm rubidium, with potassium–rubidium ratios of less than 150:1. Extreme magmatic differentiation may give rise to pegmatites with even smaller ratios, but unlike caesium, separate rubidium minerals are not formed. Rocks are also known with unusually high potassium–rubidium ratios of up to 400:1, and an explanation for this apparent anomally has been given by Taylor.[3]

Caesium. Most silicate rocks contain only a few ppm of caesium. The large size of the ion restricts the extent to which it can substitute for other elements in silicate structures, although some replacement of potassium does occur. Caesium tends therefore to be concentrated in the later rocks, particularly in the late-stage granites for which values of 100–300 ppm may be obtained, and in granitic pegmatites where pollucite, a caesium aluminium silicate with 30–40 per cent Cs_2O, may crystallise.

Determination in Silicate Rocks and Minerals

GRAVIMETRIC METHODS

Gravimetric methods have long been used for the determination of all five of the alkali metals. These methods required a most careful separation of the total alkali metals from silica, aluminium, calcium and other elements present. The most frequently used procedure for this, described originally by Lawrence Smith,[5] involved decomposing the rock sample by ignition with ammonium chloride and calcium carbonate. The alkali metals were recovered by leaching with water, and were then separated from the small amount of calcium taken into solution. Sulphates were converted into chlorides and the introduced ammonium salts were removed by volatilisation. Some authors[6] considered that special precautions were necessary to ensure a complete recovery of lithium with the remaining alkali metals. The mixed chloride residue obtained after the expulsion of ammonium salts was carefully ignited and weighed before the separation of the individual alkali metals.

An alternative procedure for decomposing the silicate rock material and recovering the alkali metals as chlorides was based upon evaporation of the sample with hydrofluoric acid and precipitation of iron, aluminium, calcium and other elements with ammonia and ammonium carbonate, a method derived from work by Berzelius.[7]

The separation of lithium was based upon the solubility of lithium chloride in organic solvents such as isobutanol, pentanol, pyridine or ether–ethanol mixtures. The determination was often completed by converting the lithium chloride to sulphate prior to weighing. This gravimetric procedure for lithium is not sufficiently sensitive for application to the majority of silicate rocks, and lithium could only be reported in those samples rich in this element.

Sodium and potassium were usually determined following the precipitation of potassium with chloroplatinic acid, perchloric acid or sodium cobaltinitrite. The insoluble potassium salts were collected and weighed directly, although a number of indirect procedures were also in common use. Sodium was generally determined by difference, although some analysts preferred to precipitate sodium as a triple acetate of uranium and zinc or other divalent metal. In all these precipitation procedures, corrections for the solubility of the potassium or sodium salts were necessary, and none of the weighing forms could be considered ideal.

Even when present in quantity, the separation of rubidium and caesium from potassium and from each other was seldom attempted, and on these few occasions the results obtained were not always very reliable.

FLAME PHOTOMETRY

All the earlier gravimetric procedures for these five alkali metals have now been rendered obsolete by the development of new physical methods, such as flame photometry and atomic absorption spectroscopy. These techniques are of particular value for this group of five elements.

Flame photometry is based upon the measurement of light emitted from a flame into which the sample solution is continuously sprayed. The emission spectra of the alkali metals are all very simple, consisting of a prominent line or doublet known as the resonance line(s), corresponding to the transition between the lowest excited state and the

TABLE 10. FLAME EMISSION SPECTRA OF THE ALKALI METALS

Resonance line(s), nm			Other useful lines, nm		
Lithium	670·8		323·3	610·4	
Sodium	589·0	589·6	330·2	818	819
Potassium	766·5	769·9	404·4		
Rubidium	780·0	794·8	420·2	421·6	
Caesium	852·1	894·4	455·5		

ground state, together with weaker lines relating to other transitions (Table 10).

Lithium and sodium are only slightly ionised in the flames normally used, in contrast to the remaining alkalis, where the degree of ionisation increases in the order K, Rb, Cs, until with caesium a large part of the metal in the flame is in the ionised state. This accounts very largely for poor sensitivity to caesium, and the general decrease of sensitivity from sodium to caesium (Table 11).

TABLE 11. DETECTION LIMITS OF THE ALKALI METALS

Element	Wavelength, nm	Detection limit, ppm
Li	670·8	0·0001
Na	589·0	0·0001
K	766·5	0·001
Rb	780·0	0·05
Ca	852·1	1·0

Any figures quoted for the sensitivities of the alkali metals can be misleading, because the useful sensitivity depends not only upon the intrinsic flame emission for given conditions of flame performance but also upon the presence or absence of other elements, the acid and anion concentration used, as well as the response of the photocell at the particular wavelength. The figures given are taken from a manufacturer's handbook[8] and can be used as a rough guide to the sensitivity likely to be obtained from a good instrument.

One way of increasing the sensitivity to a particular element in a given solution is to introduce a second, easily ionised element such as another alkali metal into the flame. This serves to decrease the extent of ionisation of the required element and thus increase the proportion of atoms available for the transition which gives rise to the resonance lines. At high alkali concentrations some loss of emission occurs by self absorption. This effect has been noted particularly with lithium and sodium,[9] and can be minimised by working at an increased dilution. Straight-line calibration curves are only obtained at low alkali concentrations, but these departures from linearity are not a serious difficulty. The concentration range over which the calibration curve is a straight line decreases in the order Li, Na, K, Rb and Cs.

To obtain the best results by flame photometry it is necessary to standardise the conditions used for the operation of the instrument (gas pressure, air pressure, fuel-to-air ratio), as well as instrument settings (slit width, resolution) and the conditions used to prepare the sample solution (acid and salt concentration). The effect of other elements upon alkali flame emission can often be minimised by including a "radiation buffer"[10] such as ammonium sulphate in the sample solution. No simple way has yet been found of accurately determining small quantities of one alkali metal in the presence of a very great excess of another. Fortunately very many silicate rocks contain both sodium and potassium in somewhat similar amounts and acceptable results for these two can be obtained by quite a simple method.

Interferences with the flame photometric determination of the alkali metals, given in outline in the preceding paragraphs, are summarized in Table 12. For a fuller treatment of these interferences, reference should be made to recent texts dealing with flame photometry[9] and with the alkali metals.[11, 12]

ATOMIC ABSORPTION SPECTROSCOPY

As noted above and in Table 12, flame emission methods are subject to a number of interferences, direct and indirect, which must be taken into account if accurate results are to be obtained. Some of the errors can be avoided by using atomic absorption spectroscopy, although interference effects associated with the use of "atomisers" (also called "nebulisers") for spraying solutions into a flame obviously remain.

For the alkali metals the hollow cathode lamps usually employed for

TABLE 12. INTERFERENCE EFFECTS IN THE DETERMINATION OF THE ALKALI METALS

Interference	Caused by	Remedy
Continuous or background emission	Spectral emission by solvent, alkalis and other elements	(1) Make a background correction, or (2) Use recording instrument with motorised wavelength drive.
Radiation interference	Enhancement or depression by alkalis and other elements present in flame	(1) Use standards containing these other elements, (2) use a "radiation buffer" or (3) separate individual alkali metals first.
Spectrum interference	Close proximity of other spectral lines	(1) Use recording instrument with motorised wavelength drive or (2) separate the elements concerned.
Difference in solution utilisation or in flame characteristics	Changes in surface tension, viscosity, acid or salt concentration, air or fuel gas pressure	Standardise the conditions used to prepare the solution and to measure the flame emission.

this technique are replaced by discharge lamps which give stable, line-emission sources. As with spectrophotometry, there is a log relation between the light absorbed and the concentration of the absorbing species in the flame. Calibration graphs for the alkali metals are straight lines, or close approximations to it over the concentration ranges used.

As with flame photometry, the sensitivity is greatest when the resonance line is used. Wavelengths that have been reported for the determination of individual alkali metals are given in Table 13, together with the "minimum detectable concentrations" noted by Billings and Adams[13] using a double-beam, modulated spectrometer. These limiting concentrations vary from one instrument to the next, and are dependent upon burner design, burner height, nature of fuel gas, fuel-to-air ratio and other similar factors.

A number of papers, in addition to that by Billings and Adams[13] referred to above, have described the determination of sodium and potassium in silicate rocks by atomic absorption spectroscopy. Recent papers have described also the determination of lithium in silicate rocks[14] and rubidium in rocks and minerals.[15] As with flame photometric methods it is necessary to add potassium to the solutions before determining rubidium. Vosters and Deutsch[15] also recommended adding lanthanum as a flame buffer, and reported values of between 0·5 and 3 ppm rubidium obtained without recourse to chemical separation. The solutions prepared for the flame photometric determination of the alkali metals can usually also be used for their determination by atomic absorption spectroscopy.

TABLE 13. WAVELENGTHS USED FOR ATOMIC ABSORPTION DETERMINATION OF THE ALKALI METALS

Element	Wavelength, nm	Minimum detectable concentration,[12] ppm
Li	670·8	
Na	589·0	0·01
	330·2	10
K	766·5	0·01
	404·4	50
Rb	780·0	0·1
Cs	852·1*	

* Obtainable with Unicam SP 90 B.

CHROMATOGRAPHIC SEPARATION OF THE ALKALI METALS

One of the most successful ways of separating mixtures of the alkali metals is by ion-exchange chromatography. Resins such as Dowex 50[16] or Amberlite 120[17] can be used for this separation, which requires the use of a cation-exchange resin as all five elements form well-characterised positive ions in solution. The separation of sodium from potassium is well established,[18] and the successful separation of rubidium and caesium has also been reported.[19] However, none of these procedures is ideal for the routine separation and determination of all five alkali metals.

Ammonium phosphomolybdate can be used as an inorganic ion-exchanger, in which the ammonium ions of the complex can be replaced by alkali metals. This material, mixed with asbestos and packed into a short column, has been shown to be of particular value for the separation of rubidium and caesium.[20-22]

In the following sections details are given for the determination of individual alkali metals. The first procedure given is for the determination of sodium and potassium, involving a decomposition with hydrofluoric and perchloric acids and the use of a "radiation buffer". This procedure is simple to use and gives good results for those rocks in which both elements are present in average amounts. Other procedures are described for those rocks containing little sodium and potassium, and for determining lithium, rubidium and caesium.

Determination of Sodium and Potassium in Silicate Rocks

The method described in detail below is based upon work by the author and by Eardley and Reed.[12] The rock material is decomposed by evaporation with hydrofluoric, nitric and sulphuric acids, as both hydrochloric and perchloric acids have been found to have a depressant effect upon the alkali flame emission. In most cases the small amount of residue that remains after this treatment can be ignored, although for really accurate work this residue should be collected and analysed separately (note 1). Aluminium sulphate solution is added to eliminate the spectral interference from calcium, and caesium sulphate solution as an ionisation buffer. The procedure given has been devised for use with a Unicam SP 900 flame spectrophotometer, but it can readily be adapted for use with other instruments that utilise a low-temperature, air–propane flame burner.

Method

Reagents: *Sodium sulphate*, anhydrous.
Potassium sulphate.
Aluminium sulphate solution, clean pure analytical-grade aluminium by washing with diluted hydrochloric acid, water, ethanol and ether. Weigh 1·06 g of the metal into a 400-ml beaker, add 20 ml of water, 10 ml of concentrated sulphuric acid and 20 ml

of concentrated nitric acid. Allow the metal to react in the cold, then gradually raise the temperature until decomposition is complete. Heat the solution to expel nitric acid and oxides of nitrogen, then to fumes of sulphuric acid. Allow to cool, dissolve the melt in water and dilute to 1 litre with water. This solution contains 2000 ppm Al_2O_3.

Caesium sulphate solution, dissolve 0·205g of pure caesium sulphate, Cs_2SO_4, in 500 ml of water.

Procedure. Accurately weigh approximately 0·1g of the finely powdered silicate rock material into a platinum dish (note 1), moisten with a little water and add 10 ml of concentrated nitric acid, 5 ml of concentrated sulphuric acid and 10 ml of hydrofluoric acid. Place the dish on a cool hotplate and gradually raise the temperature until the sulphuric acid just starts to fume. Remove the dish from the hot plate, cool, rinse down with a little water and add a further 5 ml of hydrofluoric acid. Replace the dish on the hotplate which meantime had been allowed to cool somewhat, and again increase the temperature slowly, but this time continue until the contents of the dish are fuming well. Again allow to cool, rinse down with water, replace on the hotplate and evaporate, this time until all fumes of sulphuric acid have ceased and a dry residue is obtained.

Add 1 ml of concentrated nitric acid to the cool residue, followed by 25 to 30 ml of water. Warm the dish to detach the residue and rinse the contents into a 400-ml beaker. Digest until all soluble material has passed into solution, and collect any small residue on a small close-textured filter paper, such as Whatman No. 42, and wash with a little water. Discard the residue (note 2) and dilute the filtrate and washings to volume with water in a 250-ml volumetric flask.

Transfer 25 ml of this solution by pipette to a 50-ml volumetric flask, add 5 ml of the caesium sulphate solution and 5 ml of the aluminium sulphate solution, dilute to volume with water and mix well (note 3). Prepare also a reagent blank solution in the same way as this sample solution, using the same quantities of reagents but omitting the powdered rock material.

Set the flame spectrophotometer base line whilst spraying water, and the sensitivity at 10 ppm full-scale deflection spraying the 10 ppm standard solution, with the wavelength control set at 589 nm for sodium or 766·5 nm for potassium. The adjustment of sensitivity will affect the base line position and it may be necessary to reset the controls alternatively two or three times. Once this adjustment is complete, spray also the reagent blank solution and the sample solution, noting the flame emission recorded at both wavelengths, and hence calculate the alkali content of the sample material from the calibration graphs.

Where the flame spectrophotometer is fitted with a wavelength drive motor and a pen recorder, the flame response for either element can be obtained by scanning the spectrum. The reagent blank solution and the complete set of standard solutions should similarly be examined. For sodium scan from about 560 nm to 620 nm and for potassium from 730 nm to 820 nm.

With many silicate rocks, the rock solution prepared as described above can be used to determine the lithium content of the rock. The sample solution contained in the 250-ml volumetric flask is used directly and the flame emission measured at 670·8 nm. As noted below, this emission is measured on the trailing edge of the potassium emission. It is therefore preferable to sweep through a wavelength range from about 600 to 750 nm, noting the response on a pen-recorder chart as shown in Fig. 9.

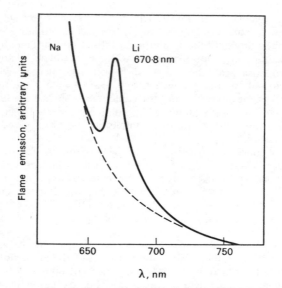

FIG. 9. Recorder trace showing lithium in a nepheline-syenite (sample R204, Li = 0·005 per cent).

Calibration. Dry small quantities of both sodium and potassium sulphates in an electric oven and allow to cool in a desiccator. Weigh exactly 0·2291 g of the anhydrous sodium sulphate and 0·1850 g of the potassium sulphate into the same beaker, dissolve in water and dilute to volume in a litre volumetric flask. This solution contains 100 μg Na_2O and 100 μg K_2O per ml. Prepare a 20 ppm standard by diluting 50 ml of this stock solution, measured with a pipette, and 1 ml of concentrated nitric acid to 250 ml with water and mixing well. Transfer 25 ml of this solution by pipette to a 50-ml volumetric flask, add 5 ml of the aluminium sulphate solution and 5 ml of the caesium sulphate solution and dilute to volume with water. This gives a 10 ppm standard for setting the sensitivity control of the flame spectrophotometer.

Prepare also a series of solutions each containing 5 ml of the aluminium sulphate solution, 5 ml of the caesium sulphate solution and a concentration of 2 ml concentrated nitric acid per litre in a volume of 50 ml, but with 2, 4, 6, and 8 ppm of each element. These, together with the 10 ppm standard, are used to prepare the calibration graph.

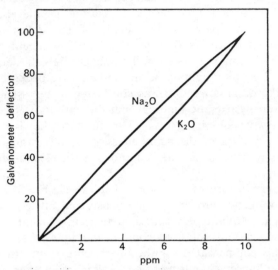

FIG. 10. Calibration graphs for sodium and potassium using an SP 900 flame photometer: sodium at 589 nm, potassium at 766·5 nm.

Measure the flame emission of each of these solutions after first setting the base line of the instrument and the overall sensitivity as described above. Plot the curves for both sodium and potassium. Both curves are slightly bowed, but in opposite directions (Fig. 10).

Notes: 1. The weight of sample material taken for the analysis and the volume of the final solution may be varied in relation to the alkali content of the sample, but the final solution for measurement should be of the correct acid concentration and contain the recommended quantities of aluminium and caesium sulphates.

2. This small residue contains much of the zircon and some of the tourmaline present in the rock, together with varying proportions of other minerals that are difficult to decompose with hydrofluoric acid. These minerals seldom contain more than trace amounts of alkali metals and the error involved in neglecting these small amounts is often within the limits of the experimental error of the method. If the amount of residue indicates that an appreciable part of the sample has not been attacked, an alternative procedure, such as fusion with lithium borate, should be used for the sample decomposition.

3. Attempts should not be made to simplify the procedure by making a mixed stock solution containing aluminium and caesium sulphates, as this leads to a slow precipitation of the sparingly soluble caesium aluminium sulphate.

Small Amounts of Alkali Metals in Ultrabasic Rocks

There are a number of ultrabasic rocks that contain only small amounts of sodium and potassium. The importance of these rocks is out of all proportion to the extent to which they occur. If reliable results are to be obtained for the alkali metals at these levels, particular attention must be paid to the details of the determination. It has been pointed out[23] that at the high instrument sensitivities required for determining these small amounts of alkali metals, a large percentage of the total light emission can be due to background radiation. A background correction was considered to be essential for the meteoritic and pyroxenitic material that the authors were examining. The simplest way of obtaining this background correction is undoubtedly to record the spectra of the flame emission over the ranges involved using an instrument fitted with a motorised wavelength drive and a pen recorder.

In their examination of a number of ultrabasic rocks (serpentines, peridotites, dunites, pyroxenites, talc schists and other related rocks) Hamilton and Mountjoy[24] calibrated their instrument response by adding small amounts of both sodium and potassium to the rock solution. This compensated for the interference from other elements and for the enhancement effect of one alkali upon another.

Procedure. Prepare the rock solution from 1 g of the rock material by evaporation to dryness with concentrated hydrofluoric acid and a mixture of concentrated sulphuric and perchloric acids, followed by solution of the moist residue in 20 ml of $0 \cdot 5$ N hydrochloric acid. Filter off and discard any residue. Dilute the filtrate to 50 ml with water and take two 5-ml aliquots for analysis. Dilute one aliquot with 5 ml of water, and the other with 5 ml of water containing 5 ppm each of sodium and potassium. Spray each of these solutions into the flame photometer and complete the determination in the usual way.

The Determination of Potassium in Micas

The need for very accurate potassium determinations has been underlined by the simple way in which geological ages can be obtained from an accurate knowledge of the $A^{40}:K^{40}$ ratios. The small amounts of argon present can readily be recovered and determined isotopically with a mass spectrometer. Potassium can also readily be determined by flame spectroscopy, but the accuracy and precision obtained by the usual procedures such as the one given in detail above, athough adequate for most petrographic purposes, do not match that obtainable for argon.

A flame photometric procedure for this accurate determination has been given by Abbey and Maxwell;[25] the sample material is decomposed with hydrofluoric acid and sulphuric acids, and the excess hydrofluoric acid removed by fuming. After the addition of a known excess of magnesium sulphate, the excess sulphuric acid is in turn removed by evaporation to dryness. The sulphates are ignited, the cooled residue is leached with warm water, and the resulting suspension filtered, giving a neutral sulphate solution containing the alkali metals, magnesium and little else. The approximate relative proportions of the alkali metals are then determined and a neutral sulphate solution prepared approximating in composition to that of the sample solution. The final potassium determination is made with a flame photometer using this solution with added potassium as standard.

Procedure. Accurately weigh duplicate $0 \cdot 1$ g samples of the finely ground rock material (see p. 18 for the preparation of micas for analysis) into 100-ml platinum dishes and add 5 ml of water, 2 ml of 20 N sulphuric acid and 5 ml of hydrofluoric acid. Cover the dishes and heat on a steam bath to decompose the samples. Rinse and remove the covers, add 2–3 ml of concentrated hydrofluoric acid and evaporate to dryness on a steam bath. Rotate each dish cautiously over a burner until copious fumes of sulphuric acid are evolved, and allow to cool. Rinse down the sides of the dish with a little water and warm on the steam bath until the solution is clear. (Any undecomposed sample will become visible at this point, and may be eliminated by further treatment with hydrofluoric acid.)

Add sufficient magnesium sulphate solution to provide 60 mg of magnesium, mix by swirling and evaporate to dryness. Heat on a sand bath until no further evolution of fumes of sulphur trioxide occurs, then heat cautiously over a burner to the reappearance of sulphur trioxide fumes and finally heat strongly over a full burner until all fuming ceases. Allow to cool. Moisten the residue with about 10 ml of water, cover and warm on the steam bath. Rinse and remove the cover, break up the residue and evaporate the slurry to dryness. Repeat the ignition to expel any sulphur trioxide trapped in the residue from the first ignition. Allow to cool.

Add 25 ml of water, cover and warm for at least 30 minutes. Decant the solution onto a close textured filter paper, collecting the filtrate in a 200-ml volumetric flask. Repeat the leaching and decanting twice, using 10-ml portions of water each time, and 10 minute leaching periods. Finally rinse the residue onto the paper and wash it three or four times with water. Cool the filtrate, dilute to volume with water and store in a polyethylene bottle until ready for spraying.

The Determination of Lithium

The simple method described above for sodium and potassium can be extended to include also the lithium present in the rock material. A somewhat greater instrument sensitivity-setting is required, but this is well within the range of most modern instruments. The solution prepared as described above for sodium and potassium, can be used directly for the lithium measurement, but for rocks containing only a few ppm of lithium, a more concentrated solution is desirable.

The addition of aluminium sulphate, as proposed for the determination of sodium and potassium (see above) for the elimination of spectral interference from calcium, can be used to prevent similar interference with the determination of lithium.

For basic rocks containing high levels of calcium Sulcek and Rubeska[26] have recommended the use of ion exchange chromatography with 0·5 M methanolic hydrochloric acid as eluate to separate the low levels of lithium from both calcium and strontium. As little as 5 ppm Li can be determined in this way.

Of all the alkali metals, lithium has the least effect on the flame response of the others, and is least affected by them.[27] For most purposes this small effect upon the lithium response can be ignored, but for accurate determinations, the amounts of sodium and potassium present in the rock solution can be measured and similar amounts added to the standard solutions used for the lithium calibration.

The most sensitive lithium line is the resonance line at 670·8 nm. At the lithium concentrations encountered in most silicate rocks, this suffers from interference from the sodium doublet at 589 nm. The simplest way of observing the lithium response is by using a flame photometer fitted with a motorised wavelength drive and a pen recorder. A typical recorder trace for the range 600 nm to 750 nm is shown in Fig. 9. The sample material was a nepheline syenite with 50 ppm lithium, and the lithium response at 670·8 nm is clearly shown above the sloping background emission from sodium.

A procedure for determining lithium, together with rubidium and caesium, as described by Abbey[28] is given below.

Determination of Lithium by Atomic Absorption Spectroscopy

The use of atomic absorption spectroscopy for the determination of lithium in silicate rocks has been described by a number of authors including O'Gorman and Suhr,[29] Stone and Chesher[30] and Zelyukova et al.[31] The rock solution for the determination can be prepared by either acid digestion (HF, H_2SO_4, etc.) or by borate fusion. The procedure given below, adapted from that given by O'Gorman and Suhr,[29] uses a fusion with sodium borate.

Procedure. Accurately weigh approximately $0 \cdot 1$ g of the finely powdered silicate rock material into a graphite fusion crucible, add $0 \cdot 5$ g of anhydrous sodium borate ($Na_2B_4O_7$), and fuse for 10 minutes in an electric muffle furnace at a temperature of 1000°. Pour the molten bead directly into a beaker containing exactly 50 ml of a 3 per cent nitric acid solution. Add a magnetic stirring bar and stir until dissolution is complete. Use this solution without further treatment for direct nebulisation into the flame of the atomic absorption spectrophotometer in the usual way.

To prepare standard solutions use portions of a silicate rock known to be essentially lithium-free (e.g. USGS Peridotite PCC-1) and add aliquots of a lithium standard solution. Two ranges of standards are recommended equivalent to 0–200 ppm and 0–2000 ppm Li in the silicate rock.

The Determination of Rubidium and Caesium

In theory there is no reason why the methods described above for the determination of sodium and potassium should not be used for rubidium and caesium. However since these elements are seldom present in more than trace amounts, and since their flame response is very dependent on other elements present in the solution, a modified technique is preferred.

Horstman[32] has described a method based upon the precipitation of oxide elements (iron, aluminium, etc.) with calcium carbonate, and of calcium by precipitation as sulphate in aqueous alcoholic solution. The sodium and potassium contents of the rock solution (and lithium where this is present in quantity) are determined, and similar amounts added to the rubidium and caesium standards used to calibrate the flame spectrophotometer. This is given in detail below.

Method

Procedure. Accurately weigh a sample of $0 \cdot 5$–2 g of the finely powdered silicate rock into a platinum dish, moisten with water and add sufficient

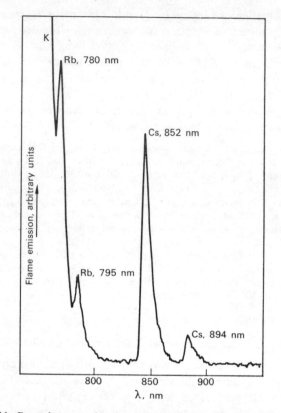

FIG. 11. Recorder trace showing rubidium and caesium in pollucite.

sulphuric and hydrofluoric acids to give complete decomposition of the sample. Add 2–3 drops of concentrated nitric acid and evaporate on a hot plate to fumes of sulphuric acid. Allow the dish to cool, wash down with a little water and again evaporate to fumes of sulphuric acid. Allow to cool, add 5 ml of water and again evaporate, this time until all excess sulphuric acid is removed, leaving only a moist residue. Allow the dish to cool and dissolve this residue, by warming if necessary, in 25 ml of water. Transfer the solution with any unattacked mineral grains (and sulphate precipitate if any) to a 150-ml beaker and dilute with water to a total volume of 70–80 ml. Add a few drops of bromothymol blue indicator solution and neutralise by adding solid calcium carbonate. Allow the precipitate to stand overnight.

Heat the solution and precipitate to boiling and filter on to a medium-textured filter paper. Wash the residue well with hot water until the volume of combined filtrate and washings reaches about 150 ml. Discard the residue and evaporate the solution to a volume of 50 ml, cool and add 50 ml of ethanol to precipitate calcium sulphate. Allow the precipitate to stand overnight and then collect on a close-textured filter paper. Wash the precipitate with a mixture of equal volumes of water and ethanol until the volume of the combined filtrate and washings again reaches 150 ml. Discard the sulphate residue and evaporate the solution to dryness on a steam bath. Dissolve the dry residue in a little water and dilute to volume in a 50-ml volumetric flask.

Using a flame spectrophotometer set according to the manufacturers instructions, determine the sodium and potassium content of the solution and also the approximate values of lithium, rubidium and caesium. For rocks containing only traces of the rarer alkali metals prepare a series of calibration solutions for each of these metals spanning the approximate values and containing the same amounts of sodium and potassium as the rock solution. For those rocks and minerals containing appreciable amounts of rubidium or caesium add each of these elements to the standard solutions used to calibrate for the other.

The flame photometer trace shown in Fig. 11 was obtained from a sample of pollucite, decomposed as described above, and illustrates the difficulty of determining small amounts of rubidium in the presence of potassium.

The determination of alkali metals by direct injection of the powdered silicate rock material into a flame has been described by Lebedev.[33] More recently this has been used with atomic absorption spectrophotometry by Govindaraju et al.[34] for rubidium and by Langmyhr and Thomassen[35] for rubidium and caesium. This is a method that should be capable of further development but its application to a wider range of rocks and minerals needs further study.

The method given in detail below is that due to Abbey.[28] The rock sample is decomposed by a conventional acid digestion procedure and there are no chemical separations. Following the addition of potassium to suppress ionisation, lithium, rubidium and caesium are determined by direct emission or absorption (lithium and rubidium only) in an air–acetylene flame. When required a standard addition technique can be used to overcome matrix effects. Aliquots of the same sample solution can be used for the determination of strontium and barium by atomic absorption spectroscopy with a nitrous oxide–acetylene flame.

Procedure. Accurately weigh approximately 0·5 g of the finely powdered rock material in duplicate into 100-ml platinum dishes and add to each 5 ml of concentrated nitric acid, 2 ml of concentrated perchloric acid and 5 ml of hydrofluoric acid. Transfer to a hot plate and evaporate to copious fumes of perchloric acid. Allow to cool, rinse down the walls with a little water and evaporate on the hot plate once more, this time to complete dryness. Allow to cool, add 2 ml of concentrated hydrochloric acid, again rinse down the walls with a little water and evaporate to complete dryness (note 1).

Using a safety pipette, add 2 ml of concentrated hydrochloric acid, rinse down with a little water, warm to dissolve the chlorides and transfer with water to a 50-ml volumetric flask. To one of the duplicates add alkali solutions containing 25 μg lithium, 250 μg rubidium and 2·5 μg caesium (note 2). Add 1 ml of potassium buffer (76·3 g potassium chloride per litre) to both spiked and unspiked samples and dilute each to volume with water (note 3).

Determine the flame emission at 670·8 nm for lithium, 780·0 nm for rubidium and 852·1 nm for caesium, preferably by recording over a wavelength range containing the complete emission peak. An air-acetylene flame should be used. Lithium and rubium can be determined in the same solution by atomic absorption spectroscopy, also using an air-acetylene flame.

Notes: 1. If barium is to be determined in the same sample solution, the residue should be collected, fused with a little anhydrous sodium carbonate and extracted with water. Discard the aqueous extract and acidify the residue with a very little diluted hydrochloric acid. This can then be added to the main rock solution before its final evaporation to dryness.

2. In the final 50-ml volume, these quantities correspond to an additional 50 ppm lithium, 500 ppm rubidium and 5·0 ppm caesium.

3. If barium and strontium are to be determined in this solution, pipette 10 ml into a clean 50-ml volumetric flask, add 2 ml of potassium buffer solution, 10 ml of a lanthanum buffer solution (50 g La_2O_3 per litre), dilute to volume with water, mix well and determine the barium and strontium by atomic absorption using a nitrous oxide–acetylene flame.

References

1. HEIER K. S. and ADAMS J. A. S., *Phys. Chem. Earth* (1964) **5**, 253.
2. GUPPY E. M., *Chemical Analyses of Igneous Rocks, Metamorphic Rocks and Minerals*, Mem. Geol. Surv. Gt. Brit., H.M.S.O., 1956.
3. TAYLOR S. R., *Phys. Chem. Earth* (1965) **6**, 133.
4. GOLDSCHMIDT V. M., *Geochemistry*, Oxford, 1954, p. 163.
5. SMITH J. L., *Amer. J. Sci.*, 2nd Ser. (1871) **50**, 269.
6. HERING H., *Anal. Chim. Acta* (1952) **6**, 340.
7. BERZELIUS J. J., *Lehrbuch der Chemie*, 3rd ed. (1841) **10**, 46.
8. UNICAM INSTRUMENTS LTD., *S.P. 900, Operating, Maintenance and Servicing Manual*, Cambridge.
9. DEAN J. A., *Flame Photometry*, McGraw-Hill, New York, 1960.
10. WEST P. W., FOLSE P. and MONTGOMERY D., *Analyt. Chem.* (1950) **22**, 667.
11. KALLMANN S., "Alkali Metals" in *Treatise on Analytical Chemistry*, Ed. KOLTHOFF I. M. and ELVING P. J., Pt. II, Vol. I, Interscience, New York, 1961.
12. EARDLEY R. P. and REED R. A., *Analyst* (1971) **96**, 699.

13. BILLINGS G. K. and ADAMS J. A. S., *Atomic Absorption Newsletter*, Perkin Elmer Corp., 1964, Aug. No. 23.
14. OHRDORF R., *Geochim. Cosmochim. Acta* (1968) **32**, 191.
15. VOSTERS M. and DEUTSCH S., *Earth Planet. Sci. Lett.* (1967) **2**, 449.
16. BEUKENKAMP J. and RIEMAN W., *Analyt. Chem.* (1950) **22**, 582.
17. REICHEN L. E., *Analyt. Chem.* (1958) **30**, 1948.
18. ELLINGTON F. and STANLEY L., *Analyst* (1955) **80**, 313.
19. COHN W. E. and KOHN H. W., *J. Amer. Chem. Soc.* (1948) **70**, 1986.
20. SMIT J. VAN R., *Nature* (1958) **181**, 1530.
21. SMIT J. VAN R., JACOBS J. J. and ROBB W., *J. Inorg. Nucl. Chem.* (1959) **12**, 95.
22. SMIT J. VAN R., *U.K. Atom. Energ. Auth. Rept.* AERE-R 3700, 1963.
23. EASTON A. J. and LOVERING J. F., *Anal. Chim. Acta* (1964) **30**, 543.
24. HAMILTON W. and MOUNTJOY W., *Geochim. Cosmochim. Acta* (1965) **29**, 661.
25. ABBEY S. and MAXWELL J. A., *Chem. in Canada* (1960) **12**, 37.
26. SULCEK Z. and RUBESKA J., *Colln. Czech. Chem. Comm.* (1969) **34**, 2048.
27. GROVE E. L., SCOTT C. W. and JONES F., *Talanta* (1965) **12**, 327.
28. ABBEY S., *Geol. Surv. Canada* Paper No. 71–50 (1972).
29. O'GORMAN J. V. and SUHR N. H., *Analyst* (1971) **96**, 335.
30. STONE M. and CHESHER S. E., *Analyst* (1969) **94**, 1063.
31. ZELYUKOVA YU V., NIKONOVA M. P. and POLUEKTOV N. S., *Zhur. Anal. Khim.* (1966) **21**, 1407.
32. HORSTMAN E. L., *Analyt Chem.* (1956) **28**, 1417.
33. LEBEDEV V. I., *Zhur. Anal. Khim.* (1969) **21**, 337.
34. GOVINDARAJU K., MEVELLE G. and CHOUARD C., *Chem. Geol.* (1971) **8**, 131.
35. LANGMYHR F. J. and THOMASSEN Y., *Z. Anal. Chem.* (1973) **264**, 122.

CHAPTER 8

ALUMINIUM

Occurrence

Of all the elements present in the igneous rocks forming the crust of the earth, only oxygen and silicon are more abundant than aluminium. Dunites and peridotites, the first rocks of the magmatic sequence to crystallise, contain very little aluminium which is thus enriched in the fraction remaining, and then appears to a much greater extent in the main stage of crystallisation. The aluminium content then decreases in successive stages as differentiation proceeds. This is shown diagrammatically in Fig. 12.

Silicate minerals containing aluminium include the micas, felspars and felspathoids, and to a lesser extent many of the amphiboles and pyroxenes. Some of the more important minerals of aluminium do not crystallise in the main stages of silicate differentiation, but appear with rare element concentration at the pegmatite and other late stages of rock emplacement. These include beryl, spodumene, topaz, amblygonite and cryolite. Andalusite, sillimanite and kyanite, industrially important aluminosilicates found in metamorphic rocks, may contain up to 60 per cent Al_2O_3. Rarer aluminium minerals include corundum, chrysoberyl, turquoise, alum and alunite.

The main sources of the metal aluminium are the bauxite and laterite ores containing diaspore, boehmite and gibbsite—hydrated oxide minerals usually contaminated with hydrated ferric oxide. Kaolin, an alteration product of felspar, is also produced commercially on a large scale.

The Determination of Aluminium in Silicate Rocks

GRAVIMETRIC METHODS

Determination by difference. In the classical procedure for determining aluminium in silicate rocks iron, aluminium and other elements of

94

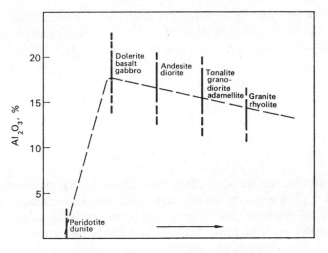

Magmatic differentiation

FIG. 12. Aluminium content of silicate rocks.

the ammonia group were precipitated together and weighed as "mixed oxides". The elements other than aluminium present in this residue were then determined separately, leaving the aluminium content to be obtained by difference. This procedure is dealt with more fully on p. 40. It involved the accurate determination of iron, titanium, vanadium, chromium, phosphate and that part of the manganese (and nickel if present in more than trace amount) precipitated with the elements of the ammonia group.

The determination of aluminium by difference is unsatisfactory in that a small amount of aluminium usually escapes precipitation by ammonia and can be recovered from the filtrate, and that any error in determining the remaining elements of the ammonia group will be accumulative and be reflected in a corresponding error in the aluminium content. Traces of silica (remaining in the solution after the dehydration and removal of the main silica fraction) will be carried down in the ammonia precipitate and, if not recovered and determined, will be counted as alumina.

Determination with 8-hydroxyquinoline (I). In dilute acetic acid solution aluminium forms an insoluble yellow complex with 8-hydroxyquinoline that can be used for the gravimetric determination of aluminium. The complex can be dried to constant weight at a temperature of 130–140°, when it has the composition $Al(C_9H_6ON)_3$ with $11 \cdot 10$ per cent Al_2O_3.

 I 8-Hydroxyquinoline

The chief advantage of 8-hydroxyquinoline as a reagent for aluminium is that it gives a direct, positive determination. Unfortunately a large number of elements form similar complexes with the reagent, and a separation procedure must therefore first be applied. Even when great care is taken with this precipitation of aluminium, small amounts of excess reagent tend to be occluded in the precipitate, giving results that are slightly high.

TITRIMETRIC DETERMINATION

8-Hydroxyquinoline can also form the basis of an indirect titrimetric determination of aluminium using bromination of the reagent. The precipitated aluminium complex is dissolved in dilute mineral acid and titrated with a standard solution of potassium bromate. Methyl red, which is destroyed by an excess of the oxidising reagent, is used to indicate the approximate end point. This leaves a slight excess of bromate which can then be measured by adding potassium iodide and continuing the titration with standard sodium thiosulphate solution until the blue colour given with starch just disappears. As with the gravimetric method based upon 8-hydroxyquinoline, a prior separation from interfering elements is necessary, and high values will result from the small amount of co-precipitation of the reagent.

Determination with EDTA and CyDTA. Aluminium forms a very stable complex with EDTA, but the reaction is extremely slow at room temperatures; complex formation is achieved however within a few minutes at the boiling point. For this reason and also because suitable indicators are lacking, most of the titrimetric EDTA methods for aluminium are based upon adding excess reagent, boiling the solution to complete complex formation, and then titrating the excess reagent with another metal ion for which an indicator is available. A variety of metals have been suggested for this, including zinc with xylenol orange as indicator,[1] zinc with dithizone,[2] thorium with alizarin red S,[3] lead with PAR[4] and copper with pyrocatechol violet.[5]

Some improvement can be obtained by replacing EDTA with CyDTA (cyclohexanediaminetetraacetic acid) which reacts much more readily with aluminium;[6] however, as suitable indicators are lacking, a back-titration of the excess complexone is still the preferred method.

It is usually necessary to separate aluminium from other elements that react with EDTA and CyDTA, including iron and titanium, as well as vanadium, manganese, nickel and chromium, all of which can reach minor-element proportion in some silicate rocks. Evans[7] has proposed a procedure involving two CyDTA back-titrations that do not require a prior separation. The first titration gives the sum of iron, aluminium and titanium. In the second, iron alone is titrated—aluminium and titanium being masked with fluoride ion. The titanium content is then determined photometrically in a separate aliquot of the rock solution, enabling the aluminium content to be calculated by difference. Any nickel present in the sample will be reported as iron, and any chromium or zirconium reported as aluminium.

PHOTOMETRIC DETERMINATION

Although numerous reagents have been suggested for photometric determination, none is specific or even selective for aluminium. Procedures based upon the formation of coloured lakes with certain organic compounds such as aluminon[8] (aurine tricarboxylic acid), eriochrome cyanine R,[9] or alizarin red S,[10, 11] although advocated for several years, are subject to interference from iron and other elements when applied to silicate rocks, and are not as precise as one would wish.

Determination with 8-hydroxyquinoline. The complex formed between aluminium and 8-hydroxyquinoline, previously described for both gravimetric and titrimetric determination of aluminium, can also be used for a photometric determination. The yellow solution has a maximum optical density at a wavelength of about 400 nm, and the Beer–Lambert Law is valid up to at least 200 μg Al_2O_3 per 25 ml of chloroform. Some fading occurs, particularly when solutions are exposed to direct sunlight, and Riley[12] has recommended storage in a dark cupboard. The stability depends to some extent upon the quality of the chloroform used.[13]

Riley[12] has described the application of this photometric method to silicate rocks, using 2:2'-dipyridyl to complex iron present in the solution, and applying a correction for titanium. Manganese does not interfere. Low results can be obtained if fluoride ions are not completely removed, although small traces of fluoride can be complexed by adding beryllium.

Determination with pyrocatechol violet (also known as catechol violet and catecholsulphonphthalein-II). Aluminium forms a blue-coloured complex with pyrocatechol violet that has been used by Wilson and Sergeant[14] for the determination of aluminium in silicate rocks and minerals. The recommended pH is $6 \cdot 1$–$6 \cdot 2$, obtained with an ammonium acetate–acetic acid buffer. The colour development occurs over a period of about 1 hour, after which the colour is virtually constant. The Beer–Lambert Law is valid up to 80 μg Al_2O_3 per 100 ml of solution, and maximum values of optical density are obtained at a wavelength of 580 nm.

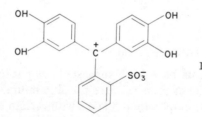

II Pyrocatechol violet

Many elements interfere with the determination, but as a routine method where extreme accuracy is not important, the addition of a solution of 1:10-phenanthroline will serve to complex iron, leaving titanium as the only major interference for which a correction is necessary. In a more rigorous procedure, given in detail below, iron, titanium, vanadium and zirconium can be removed by a cupferron extraction.

Other photometric reagents. Although a large number of other reagents have been suggested for the photometric determination of aluminium, only a very few appear to have been applied to the analysis of silicate rocks. Reagents used for other purposes that might find application in this field include chrome azurol S (Colour Index Mordent Blue 29),[15, 16] xylenol orange,[17] stilbazo[18] [4,4'-bis(dihydroxyphenylazo) stillbene-2,2'-disulphonic acid] and pyrogallol red.[19]

ATOMIC ABSORPTION SPECTROSCOPY

One of the earliest methods using this technique involved the extraction of aluminium cupferrate into isobutylmethyl ketone solution, which was then sprayed into a fuel-rich oxyacetylene flame.[20] This procedure is not very sensitive, and in more recent papers nitrous oxide (high temperature) burners have been suggested.[21, 22] Inter-element effects and compound formation make this element one of the most difficult to determine by this technique. Further work appears to be necessary before reliable results can be obtained for silicate rocks and minerals.

The Separation of Aluminium

EXTRACTION PROCEDURES

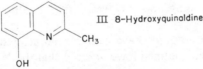

III 8-Hydroxyquinaldine

8-Hydroxyquinaldine(III), unlike 8-hydroxyquinoline, does not form an insoluble aluminium complex in dilute acetic acid solution, but can be used to precipitate iron, titanium and other metal ions. This compound has been suggested by Hynek[23] for the separation of aluminium from those elements that interfere with the gravimetric determination with 8-hydroxyquinoline. However, it has been observed that in the presence of much iron, small amounts of aluminium are lost from the solution by co-precipitation. Riley and Williams[24] have used extraction with 8-hydroxyquinaldine at pH 10 to remove iron, chromium, nickel and vanadium from solutions. Titanium is not removed at this pH, but is extracted in a second operation at pH 4, prior to forming the aluminium complex with 8-hydroxyquinoline at pH 4·5—this low pH being selected to prevent complex formation with beryllium and manganese which are also not removed with 8-hydroxyquinaldine. Zirconium is not extracted, but is not normally present in silicate rocks in amounts sufficient to interfere with the aluminium determination. If present in larger amounts it can be removed as the mauve-coloured lake with quinalizarin sulphonic acid at pH 4·5, which is not extracted with the aluminium into the chloroform solution.

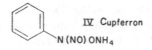

IV Cupferron

N (NO) ONH₄

N-Nitrosophenylhydroxylamine(IV) (used as the ammonium salt, with the trivial name of "cupferron") forms insoluble complexes with ferric iron, titanium and vanadium. These can be removed from aqueous acid solution by extraction with chloroform.[25] Aluminium remains in the aqueous phase and can, after extraction of the excess reagent and removal of residual chloroform, be determined gravimetrically, titrimetrically or photometrically with 8-hydroxyquinoline. Solutions of cupferron in chloroform are unstable and must be kept ice-cold or, as preferred, prepared from the solid reagent *in situ*. Even the solid reagent deteriorates on keeping, and only fresh, good-quality cupferron should be used.

A number of procedures have been recommended using cupferron, and the separations obtained have differed slightly. Vlacil and Zatka[26] describe a two-stage cupferron separation, first from a 2 per cent sulphuric acid solution and then from a solution buffered to pH 5–5·6, for the removal of iron, titanium and other elements, whereas, as noted above, titanium is not extracted if a pH of about 10 is used.[24] Kiss[27] has recommended a procedure based upon removing iron by extraction into isobutylmethyl ketone from 6 N hydrochloric acid, before removing titanium with cupferron. This enables the determination of iron to be made in the same portion of rock material. Miller and Chalmers[28] extracted the cupferrates of iron, titanium and vanadium into *o*-dichlorobenzene solution before extracting aluminium and beryllium into diethyl ether as complexes with acetylacetone. Manganese and nickel are not extracted with cupferron, and some doubt exists as to the extraction of chromium, which can however be removed by volatilisation as chromyl chloride.[29]

PRECIPITATION WITH SODIUM HYDROXIDE

Pritchard[30] has used a separation based upon precipitation with sodium hydroxide in the presence of CyDTA. Aluminium is said to be retained in solution, and iron, titanium, magnesium, manganese and calcium all precipitated. This procedure was investigated by Mercy and Saunders[31] who recorded loss of aluminium by adsorption on the hydroxide precipitate.

ION-EXCHANGE SEPARATION

Aluminium is not retained when hydrochloric acid solutions of silicate or other rocks are allowed to pass through a column of strongly basic anion exchange resin. This enables a good separation to be made from iron, cobalt and zinc which are retained from solutions 9 M in hydrochloric acid. Titanium and manganese accompany the aluminium in the eluate from the column. This separation is useful where a separate determination of iron is to be made in the same sample portion.

ELECTROLYSIS WITH A MERCURY CATHODE

Iron, chromium, cobalt, nickel, zinc and a number of other elements can easily and simply be removed from dilute sulphuric acid solution by electrolysis using a mercury pool cathode.[32] Aluminium remains in the dilute acid solution. This separation procedure has not found wide application to silicate or other rocks, as titanium, vanadium, zirconium and phosphorus are similarly not removed at the cathode, but remain in solution with the aluminium.

The Gravimetric Determination with 8-Hydroxyquinoline

In the procedure given in detail below, iron, titanium and other elements are removed by an extraction of their cupferrates into chloroform, followed by a direct precipitation of aluminium in the aqueous phase with 8-hydroxyquinoline. The precipitate is collected, dried to constant weight in an electric oven, and weighed as $Al(C_9H_6ON)_3$.

Method

Reagents: *Cupferron,* the ammonium salt of *N*-nitrosophenylhydroxylamine. Use only fresh, good quality reagent.
Chloroform.
8-Hydroxyquinoline solution, dissolve 2·5 g of the reagent in 100 ml of 2 N acetic acid.
Ammonium acetate.
Bromocresol purple indicator solution, dissolve 0·1 g in 100 ml of ethanol.

Procedure. Accurately weigh approximately 0·5 g of the finely powdered silicate rock material into a platinum dish, moisten with water and add 1 ml of concentrated nitric acid, 10 ml of 20 N sulphuric acid and 10 ml of concentrated hydrofluoric acid. Transfer the dish to a hot plate and evaporate to fumes of sulphuric acid. Allow to cool, rinse down the sides of the dish with a little water, add 5 ml of water and 5 ml of concentrated hydrofluoric acid and again evaporate to fumes of sulphuric acid. Allow to cool, dilute with 5 ml of water and again evaporate to fumes. Repeat the evaporation once more to ensure expulsion of residual fluorine and to remove most of the remaining sulphuric acid.

Transfer the moist residue to a 250-ml beaker using about 100 ml of 2 N hydrochloric acid and heat on a hot plate until all soluble material has dissolved. Collect any insoluble material on a small paper, wash with a little water and transfer to a small platinum crucible. Dry and ignite the residue and fuse with 0·5 g of a flux containing anhydrous sodium carbonate and borax glass (note 1). Allow to cool, dissolve the melt in 2 N hydrochloric acid and add to the main rock solution.

Transfer the solution to a 250-ml volumetric flask and dilute to volume with 2 N hydrochloric acid. Pipette 100 ml of this solution into a 250-ml beaker, cover with a clock glass and allow to stand in a refrigerator until a temperature of less than about 8° is reached, and then transfer to a 500-ml separating funnel with about 100 ml of ice cold water (note 2). Extract iron, titanium, etc., by adding about 0·5 g of solid cupferron and 20 ml of chloroform. Stopper the funnel, invert and release the pressure by opening the tap momentarily. Close the tap and shake the funnel for about 1 minute. The organic layer becomes coloured with the iron cupferrate, and can be run off and discarded. Repeat the extraction with 0·25 g portions of cupferron and 20 ml aliquots of chloroform until the extracts are no longer coloured, and then with pure chloroform two or three times to remove all excess cupferron from the aqueous phase. Discard the organic extracts and run the aqueous solution into a 400-ml beaker. Rinse the funnel with a little N hydrochloric acid and add to the main solution.

Heat to boiling to remove all traces of chloroform remaining in the solution and add 8 g of ammonium acetate, a few drops of bromocresol purple indicator solution and aqueous ammonia drop by drop until the yellow colour imparted by the indicator is just replaced by purple. Heat the solution to about 60° and precipitate the aluminium by slowly adding 20 ml of the 8-hydroxyquinoline solution (note 3). Bring the solution just to boiling and allow to stand for 30 minutes.

Collect the precipitate on a weighed, sintered glass crucible of medium porosity, wash with cold water, dry in an electric oven set at a temperature of 130–140° and weigh as $Al(C_9H_6ON)_3$. This precipitate contains $11·10$ per cent Al_2O_3.

Notes: 1. Any corundum present in the rock material is particularly resistant to decomposition unless borax or boric acid is added. For most rocks $0·5$ g of a flux containing 10 per cent borax glass in anhydrous sodium carbonate will be sufficient.

2. Failure to keep the temperature down will result in the formation of a gummy organic product on the addition of cupferron.

3. Sufficient reagent should be added to provide 15–20 per cent excess over that required as calculated from the composition of the aluminium precipitate. 1 ml of a $2·5$ per cent solution of 8-hydroxyquinoline will precipitate $2·9$ mg of Al_2O_3.

4. The precipitated complex of aluminium should be pure yellow in colour, but may be tinged green by the presence of a trace of iron remaining after the cupferron extraction. The residue can be removed from the sintered glass crucible with 6 N hydrochloric acid, and the crucible cleaned with concentrated nitric acid before the next determination.

5. Some analysts prefer to remove the bulk of mineral acid from the solution after the extraction of the cupferrates by evaporation with concentrated nitric acid with a small amount of sulphuric acid present. This serves also to destroy any organic material remaining in the solution.

The Titrimetric Determination with CyDTA

This procedure is similar to that described by Mercy and Saunders,[31] and is based upon the removal of interfering elements by a chloroform extraction of their cupferrates.

Method

Reagents: *Cyclohexanediaminetetraacetic acid solution* (CyDTA), add approximately 8 g of reagent to 100 ml of water and dissolve by adding sodium hydroxide solution drop by drop. Dilute to 1 litre with water and store in a polyethylene bottle.

Hexamethylenetetramine solution, saturated solution in water.

Xylenol orange indicator solution, dissolve $0·25$ g in 50 ml of water.

Lead nitrate solution, dissolve $6·5$–$6·7$ g of lead nitrate in 1 litre of water.

Aluminium chloride standard solution, accurately weigh about $0·25$ g of pure clean aluminium foil or wire into a small beaker and dissolve in about 40 ml of 3 N hydrochloric acid. Transfer to a 1 litre volumetric flask and dilute to volume with water. This solution contains about 500 μg Al_2O_3 per ml, the exact concentration can be calculated from the weight of aluminium taken.

Standardisation. Transfer 25 ml of the standard aluminium solution to a 500-ml conical flask, add 5 ml of N hydrochloric acid, 150 ml of water and 25 ml of CyDTA solution. Add sufficient hexamethylenetetramine buffer solution to bring the pH of the solution to a value in the range 5·0–5·5, add a few drops of indicator solution and titrate the excess CyDTA solution with the lead nitrate solution to an end-point given by the permanent appearance of a violet colour.

Also transfer 15 ml of the CyDTA solution to a separate 500-ml conical flask, add 5 ml of N hydrochloric acid, 150 ml of water, sufficient hexamethylenetetramine buffer solution to bring the pH to 5·0–5·5 and a few drops of indicator solution. Again titrate with lead nitrate solution to a violet endpoint. Calculate the equivalence of the lead solution to the CyDTA solution, and hence the equivalence of the CyDTA solution to that of the standard aluminium.

Procedure. Prepare a solution of the rock material by evaporation with hydrofluoric, nitric and sulphuric acids as described above for the method based upon precipitation with 8-hydroxyquinoline, but using only 0·1 g of the finely ground sample material. Decompose any residue by fusion, add the acid extract to the main rock solution to give a solution volume of about 30 ml and extract iron, titanium and vanadium with a chloroform solution of cupferron. Remove all organic material by extraction with chloroform containing a little acetone.

Transfer the aqueous solution to a 400-ml beaker, rinse the funnel with a little water and add the washings to the solution in the beaker. Add 5 ml of concentrated perchloric acid and 5 ml of concentrated nitric acid and evaporate to fumes of perchloric acid. As soon as fumes appear add a few drops of concentrated hydrochloric acid and continue the evaporation (note 1). After about 5 minutes fuming add a further few drops of concentrated hydrochloric acid, and repeat after fuming for a further 5 minutes. Continue the evaporation to incipient crystallisation, then allow to cool, add 150 ml of water, warm to give a clear solution and allow to cool.

Add 25 ml of the CyDTA solution (note 2), add buffer solution to bring the pH to 5·0–5·5, and a few drops of indicator solution. Titrate the excess CyDTA with the lead nitrate solution to a violet endpoint and hence calculate the aluminium content of the solution.

Notes: 1. This evaporation with concentrated hydrochloric acid serves to volatilise any chromium as chromyl chloride, and can be omitted for rocks low in chromium.

2. This provides 10–15 ml of CyDTA solution in excess for rocks containing about 10 per cent Al_2O_3, and should be increased for those rocks containing more alumina.

3. If manganese or nickel are present in more than trace amounts, these should be determined separately and corrections applied to the aluminium value as determined. 1 mg MnO is equivalent to $0 \cdot 72$ mg Al_2O_3 and 1 mg NiO to $0 \cdot 68$ mg Al_2O_3.

4. According to Mercy and Saunders,[31] vanadium is not completely removed in the cupferron extraction. Interference from this small amount can be prevented by adding 3 drops of concentrated (100-vol) hydrogen peroxide to the solution before adding the indicator prior to titrating with lead nitrate solution.

The Photometric Determination with 8-Hydroxyquinoline

This procedure is based upon that given by Riley.[12] Apart from the removal of silica, no prior separations are made.

Method

Reagents: *8-Hydroxyquinoline solution*, dissolve $1 \cdot 25$ g of the reagent in 250 ml of a pure grade of chloroform and store in a refrigerator. This solution deteriorates slowly giving a brown coloration, when it should be discarded.

Complexing reagent solution, dissolve 1 g of hydroxylamine hydrochloride, $3 \cdot 6$ g of sodium acetate trihydrate and $0 \cdot 4$ g of beryllium sulphate tetrahydrate in 50 ml of water. Add $0 \cdot 04$ g of $2:2'$-dipyridyl dissolved in 20 ml of $0 \cdot 2$ N hydrochloric acid and dilute to 100 ml with water. *Note:* before handling beryllium compounds or solutions read the introduction to Chapter 12.

Standard aluminium stock solution, dissolve $0 \cdot 106$ g of pure aluminium wire or foil in dilute hydrochloric acid, avoiding an excess, transfer to a 1 litre volumetric flask and dilute to volume with water. This solution contains 200 μg Al_2O_3 per ml.

Standard aluminium working solution, pipette 50 ml of the stock solution into a 500-ml volumetric flask and dilute to volume with water. This solution contains 20 μg Al_2O_3 per ml.

Procedure. Decompose a $0 \cdot 1$ g portion of the finely powdered silicate rock material by evaporation with hydrofluoric, nitric and perchloric acids and dilute with water to 500 ml as outlined previously for the determination of the alkali metals, p. 82.

Pipette 5 ml of this rock solution into a 75- or 100-ml separating funnel, add 5 ml of water and 10 ml of the complexing reagent solution (measured with an "automatic pipette" or similar device and NOT by mouth-operated pipette— see introduction to Chapter 12), and allow to stand for a few minutes. Then add 20 ml of the 8-hydroxyquinoline reagent solution. Stopper tightly, invert the funnel and release the pressure by momentarily opening the tap. Close the tap and shake for 5 to 8 minutes, releasing the pressure at intervals. Allow the layers to separate and run the organic layer through a small wad of filter paper inserted in the stem of the funnel into a 25-ml volumetric flask. Rinse the separating funnel three times with 3 or 4 ml of chloroform and add these washings to the solution in the 25-ml flask. Dilute to volume with chloroform.

Measure the optical density in a 1-cm cell, preferably stoppered, using chloroform as the reference solution, with the spectrophotometer set at a wavelength of 410 nm. Measure also the optical density of a reagent blank extract, prepared in the same way as the sample extract, but omitting the rock material.

Calibration. Transfer aliquots of 0–10 ml of the standard aluminium solution containing 0–200 μg Al_2O_3 to separate 75- or 100-ml separating funnels and dilute each solution to 10 ml with water if necessary. Add complexing reagent and 8-hydroxyquinoline solution and extract the aluminium complex as described above. Measure the optical density of each extract relative to chloroform in 1-cm cells at a wavelength of 410 nm, as for the sample extract, and plot the relation of optical density to aluminium concentration.

Notes: 1. The aluminium complex is pure yellow in colour, and gives a pure yellow solution in chloroform. If the pH of the aqueous solution is too low, not all the iron will be reduced by the hydroxylamine and a green coloured extract will be obtained. This can be avoided by increasing the buffer concentration to bring the pH to a value between 4·9 and 5·0.

2. A small correction is necessary for any titanium present in the sample material. This can be determined by adding 10 μg TiO_2 (equivalent to 1 per cent TiO_2 in the sample), to 100 μg Al_2O_3 and when extracting with 8-hydroxyquinoline as for the calibration solutions, measuring the increased optical density due to titanium alone.

The Photometric Determination with Pyrocatechol Violet

This procedure, due to Wilson and Sergeant,[14] is also based upon the use of cupferron for the removal of interfering elements.

Method

Reagents: *Pyrocatechol violet solution,* dissolve 0·075 g of reagent in 50 ml of water.

Hydroxylamine hydrochloride solution, dissolve 10 g of reagent in 100 ml of water.

Buffer solution, dissolve 50 g of ammonium acetate in 450 ml of water, bring the pH to 6·2 by adding acetic acid (use a pH meter) and dilute to 500 ml with water.

Cupferron.

Chloroform.

Standard aluminium working solution, dilute the stock solution obtained as described above, with water to give a working solution containing 4 μg Al_2O_3 per ml.

Procedure. Decompose 0·1 g of the finely powdered silicate rock material by evaporation with hydrofluoric and sulphuric acids as previously described and extract iron, titanium, vanadium and zirconium with cupferron. Transfer the aqueous solution to a 250-ml volumetric flask and dilute to volume with water. Pipette an aliquot of this solution containing not more than 40 μg of aluminium to a 100-ml beaker and dilute to 20 ml with water. Add 2 ml of hydroxylamine hydrochloride solution, 2 ml of pyrocatechol violet solution and 5 ml of the buffer solution. Mix well and bring the pH to a value of 6·1–6·2 by careful addition of ammonia (use a pH meter), taking care not to allow the solution to become distinctly alkaline at any time.

Rinse the solution into a 100-ml volumetric flask with a little water, add 50 ml of the buffer solution and dilute to volume with water. Set aside for 2 hours and then measure the optical density in 1-cm cells with the spectrophotometer set at a wavelength of 580 nm. Determine also the optical density of a reagent blank solution similarly prepared but omitting the sample material.

Calibration. Transfer aliquots of 0–20 ml of the standard aluminium solution containing 0 to 80 μg Al_2O_3 to separate 100-ml beakers and dilute each solution to 20 ml as necessary. Add hydroxylamine hydrochloride, pyrocatechol violet and buffer solutions and continue as described above for the sample solution. Plot the relation of optical density to aluminium concentration.

References

1. CLULEY H. J., *Glass Technol.* (1961) **2**, 71.
2. VOINOVITCH I. A. and LEFRANC-KOUBA A., *Chim. Anal.* (1960) **42**, 543.
3. TER HAAR K. and BAZEN J., *Anal. Chim. Acta* (1954) **10**, 23.
4. LANGMYHR F. J. and KRISTIANSEN H., *Anal. Chim. Acta* (1959) **20**, 524.
5. KRIZ M., *Sklar a Keramik* (1956) **6**, 140.
6. PRIBIL R. and VESELY V., *Talanta* (1962) **9**, 23.
7. EVANS W. H., *Analyst* (1967) **92**, 685.
8. COREY R. B. and JACKSON M. L., *Analyt. Chem.* (1953) **25**, 624.
9. GLEMSER O., RAUELF E. and GRISEN K., *Zeit. Anal. Chem.* (1954) **141**, 82.
10. BEHR A., BLANCHET M. L. and MALAPRADE C., *Chim. Anal.* (1960) **42**, 501.
11. SHAPIRO L. and BRANNOCK W. W., *U.S. Geol. Surv. Bull.* 1144-A, 1962.
12. RILEY J. P., *Anal. Chim. Acta* (1958) **19**, 413.
13. LINNELL R. H. and RAAB F. H., *Analyt. Chem.* (1961) **33**, 154.
14. WILSON A. D. and SERGEANT G. A., *Analyst* (1963) **88**, 109.
15. BUCK L., *Chim. Anal.* (1965) **47**, 10.
16. MATSUO T. and SUGAWARA M., *Japan Analyst* (1960) **9**, 706.
17. OTOMO M., *Bull. Chem. Soc. Japan* (1963) **36**, 809.
18. LANDI M. F. and BRAICOVICH L., *Metallurg. Ital.* (1962) **54**, 389.
19. TANAKA T., NAKAGAWA Y. and HONDA S., *Japan Analyst* (1961) **10**, 1148.
20. CHAKRABARTI C. L., LYLES G. R. and DOWLING F. B., *Anal. Chim. Acta* (1963) **29**, 489.
21. CAPACHO-DELAGADO L. and MANNING D. C., *Analyst* (1967) **92**, 553.

22. RAMAKRISHNA T. V., WEST P. W. and ROBINSON J. W., *Anal. Chim. Acta* (1967) **39, 81.**
23. HYNEK R. J., *ASTM Spec. Tech. Publ.* (1959) No. 238, p. 36.
24. RILEY J. P. and WILLIAMS H. P., *Mikrochim. Acta* (1959) 825.
25. MILNER G. W. C. and WOODHEAD J. L., *Anal. Chim. Acta* (1955) **12,** 127.
26. VLACIL F. and ZATKA V., *Chem. Prumysl* (1961) **11,** 139.
27. KISS E., *Anal. Chim. Acta* (1967) **39,** 223.
28. MILLER C. C. and CHALMERS R. A., *Analyst* (1953) **78,** 686.
29. VAN LOON J. C., *Talanta* (1966) **13,** 1555.
30. PRITCHARD D. T., *Anal. Chim. Acta* (1965) **32,** 184.
31. MERCY E. L. P. and SAUNDERS M. J., *Earth Planet. Sci. Lett.* (1966) **1,** 169.
32. MAXWELL J. A. and GRAHAM R. P., *Chem. Revs.* (1950) **46,** 471.

CHAPTER 9

ANTIMONY

Occurrence

The occurrence of antimony in silicate rocks was discussed by Onishi and Sandell[1] and by Fleisher.[2] Most of the igneous rocks examined by these authors were found to contain 0·1 to 0·2 ppm Sb—values that are close to the limit of determination of the spectrophotometric methods used. More recent determinations by Lombard et al.,[3] Steinnes[4] and others in a number of standard rocks by neutron activation analysis gave a range of antimony values from 0·04 ppm in G-2 to 4·6 ppm in AGV-1. These values are well below those that can be determined by atomic absorption spectroscopy.[5] Antimony shows no marked preference for silicic or subsilicic rocks, resembling arsenic in this respect.

Shales are strongly variable in antimony, some contain no more than igneous rocks, others a few ppm. It is possible that antimony, like certain other elements, is concentrated in those shales that are rich in sulphides and carbonaceous matter. Limestones and sandstones are probably low in antimony.

The Determination of Antimony in Silicate Rocks

In their examination of silicate rocks. Onishi and Sandell[6] used rhodamine B (XIV, p. 248) as a photometric reagent. Special care was required to ensure that a complete separation of antimony was obtained, particularly from iron, gallium, thallium, tungsten and gold, which all give sensitive colour reactions with rhodamine B.

` In this method the rock material was obtained in sulphuric acid solution. Copper sulphate was added and the antimony and copper precipitated together with hydrogen sulphide. The separation is far from complete and additional separation stages were required to remove the final traces of iron, gallium, thallium and gold. This method is limited more by poor reproducibility at the ppm level than by sensitivity. The authors claim a limit of determination of $0 \cdot 1 - 0 \cdot 2$ ppm Sb—the concentration that they observed in many igneous rocks.

Rhodamine B was also used by Ward and Lakin[7] to determine antimony in soils and rocks. Their procedure incorporated an extraction into isopropyl ether, followed by the formation of the coloured complex with rhodamine B in the etherial solution. Iron(III), arsenic, gold, tin and thallium are also extracted into isopropyl ether and could give rise to serious errors in antimony determinations, particularly at the ppm level. The procedure as described was devised primarily for soil samples containing relatively large amounts of antimony, and is not directly applicable to rock material containing much less than $0 \cdot 5$ ppm Sb.

The reaction between chloroantimonate ions and brilliant green has also been used to determine antimony.[8] Soil and rock samples can be decomposed either by fusion with potassium hydrogen sulphate or by heating with ammonium chloride. The complex with brilliant green is formed in hydrochloric acid solution containing sodium hexametaphosphate, and is extracted into toluene for photometric measurement. The lower limit of determination appears to be about $0 \cdot 4$ ppm Sb.

PHOTOMETRIC DETERMINATION

The procedure given here is based upon that described by Schnepfe.[9] Antimony in the sample solution is reduced and volatilised as stibine, SbH_3. This is absorbed in mercuric chloride solution, oxidised with ceric sulphate and extracted as rhodamine B chloroantimonate. Arsenic distils as arsine, AsH_3, but the recovery of 1 μg Sb in the presence of 1000 μg As suggests that arsenic is unlikely to interfere in the determination. If it is present in high concentration in the rock sample, additional oxidant will be required.

A variety of procedures were used by Schnepfe[9] for sample decomposition including fusion with a mixture of potassium and sodium carbonates, potassium hydroxide, potassium hydroxide followed by sodium peroxide and potassium pyrosulphate. Only fusion with potassium pyrosulphate was successful in recovering antimony added to silicate rocks, and this procedure is used below. The decomposition of the rock material is far from complete using this fusion and the possibility therefore exists that some of the antimony present in the silicate rock may not be recovered.

Method

Apparatus. For the reduction of antimony, a hydride generator is required (also known as an arsine generator). A simple form of this is shown in Fig. 13.

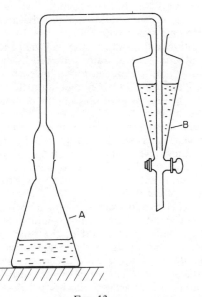

FIG. 13

Reagents: *Tartaric acid solution,* dissolve 10 g in water and dilute to 100 ml.
Zinc shot, low in antimony.
Cadmium sulphate solution, dissolve 5 g in water and dilute to 100 ml.
Mercuric chloride solution, dissolve 5 g of mercuric chloride in 1 litre of 6 M hydrochloric acid.
Ceric sulphate solution, dissolve 10·6 g of tetrasulphateoceric acid, $H_4Ce(SO_4)_4$ in 80 ml of water, add 5·56 ml of 20 N sulphuric acid and dilute to 100 ml.
Rhodamine B solution, dissolve 0·2 g in 100 ml of water.
Rhodamine B-hydroxylamine hydrochloride solution, dilute 50 ml of the rhodamine B solution with 50 ml of water, add 1 g of hydroxylamine hydrochloride. Prepare freshly each day.

Standard antimony, stock solution, 100 μg Sb per ml. Dissolve 0·274 g of potassium antimonyl tartrate hemihydrate in 1 litre of N sulphuric acid. Prepare working standards by dilution of this stock solution with water, as required.

Procedure. Weigh approximately 1·5 g of potassium pyrosulphate into a fused silica crucible and fuse gently to remove all moisture and give a quiescent melt. Allow to cool and accurately weigh approximately 0·2 g of the finely powdered silicate rock material into the crucible. Cover the crucible with a silica lid and fuse gently for 10 to 15 minutes. Allow to cool. Add 14 ml of water and 1 ml of tartaric acid solution and dissolve all soluble constituents of the melt by heating on a steam bath.

Transfer the solution and residue to the hydride generator, Fig. 13, rinsing the crucible and lid with 4 ml of water followed by 40 ml of 20 N sulphuric acid. Add the washings to the generator followed by 1 ml of the cadmium sulphate solution. Place a small filter funnel in the neck of the flask and bring the liquid gently to the boil. Do not prolong the boiling, but cool under a tap to bring to a temperature of 20° to 25°.

Add 15 g of zinc shot to the generator and immediately complete the assembly of the apparatus as shown in Fig. 13, with the delivery tube in the 60-ml separating funnel which contains 20 ml of the mercuric chloride solution. Allow to stand for 2 hours, then remove the generator and delivery tube, rinsing any solution adhering to the tube back into the separating funnel. Discard the contents of the flask.

Using a safety pipette, add exactly 10 ml of benzene to the separating funnel followed by 0·5 ml of the cerium solution. Stopper the funnel and shake for about a minute. Add 10 ml of the rhodamine B-hydroxylamine hydrochloride solution, again stopper and shake for 1 minute. Allow the phases to separate, drain off and discard the lower aqueous phase. Wash the benzene layer by shaking twice with 10 ml of 4 M hydrochloric acid and discard the washings. Drain the benzene layer into a small glass beaker and transfer it to a 15-ml centrifuge tube. Centrifuge to give a clear solution and then measure the optical density of the solution relative to benzene at a wavelength of 550 nm.

Prepare working curves for 0 to 1·0 μg Sb and 0 to 10 μg Sb using aliquots of a working standard solution. Transfer each aliquot to the hydride generator, add 1·5 g of potassium pyrosulphate, 1 ml of tartaric acid solution, 18 ml of water, 1 ml of cadmium solution and 40 ml of 20 N sulphuric acid. Then proceed as described above.

POLAROGRAPHIC DETERMINATION

Both arsenic and antimony form well-developed polarographic reduction waves in dilute sulphuric acid solution containing potassium chloride and methylene blue. These waves have been used by Morachevskii and Kalinin[10] to determine these two elements in silicate materials. Arsenic gives two reduction waves at $E_{\frac{1}{2}} = -0\cdot42$ V and $-0\cdot65$ V, although only the second can be used for the quantitative determination of arsenic; whilst antimony gives a single wave at $E_{\frac{1}{2}} = -0\cdot14$ V, all relative to a saturated calomel electrode. In the proportions in which they occur in silicate rocks it is not necessary to separate arsenic from antimony, although they must be separated together from other constituents of silicate rocks. This can be done conveniently by distillation of the halides. The lower limit of determination is of the order of $0\cdot1$ ppm Sb.

Method

Reagents: *Hydrazine sulphate.*

Acid bromide solution, dissolve 5 g of potassium bromide in 100 ml of water and add with cooling 200 ml of concentrated sulphuric acid.

Supporting electrolyte, $0\cdot05$ M sulphuric acid, $0\cdot5$ M potassium chloride and 10^{-5} M methylene blue.

Standard antimony, stock solution, dissolve $0\cdot0120$ g of antimony trioxide in 55 ml of concentrated hydrochloric acid and dilute with water to 100 ml. This solution contains 100 μg Sb per ml.

Standard antimony, working solution, dilute 10 ml of the stock solution to 100 ml with 6 M hydrochloric acid. This solution contains 10 μg Sb per ml.

Procedure. Accurately weigh approximately 5 g of the finely powdered rock material into a platinum basin and decompose by evaporation to copious fumes with 5 ml of concentrated nitric acid, 5 ml of 20 N sulphuric acid and 25 ml of concentrated hydrofluoric acid. Allow the basin to cool, break up any crust that has formed and, using a small amount of water, transfer the contents of the basin to a 100-ml distilling flask (Fig. 14). Add $0\cdot5$ g of potassium bromide, 20 ml of concentrated sulphuric acid, 10 ml of concentrated hydrochloric acid and $0\cdot5$ g of hydrazine sulphate. Use a 50-ml conical flask containing 10 ml of water as a receiver.

Gently heat the distilling flask until a temperature of 80° is reached, then add drop by drop 30 ml of the acid bromide solution. Continue heating until a distillant temperature of 190–200° is reached, and distil at this temperature for about 1 hour. Add 2 ml of concentrated nitric acid to the distillate and evaporate on a steam bath to a volume of 1–2 ml. Add 1 ml of 6 N sulphuric acid and continue the evaporation, this time to fumes of sulphuric acid. Allow to cool, add $0\cdot1$ g of hydrazine sulphate and evaporate to complete dryness. Dissolve the residue in 1 ml of the supporting electrolyte, transfer to a polarographic cell, de-aerate the solution with carbon dioxide or nitrogen in the usual way and determine the arsenic and antimony in the sample by polarographic reduction, measuring the waves at $-0\cdot14$ V (antimony) and $-0\cdot65$ V (arsenic), relative to a saturated calomel electrode.

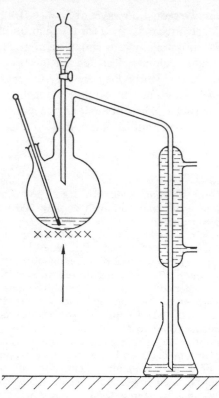

FIG. 14. Apparatus for the distillation of arsenic and antimony.

Calibrate the polarograph by taking aliquots of the standard antimony (and arsenic) solutions, evaporating with sulphuric acid, etc., as described above, and measuring the reduction waves at $-0 \cdot 14$ V and $-0 \cdot 65$ V.

References

1. ONISHI H. and SANDELL E. B., *Geochim. Cosmochim. Acta* (1955) **8**, 213.
2. FLEISCHER M., *Econ. Geol.* (1955) **50**, 970.
3. LOMBARD S. M., MARLOW K. W. and TANNER J. T., *Anal. Chim. Acta* (1971) **55**, 13.
4. STEINNES E., *Analyst* (1972) **97**, 241.
5. STRESKO V. and MARTINY E., *Atomic Absorpt. Newsl.* (1972) **11**, 4.
6. ONISHI H. and SANDELL E. B., *Anal. Chim. Acta* (1954) **11**, 444.
7. WARD F. N. and LAKIN H. W., *Analyt. Chem.* (1954) **26**, 1168.
8. STANTON R. E. and McDONALD A. J., *Analyst* (1962) **87**, 299.
9. SCHNEPFE M. M., *Talanta* (1973) **20**, 175.
10. MORACHEVSKII YU. V. and KALININ A. I., *Zavod. Lab.* (1961) **27**, 272.

CHAPTER 10

ARSENIC

Occurrence

Much of the early data on the occurrence and distribution of arsenic in silicate rocks and minerals is clearly very unreliable. The first paper of real interest was probably that of Onishi and Sandell,[1] who recovered arsenic from silicates by distillation as arsine, AsH_3. Table 14, adapted from the data given in their paper, shows only small variations in the arsenic content of the various rock types. These differences are too small to establish any pattern with certainty. It is clear, however, that arsenic does not exclusively or even principally follow any of the major constituents of silicate rocks.

TABLE 14. THE ARSENIC CONTENT OF SOME IGNEOUS ROCKS

Rock type	As, ppm	
	mean	range
Rhyolites, felsites, liparites	3·5	0·7– 7·5
Silicic volcanic glasses	5·9	2·0–12·2
Granitic rocks	1·7*	0·0– 8·5*
Intermediate rocks	2·4	0·5–13·4
Basalts and diabases	1·9	0·6– 9·0
Gabbros	1·4	0·3– 5·6
Peridotites, dunites, etc.	1·0	0·3– 3·0
Serpentines	2·8	0·8–15·8

* Excluding 0·26 per cent As in a granite from Sterkwater, S.A.

Shales contain rather more arsenic than igneous rocks, whilst those shales containing sulphide minerals and carbonaceous matter tend to be even higher in arsenic. A value of 10 ± 5 ppm was suggested by Onishi and Sandell as the likely abundance of arsenic in shales, and the range noted in twenty-nine individually analysed samples, mostly from the U.S.A., was $0 \cdot 8$–59 ppm. Except for those specimens containing pyrite or carbonaceous matter, sandstones and limestones contain only small amounts of arsenic, usually in the range 1–2 ppm.[1]

The occasional occurrence of both igneous and sedimentary rocks with what appear to be unusually large amounts of arsenic can often be attributed to the presence of arsenopyrite or other sulphide mineral containing arsenic. The association of arsenic with cobalt has also been noted.[2]

The Determination of Arsenic in Silicate Rocks

THE SEPARATION OF ARSENIC

Distillation as arsine, AsH_3, although traditionally used for the recovery of arsenic, and used by Onishi and Sandell[3] in their examination of silicate rocks, has the disadvantage that the reagents used for the reaction may contain similar amounts of arsenic to the rock samples. Another distillation procedure commonly used is that of arsenious chloride from aqueous hydrochloric acid solution at a temperature not exceeding 108°. At this temperature neither antimony nor tin will distil. Any germanium present in the sample material will accompany the arsenic, but is not likely to interfere. Small amounts of selenium may also be recovered in the distillate.[4] Where arsenic and antimony can be determined in the same solution, as for example by polarography,[5] a higher temperature can be used to obtain both elements in the distillate. Details of this procedure are given in the preceding section (antimony, p. 112).

The extraction of arsenious iodide with carbon tetrachloride from solutions 9–12 M in hydrochloric acid[6] does not appear to have been applied to silicate or carbonate rocks, but could provide an alternative separation procedure to distillation or a co-crystallisation technique, as for example, that described by Portman and Riley.[7]

In the initial stages of rock decomposition, loss of arsenic as the volatile arsenious fluoride may occur if hydrofluoric acid is used. An alkali fusion is preferred, although Rader and Grimaldi[4] have successfully used both fusion and acid digestion techniques to determine arsenic in marine shales.

PHOTOMETRIC DETERMINATION

A variety of procedures are in common use for the photometric determination of arsenic, many of them based upon the formation of an arsenomolybdate complex which is then reduced to a molybdenum blue. Silica and phosphorus, which form similar yellow-coloured complexes that can be reduced to molybdenum blues, will interfere unless removed in the earlier stages of the analysis. For this reason it is recommended[7] that volumetric flasks used should be filled with concentrated sulphuric acid and allowed to stand overnight.

Stannous chloride has been used[8, 9] to reduce the arsenomolybdate, but the blue colour produced is relatively unstable. Hydrazine[10] and ascorbic acid[11] have been used to give stable blues, but the reduction is slow and requires several hours for complete colour development. Portman and Riley[7] have suggested using a mixed colour-reagent containing ammonium molybdate, ascorbic acid and an antimony salt, to produce an arsenic–molybdenum blue by a reaction analogous to that described for the determination of phosphorus.[12]

It is necessary to effect a compromise between a satisfactory sensitivity to arsenic (coupled with a low reagent blank) obtained at high acid concentrations, and rapid colour development which is promoted by low acidities. A major factor in the formation of molybdenum blue is the relative proportion of acid and molybdenum present—the amount of reducing agent being only of minor importance provided that it is above a critical level.[7]

The absorption spectrum of a molybdenum blue solution (Fig. 15) resembles that obtained by reduction of silicomolybdate or phosphomolybdate solutions, with maximum absorption occurring at a wavelength of about 835 nm. With their antimonial reagent, Portmann and Riley recommended a somewhat higher wavelength of 866 nm. Straight-line calibration graphs indicate that the Beer–Lambert Law is followed up to at least $2 \cdot 5$ ppm As (250 μg As per 100 ml of solution), covering the concentration range normally encountered in the examination of silicate, carbonate and other types of rock. The lower part of this range is shown in Fig. 16.

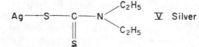

V Silver diethyldithiocarbamate

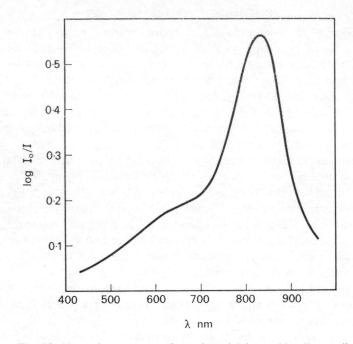

FIG. 15. Absorption spectrum of arsenic-molybdenum blue (1-cm cells,
40 μg As/25 ml).

An alternative spectrophotometric method for arsenic is that based
upon the red colour given with the silver salt of diethyldithiocarbamate
(V) in pyridine solution.[13, 14] To form the complex it is necessary to
reduce the arsenic to arsine, for which arsenic-free zinc and hydrochloric acid should be used. Antimony and germanium form similar
complexes[15] of somewhat lower sensitivity and can interfere, although
the small amounts of antimony and germanium present in silicate rocks
do not give rise to significant errors. Absorption spectra of the three
complexes are shown in Fig. 17.

A Molybdenum Blue Method Using Acid Digestion

Rader and Grimaldi[4] have reported that a simple acid digestion
technique can be used to recover arsenic from certain marine shales.

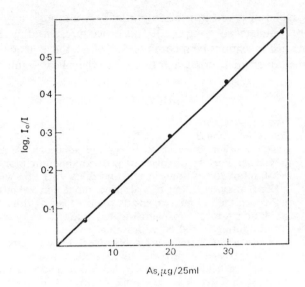

FIG. 16. Calibration graph for arsenic by a molybdenum blue method (1-cm cells, 840 nm).

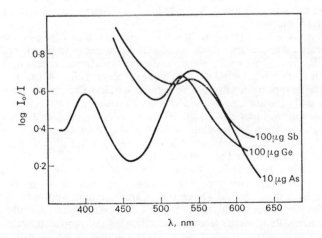

FIG. 17. Absorption spectra of arsenic, antimony and germanium complexes of diethyldithiocarbamate.

The rock material is not appreciably attacked, although all sulphides present are completely decomposed. This procedure, avoiding the use of an alkaline fusion, cannot be recommended where the arsenic is present in very small quantities, nor when present in the silicate matrix.

Method

Reagents: *Hydrazine sulphate.*
Potassium bromide.
Colour reagent, prepare two separate solutions, A and B as follows. For A, dissolve 10 g of ammonium molybdate in 250 ml of 20 N sulphuric acid and dilute to 1 litre with water. For B, dissolve 0·75 g of hydrazine sulphate in 500 ml of water. Prepare the colour reagent as required by diluting 50 ml of solution A to 450 ml with water, adding 15 ml of solution B and diluting to 500 ml with water.
Standard arsenic stock solution, dissolve 0·132 g of pure arsenious oxide in 5 ml of approximately 5 N sodium hydroxide solution, make just acid with dilute hydrochloric acid and dilute to 1 litre with water. This solution contains 0·1 mg As per ml.
Standard arsenic working solution, dilute 5 ml of the stock solution to 500 ml with water. This solution contains 1 μg As per ml.

Procedure. Accurately weigh approximately 1 g of the finely powdered rock material into a 100-ml beaker, add 10 ml of concentrated nitric acid, gently swirl the contents to moisten the whole of the sample and digest on a steam bath for 15 minutes.

Allow to cool, add 10 ml of 20 N sulphuric acid and 7 ml of concentrated perchloric acid and evaporate on a hot plate to fumes of sulphuric acid. Allow to cool. If any organic matter remains after this treatment, repeat the evaporation with further additions of nitric and perchloric acids. Allow to cool and repeat the evaporation to fumes of sulphuric acid to complete the expulsion of nitric and perchloric acids. Again allow to cool, add 15 ml of water, stir to break up any residue, and using 20 ml of concentrated hydrochloric acid transfer the solution and residue to the distilling flask (Fig. 14, p. 113). Add 0·5 g of potassium bromide and 1 g of hydrazine sulphate and assemble the apparatus. Transfer 10 ml of water to the receiver.

Distil the sample solution, keeping the tip of the delivery tube below the surface of the water in the receiver and ensuring that the temperature in the distilling flask does not rise above 108°. Continue until approximately 25 ml of distillate have been collected (35 ml in the receiver), this usually takes 30 to 35 minutes. Remove the receiver, add 10 ml of concentrated nitric acid and evaporate to dryness on a hot plate. Place the flask in an electric oven at a temperature of 130°, and leave for 30 minutes to complete the expulsion of all traces of nitric acid. Add exactly 25 ml of the colour reagent to the flask, cover and heat on a steam bath for 20 minutes to develop the colour. Remove the flask, allow to cool, and measure the optical density in 1-cm cells using the spectrophotometer set at a wavelength of 840 nm (note 1). Measure also the optical density of a reagent blank solution prepared in the same way as the sample solution, but omitting the sample material.

Calibration. Transfer of aliquots of 5–40 ml of the standard arsenic solution containing 5 to 40 μg As, to separate 100-ml conical flasks. Add 10 ml of concentrated nitric acid to each and evaporate to dryness on a hot plate. Proceed as described above for colour development and then measure the optical density of each solution. Plot the relation of optical density to arsenic concentration.

Note: 1. The direct measurement of optical density can be made if the sample solution contains less than 50 μg As. If the sample material contains more arsenic than this, transfer the solution to a 250-ml volumetric flask and dilute to volume with 0·5 N sulphuric acid. Additional colour reagent will not be required unless the arsenic content of the solution exceeds 400 μg.

A Molybdenum Blue Method Using a Fusion[4]

The flux used consists of a mixture of potassium carbonate and magnesium oxide. Sodium carbonate should not be used, as this will result in the precipitation of sodium chloride during the distillation. No provision is made for removing silica, which tends to precipitate towards the end of the distillation.

Method

Reagents: *Hydrobromic acid,* concentrated.
Fusion mixture, mix one part by weight of magnesium oxide with three parts by weight of potassium carbonate.

Procedure. Accurately weigh 0·25–0·5 g of the finely powdered rock material into a 30-ml platinum crucible and add 2–3·5 g of the fusion mixture. Mix the charge carefully and cover with an additional 0·5 g of the fusion mixture. Cover the crucible and heat in an electric furnace set at 650° for a period of 30 minutes. Gradually raise the temperature of the furnace to 900° and then heat at this temperature for a further period of 30 minutes or until all organic matter has been destroyed. Allow to cool. Place the crucible in a beaker and add 20 ml of water (but no alcohol, even if manganate has been formed in the fusion). Cover the beaker and carefully add 60 ml of concentrated hydrochloric acid, gently agitating the solution. Allow the melt to disintegrate in the cold. Remove the crucible and rinse it with a further 10 ml of concentrated hydrochloric acid.

Transfer the solution to the distilling flask, rinse the beaker with a little hydrochloric acid and add this to the flask (Fig. 14, p. 113), giving a total volume of about 100 ml. Add 2 ml of concentrated hydrobromic acid and 0·5 g of hydrazine sulphate. Distil the solution, collecting about 50 ml of distillate in a beaker containing about 50 ml of water and kept cold by immersion in an ice bath. Add 25 ml of concentrated nitric acid to the distillate and evaporate to dryness on a steam bath. Transfer the beaker to an electric oven to complete the expulsion of nitric acid and then complete the determination of arsenic in the distillate as described in the previous section.

An Arsine–Molybdenum Blue Method

In this method the silicate rock material is decomposed by an alkaline fusion and the arsenic present is reduced to arsine with zinc and hydrochloric acid. The arsine can be trapped in a Gutzeit-type apparatus, producing stains that can be matched against those produced from standard arsenic solutions as described by Carmichael and McDonald.[16] The procedure described below is a modification of that given by Onishi and Sandell[3] involving an oxidation of the arsine and subsequent reduction of arsenomolybdate to a molybdenum blue for photometric measurement.

Method

Apparatus. This is shown in Fig. 19. The plug of glass wool prevents drops of liquid from being carried over. The second plug is lead acetate wool and is used to remove hydrogen sulphide which interferes with the determination from the gas stream. Lead acetate wool is prepared by soaking cotton wool in a solution of lead acetate containing 200 g of the reagent per litre, and drying in a current of warm air.

Reagents: *Sodium hydroxide wash solution*, dissolve 1 g of reagent in 100 ml of water.
Hydrochloric acid, "arsenic-free".
Zinc metal, 20–30 mesh, "arsenic-free".
Potassium iodide solution, dissolve 15 g of reagent in 100 ml of water.
Stannous chloride solution, dissolve 40 g of stannous chloride dihydrate in 100 ml of concentrated hydrochloric acid.
Iodine solution, dissolve 0·25 g of iodine and 0·4 g of potassium iodide in water and dilute to 100 ml.
Sodium bicarbonate solution, dissolve 4·2 of the reagent in 100 ml of water.

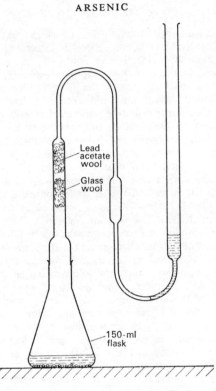

Fɪɢ. 18. Apparatus for the evolution and collection of arsine.

> *Sodium metabisulphite solution*, prepare as required by dissolving
> 0·5 g of reagent in 10 ml of water.
> *Ammonium molybdate reagent*, when required for use, mix 10 ml
> each of two solutions A and B. For A, dissolve 1·0 g of
> ammonium molybdate in 10 ml of water and add 90 ml of 6 N
> sulphuric acid. For B, dissolve 0·15 g of hydrazine sulphate
> in 100 ml of water.

Procedure.   Fuse approximately 2·5 g of sodium hydroxide in a 10-ml
nickel or zirconium crucible, and allow to cool. Accurately weigh approximately 0·5 g of the finely powdered rock material on to the solidified sodium
hydroxide and add approximately 0·25 g of sodium peroxide (note 1). Cover
the crucible and gently fuse the contents for 20 to 30 minutes or until complete
decomposition has been obtained. Care should be taken, particularly in the
early stages of the fusion as a vigorous reaction may occur if the sample
contains much sulphide. Allow to cool.

Fill the crucible with approximately 8 ml of water and warm on a hot plate or steam bath until the melt is completely disintegrated. Filter the solution through a small, medium-textured paper and wash the residue well with about ten 1–2 ml portions of the sodium hydroxide wash solution. Discard the residue and collect the filtrate and washings in the flask used for the arsenic reduction (Fig. 18). The solution should not exeed 25 ml in volume. Cool the solution and add 11 ml of concentrated hydrochloric acid drop by drop while swirling. Silica may separate from the solution, but this can be ignored. Now add 2 ml of the potassium iodide solution and 1 ml of the stannous chloride solution. Allow the solution to stand at room temperature for 15 to 30 minutes.

Pipette 1·0 ml of the iodine solution and 0·2 ml of the sodium bicarbonate solution into the absorption arm of the apparatus. Now add 2 g of zinc metal to the solution in the flask and immediately complete the assembly of the apparatus as in Fig. 18. Allow the gases evolved at room temperature to bubble gently through the iodine solution for about 30 minutes. At the end of this period the solution in the absorber should still be faintly yellow in colour, indicating an excess of iodine. Disconnect the flask and pour the contents of the absorber into a 10-ml volumetric flask. Rinse the absorber with no more than 1·0 ml of water and add this to the volumetric flask. Add a single drop of sodium metabisulphite solution to the flask to remove the excess iodine and then add 5·0 ml of the ammonium molybdate reagent solution. Heat the flask on a steam bath for 15 minutes, cool and dilute to volume with water.

Measure the optical density of the solution relative to water in 4-cm cells, using the spectrophotometer set at a wavelength of 840 nm. Determine also the optical density of a reagent blank solution prepared in the same way as the sample solution but omitting the rock material. The arsenic content of the sample can then be obtained by reference to the calibration graph (note 2).

Calibration. Transfer aliquots of 1–3 ml of the standard arsenic solution containing 1 to 3 μg arsenic to separate 10-ml volumetric flasks and add 1 ml of the iodine solution and 0·2 ml of the sodium bicarbonate solution to each. Swirl gently to mix and then add 1 drop of sodium metabisulphite solution and 5 ml of the ammonium molybdate reagent solution to each flask. Heat on a water bath for 15 minutes, cool, dilute to volume with water and measure the optical density of each solution in 4-cm cells as described above. Plot the relation of these values to arsenic concentration.

Notes: 1. The addition of sodium peroxide is not essential, but assists the decomposition, particularly in the presence of sulphide minerals.

2. It is advisable to check the recovery of arsenic by adding a suitable aliquot of the standard arsenic solution to a rock material of known arsenic content, and determining the total arsenic content by the procedure given. Onishi and Sandell[3] suggest that the recovery of arsenic from normal silicate rocks is usually better than 95 per cent.

Method Using Silver Diethyldithiocarbamate

For this method the rock material can be decomposed by fusion with sodium hydroxide as described in the previous section. The arsenic present in solution is then reduced to arsine with metallic zinc and hydrochloric acid, prior to reaction with the silver reagent solution in the apparatus shown in Fig. 18.

Method

Reagents: *Silver diethyldithiocarbamate solution,* dissolve 0·5 g of the reagent in 100 ml of pyridine.
Hydrochloric acid, zinc, potassium iodide and stannous chloride, as in the previous section.

Procedure. Prepare a solution of the silicate rock by fusion with sodium hydroxide as described in the previous section, and reduce any arsenic present to arsine with zinc and hydrochloric acid in the apparatus shown in Fig. 19, passing the gases through 3 ml of the silver diethyldithiocarbamate solution placed in the absorption arm. Allow the reduction to proceed for 45 to 60 minutes at room temperature and then remove the reaction flask. Gently tilt the absorber to and fro to dissolve any complex adhering to the walls of the tube, transfer the pyridine solution to a 1-cm cell and measure the optical density with the spectrophotometer set at a wavelength of 540 nm. Before re-using the apparatus, rinse it out thoroughly with acetone, allow to dry and repack with fresh lead acetate wool.

Calibration. Transfer aliquots of 1–10 ml of the standard arsenic solution containing 1–10 μg As to the reaction flask, add hydrochloric acid, potassium iodide solution and arsenic-free metallic zinc, as for the sample solution. Assemble the apparatus with 3 ml of the silver diethyldithiocarbamate solution in the absorption arm, and collect the arsine as described above. Measure the optical density of each solution and plot the relation of optical density to arsenic concentration as in Fig. 19.

Method Using Co-crystallisation with Thionalide

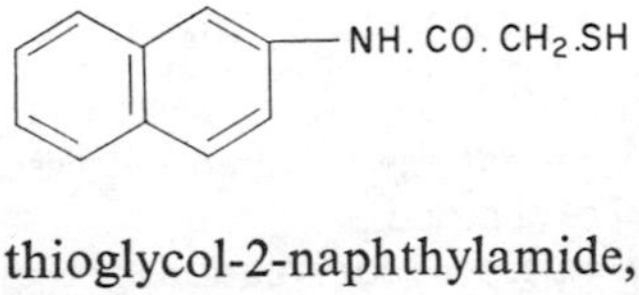

$NH.CO.CH_2.SH$

VI Thionalide

Thionalide, thioglycol-2-naphthylamide, (VI) forms insoluble complexes with a number of elements including arsenic. This complex can

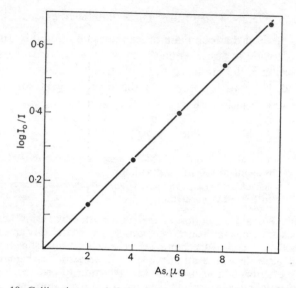

FIG. 19. Calibration graph for arsenic using silver diethyldithiocarbamate (1-cm cells, 540 nm).

be recovered from aqueous solution by the technique of co-crystallisation from an organic solvent such as acetone. This recovery of arsenic was investigated by Portmann and Riley[7] (using arsenic-74 as a radioactive tracer) who showed that as little as $0 \cdot 05$ μg As could be recovered in 97–98 per cent yield from 1 litre of sea water. This procedure is described below in its application to silicate rocks and minerals. The determination is completed using a spectrophotometric antimonial–arsenic–molybdenum blue reaction.

Method

Apparatus. *Silver crucibles*, 10-ml, for fusions with caustic alkali.

Reagents: *Sulphuric acid*, 5 N.
  Thionalide solution, prepare as required by dissolving 2 g of the reagent in 100 ml of acetone.
  Ascorbic acid solution, 5 per cent solution, prepare as required by dissolving 5 g of the solid reagent in 100 ml of water.

Ascorbic acid solution, 0·1 M solution, prepare as required by dissolving 1·76 g of the reagent in water and diluting to 100 ml. Keep this solution in a refrigerator at 0° until ready to use, and discard when it darkens in colour.

Antimony solution, dissolve 0·274 g of potassium antimony tartrate in water and dilute to 100 ml.

Ammonium molybdate solution, dissolve 4·8 g of ammonium molybdate in water and dilute to 100 ml.

Mixed reagent solution, mix together 50 ml of 5 N sulphuric acid, 15 ml of ammonium molybdate solution, 5 ml of the antimony solution and 30 ml of the 0·1 M ascorbic acid solution. Dilute with water to 125 ml. This mixed reagent solution is unstable and must be used within 1 hour of preparation.

Procedure. Accurately weigh approximately 0·25 g of the finely powdered silicate rock material into a 10-ml silver crucible and add 1·5 g of sodium hydroxide. Transfer the crucible to an electric furnace set to a temperature of 750°, and allow to fuse for about 10 minutes. Allow to cool, transfer the crucible to a small beaker and leach the melt with about 50 ml of water. Rinse and remove the crucible and leave the solution to stand overnight to allow the insoluble material to settle. Decant the solution through a hardened filter paper using suction if necessary, and wash the residue thoroughly with water.

Transfer the combined filtrate and washings to a 1-litre conical flask, dilute to 500 ml with water and add 15 ml of 5 N sulphuric acid and 1 ml of ethanol. Heat the solution for a few minutes on a hot plate to reduce any manganate formed during the fusion, add 4 ml of the 5 per cent ascorbic acid solution and heat to boiling on a hot plate to reduce any arsenic compounds to the lower valency state.

Allow to cool for a period of 10 minutes, then add a further 2 ml of the 5 per cent ascorbic acid solution and cool the beaker to room temperature. Stir continuously with a mechanical stirrer whilst adding 10 ml of 5 N sulphuric acid followed by 7 ml of thionalide solution. Continue stirring for a further period of 5 minutes to complete the coagulation of the precipitate. Allow the beaker to stand for 10 minutes and then heat to boiling on a hot plate and boil gently for 30 minutes to remove all the acetone. Allow to cool once more, stir, allow to stand overnight, collect the precipitate on a small open-textured filter paper, and wash the precipitate with water.

Transfer the precipitate and filter to a 25-ml conical flask and add 7·5 ml of concentrated nitric acid. Close the neck of the flask with a bulb stopper, as in Fig. 20, and heat gently on a hot plate until the solution obtained is pale yellow in colour—this may take 24 to 36 hours. When a viscous yellow liquid remains, blow in a current of carbon dioxide and continue heating to remove all residual traces of nitric acid. Continue the evaporation until dense white fumes are no longer evolved. Care must be taken to avoid over-heating and subsequent charring. If this does occur, add 1 ml of concentrated nitric acid and repeat the evaporation.

Cool the residue, which should then set to a pale yellow solid, add 1 ml of N sulphuric acid and warm gently until all solid material has dissolved. Transfer the solution to a 10-ml volumetric flask, add 2 ml of the mixed reagent solution and dilute to volume with water. Allow the solution to stand for not less than 30 minutes and then measure the optical density relative to water in 4-cm cells with the spectrophotometer set at a wavelength of 866 nm. The yellow colour of the solution may partially mask the molybdenum blue, but this does not interfere with the determination. Determine also the optical density of a reagent blank solution prepared in the same way as the sample solution but omitting the rock material. The reagent blank value should not exceed $0 \cdot 100$—equivalent to about 1 μg arsenic.

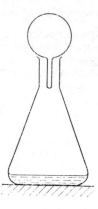

FIG. 20. Apparatus for the oxidation of thionalide.

Calibration. Transfer aliquots of 1–5 ml of the standard arsenic solution containing 1–5 μg As, to separate 10-ml volumetric flasks. Add 2 ml of the mixed reagent solution to each, allow to stand for not less than 30 minutes and then measure the optical density of each solution relative to water as described above. Plot the relation of optical density to arsenic concentration.

References

1. ONISHI H. and SANDELL E. B., *Geochim. Cosmochim. Acta* (1955) **7**, 1.
2. GOLDSCHMIDT V. M., *Geochemistry*, Oxford, 1954, p. 470.
3. ONISHI H. and SANDELL E. B., *Mikrochim. Acta* (1953) 34.
4. RADER L. F. and GRIMALDI F. S., *U.S. Geol. Surv. Prof. Paper* 391A, 1961.
5. MORACHEVSKII YU. V and KALININ A. I., *Zavod. Lab.* (1961) **27**, 272.
6. MILAEV S. M. and VOROSHINA K. P., *Zavod. Lab.* (1963) **29**, 410.
7. PORTMANN J. E. and RILEY J. P., *Anal. Chim. Acta* (1964) **31**, 509.

8. TUROG E. and MEYER A. H., *Ind. Eng. Chem., Anal. Ed.* (1929) **1**, 136.
9. HERON A. E. and ROGERS D., *Analyst* (1946) **71**, 414.
10. MORRIS H. J. and CALVERY H. O., *Ind. Eng. Chem., Anal. Ed.* (1937) **9**, 447.
11. JEAN M., *Anal. Chim. Acta* (1956) **14**, 172.
12. MURPHY J. and RILEY J. P., *Anal. Chim. Acta* (1962) **27**, 31.
13. VASAK V. and SEDIVEC V., *Chem. Listy* (1952) **46**, 341.
14. LIEDERMAN D., BOWEN J. E. and MILNER O. I., *Analyt. Chem.* (1959) **31**, 2052.
15. FOWLER E. W., *Analyst* (1963) **88**, 380.
16. CARMICHAEL I. and McDONALD A., *Geochim. Cosmochim. Acta* (1961) **25**, 190.

CHAPTER 11

BARIUM

Occurrence

Barium has been widely reported from all types of igneous rock, but particularly from acidic and alkaline silicates. Table 15, compiled from data by Engelhardt,[1] gives average values for a number of types. The average content of igneous rocks was given by Engelhardt as 480 ppm BaO, a somewhat similar figure of 430 ppm being calculated for the average value of sedimentary rocks. Marine limestones and evaporite sediments contain only small amounts of barium—1–10 ppm in most specimens. High barium values are one of the characteristic features of carbonatite rocks.

TABLE 15. THE BARIUM CONTENT OF SOME IGNEOUS ROCKS

Rock type	BaO, ppm
Pyroxenite and peridotite	3
Gabbro, anorthosite and basalt*	70
Diorite and andesite	260
Granite and liparite	480
Nepheline syenite and phonolite	580
Syenite and trachyte	1800
Leucite rocks	1000–4000

* Alkaline basalts and gabbros contain appreciably more barium.

Barium occurs in silicate rocks as the sulphate mineral, barite (baryte or barytes) and as the barium felspar, celsian. In addition barium is found in other silicate minerals where the close similarity of ionic radii (Ba^{2+}, R† $= 143$ pm; K^+, $R = 133$ pm; Ahrens[2]) enables it to substitute for potassium, particularly in the potash felspars where it appears

† R = ionic radius, pm = picometre, 10^{-12} m.

in the later crystallates. Ore minerals of barium include the sulphate barite, and a number of carbonates—witherite $BaCO_3$, barytocalcite $BaCa(CO_3)_2$ and celestobaryte $BaSr(CO_3)_2$.

The Determination of Barium in Silicate Rocks

GRAVIMETRIC METHOD

The residue obtained on decomposing silicate rocks by evaporation with concentrated hydrofluoric and sulphuric acids contains the barium present in the rock material now in the form of the insoluble sulphate. Accessory minerals that are incompletely attacked in this procedure remain with the barium sulphate and must be separated before the residue can be weighed. This method of determining barium can be combined with the determination of manganese, total iron, titanium and phosphorus. Carbonate rocks and minerals are readily decomposed with hydrochloric acid, but subsequent precipitation of the barium with dilute sulphuric acid gives a residue that may be heavily contaminated with calcium, strontium and accessory minerals.

As an alternative procedure, silicate rocks can be decomposed by fusion with sodium carbonate (as, for example, described by Pickup,[3] where the determination of barium can be combined with those of chromium, vanadium, sulphur, chlorine and zirconium) or by sinter with sodium peroxide. This alternative has the advantage that the accessory minerals are more likely to be decomposed, and the disadvantage that the collected barium sulphate is likely to be more contaminated with other elements. After leaching and filtering, the insoluble residue can be cautiously acidified with sulphuric acid to precipitate the barium as sulphate, in the presence of added calcium if there is little or none in the silicate rock material.

Gravimetric methods based upon the precipitation and weighing as barium chromate are not recommended. Other elements are likely to be co-precipitated and weighed, and the barium chromate may contain too much chromium.[4]

The insolubility of barium sulphate provides a simple and effective way of collecting barium in a small compact residue, separated from the major part of the rock matrix. Small amounts of calcium and some strontium will accompany the barium as sulphates, and any mineral unattacked or incompletely attacked will also remain behind and be collected with the barium sulphate. After dehydration with concentrated sulphuric acid, ferric and aluminium sulphates will also appear in the residue, but these will pass into the solution on prolonged digestion. One procedure commonly employed for the purification of the barium residue makes use of the solubility of barium sulphate in concentrated sulphuric acid, from which it may be recovered by dilution with water. A more recently described procedure[5] makes use of the solubility of barium sulphate in ammoniacal EDTA solution. The calcium and strontium EDTA complexes are stable over a wide pH range, in contrast to the barium complex, which is stable only in alkaline solution. Barium sulphate is therefore precipitated on acidification. The procedure given in detail below is from Wilson.[6] In both procedures the barium sulphate is collected, dried, ignited and weighed. The lower limit for this determination is about 100 ppm Ba, using a 2-g sample of the rock material.

The collection of small quantities of a barium sulphate precipitate is not an easy thing to do. The particle size is usually very small and a close-textured filter paper is essential. Even so, some small part of the precipitate may pass through the paper and give rise to low and erratic results for barium.

Method

Reagents: *Sodium carbonate wash solution*, dissolve 10 g of anhydrous reagent in 500 ml of water.

Acid EDTA solution, suspend 10 g of ethylenediaminetetraacetic acid (i.e. the free acid, not the sodium salt) in 50 ml of water, add 10 ml N sulphuric acid, neutralise and dissolve the EDTA by adding ammonia, make just acid to methyl red indicator and dilute to 100 ml with water.

Alkaline EDTA solution, suspend $0 \cdot 5$ g of ethylenediaminetetra-acetic acid (free acid) in 50 ml of water, add 10 ml of N sulphuric acid, 10 ml of concentrated ammonia solution and dilute to 100 ml with water.

Procedure. Accurately weigh approximately 2 g of the finely powdered rock material into a 3-inch platinum dish and add 1 ml of concentrated nitric acid (note 1), 15 ml of 20 N sulphuric acid and 25 ml of concentrated hydrofluoric acid. Transfer the dish to a hot plate and evaporate to fumes of sulphuric acid.

Allow to cool, rinse down the sides of the dish with a little water, add an additional 10 ml of water and again evaporate to fumes of sulphuric acid. Allow to cool and rinse the contents of the dish into a 250-ml beaker containing about 100 ml of water. Transfer the beaker to a hot plate or steam bath and digest until all soluble material has passed into solution. The ferric and aluminium sulphates may take a little while to do this. Allow the solution to cool and stand, preferably overnight. Barium sulphate will appear as a fine white precipitate, best seen by gently stirring the solution once or twice with a glass rod.

Collect the small insoluble residue on a small close-textured filter paper, after first decanting the bulk of the solution. Wash the residue well with small quantities of cold water. The filtrate and the washings may be combined and allowed to stand overnight to ensure complete collection of the barium sulphate precipitate. It may then be used for the determination of manganese, titanium, total iron and phosphorus if required. Transfer the filter and residue to a small platinum crucible, dry and ignite over a burner set to give only a faint dull red colour to the bottom of the crucible. Allow to cool, add a small quantity of anhydrous sodium carbonate (note 2) and fuse over a Bunsen burner at full heat for 30 minutes. Allow to cool. Extract the melt with hot water and rinse the solution and residue into a small beaker. Collect the residue on a small close-textured filter paper and wash it with a little hot sodium carbonate wash solution. Discard the filtrate and washings. Using a fine jet of water rinse the residue back into the beaker and dissolve in a few ml of dilute hydrochloric acid. Rinse the platinum crucible used for the fusion with a little dilute acid and add this to the solution in the beaker. Filter this solution through the paper used for the previous filtration, and wash well with small quantities of water. Collect the filtrate and washings, which should have a volume of 30 to 35 ml, in a 50-ml beaker, add a few drops of 20 N sulphuric acid, stir and set aside to stand overnight.

Collect the precipitated barium sulphate on a small close-textured filter paper, washing first with $0 \cdot 1$ N sulphuric acid (note 3) and then with a little water. Transfer the filter to a weighed platinum crucible, dry the paper, burn off, ignite over the full flame of the burner in the open crucible (note 4), and weigh as barium sulphate $BaSO_4$.

Purification of barium sulphate. Add 2 ml of 20 N sulphuric acid to the ignited residue contained in the small platinum crucible. Transfer to a hot plate and heat until all the barium sulphate has passed into solution. (Any black specks are probably particles of carbon from the ignited filter paper and may be cleared by adding a little concentrated nitric acid to the COLD sulphuric acid solution, and evaporating to fumes of sulphuric acid.) Allow to cool, rinse the solution into a small beaker, dilute to 25 ml with water and allow to stand overnight. Collect the precipitate, ignite and weigh as described above.

The alternative procedure for purification of the acid insoluble residue, based upon the stability of the EDTA complexes is as follows. Rinse the residue into a 50-ml beaker using a few ml of water and add 5 ml of the acid

EDTA solution. Warm the beaker on a hot plate or steam bath for a few minutes; any calcium sulphate present will dissolve, the solution can be filtered through the original paper, and the filtrate discarded. Again rinse the residue back into a 50-ml beaker with a little water, and 5 ml of the alkaline EDTA solution, a little macerated filter paper and heat on the steam bath to dissolve the barium sulphate (note 5). Filter the solution through a small close-textured filter to remove any insoluble material and wash well with a little hot water. Collect the filtrate and washings in a clean 50-ml beaker, add 2 drops of methyl red indicator solution, and just acidify with dilute hydrochloric acid.

Transfer the beaker to a steam bath, digest for an hour and allow to stand overnight. Collect the barium sulphate on a small close-textured filter paper, wash with water containing a few drops of N sulphuric acid, dry, ignite and weigh as barium sulphate as described above.

Notes: 1. Care must be taken to avoid loss of sample by spitting if the rock material contains much carbonate.

2. The amount of sodium carbonate required will depend upon the amount of residue remaining. For most rocks $0 \cdot 25$–$0 \cdot 5$ g will suffice, but if the weight of the residue exceeds more than about $0 \cdot 1$ g, between 6 and 10 times as much flux should be taken.

3. At room temperature barium sulphate is soluble in water to the extent of $2 \cdot 3$ mg per litre. The solubility is much less in dilute sulphuric acid solution.

4. An oxidising atmosphere is required to burn off all the carbon of the filter at as low a temperature as possible. Failure to do this may result in some reduction of the barium sulphate giving low results.

5. Ignited barium sulphate does not readily dissolve in alkaline EDTA solution, and if this method of purification is to be used the first ignition in platinum should be omitted.

DETERMINATION BY FLAME PHOTOMETRY

Although barium compounds impart a characteristic apple-green colour to a flame, the sensitivity of the reaction does not compare with the flame reactions of the alkali metals. The barium spectrum is complex (Fig. 21), but the most prominent feature is the atomic resonance line at $553 \cdot 6$ nm. A systematic study of the barium spectrum was made by Dean et al.[7] who found that the presence of alkali metals and other alkaline earth elements tended to enhance the barium emission at $553 \cdot 6$ nm, whilst aluminium, iron and titanium tended to depress the emission. The calcium emission at 554 nm interferes with the barium emission at $553 \cdot 6$ nm, and effectively prevents the use of flame photometry for the determination of barium in silicate and carbonate rocks unless the two elements are separated first.

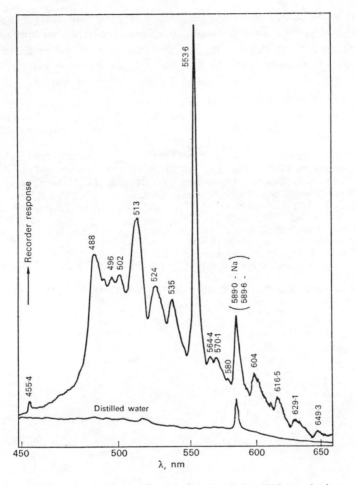

FIG. 21. Emission spectrum of a solution containing 1000 ppm barium.

Determination by Atomic Absorption Spectroscopy

In air–acetylene flames, the barium absorption at its most sensitive wavelength of 553·6 nm, is masked by calcium oxide bands. However, it has been observed by Amos and Willis[8] that there is a vast increase in sensitivity to barium by absorption, relative to calcium, when a nitrous oxide–acetylene flame is used. This is presumably due to the more effective decomposition of oxide structures in the hotter flame.

As with the determination of some other elements, the presence of aluminium in the amounts found in silicate rocks has an appreciable depressant effect upon the barium absorption. This can be eliminated by adding lanthanum to the rock solution. Ionisation suppression, which is particularly important in the determination of barium, can be achieved by adding potassium.

References

1. ENGELHARDT W. V., *Chem. Erde* (1936) **10,** 187.
2. AHRENS L. H., *Geochim. Cosmochim. Acta* (1952) **2,** 168.
3. BENNETT W. H. and PICKUP R., *Colon. Geol. Min. Res.* (1952) **3,** 171.
4. INGAMELLS C. O., SUHR N. H., TAN F. C. and ANDERSON D. H., *Anal. Chim. Acta* (1971) **53,** 345.
5. BUSEV A. I. and KISELEVA L. V., *Vestnim Moskov. Univ. Ser. Mat., Mekd., Astron., Fiz. i. Khim.* (1957) **12,** 227.
6. WILSON A. D., unpublished work.
7. DEAN J. A., BURGER J. C., RAINS T. C. and ZITTEL H. E., *Analyt. Chem.* (1961) **33,** 1722.
8. AMOS M. D. and WILLIS J. B., *Spectrochim. Acta* (1966) **22,** 1325.

CHAPTER 12

BERYLLIUM

Occurrence

Although beryllium is essentially a rare and dispersed constituent of silicate rocks and minerals, a considerable degree of enrichment can occur, particularly in the later stages of granitic emplacement and pegmatite formation. This leads eventually to the crystallisation of silicate minerals with high beryllium content and finally to the formation of independent beryllium minerals. Examples of silicate minerals rich in beryllium recorded by Solodov[1] include albite with 56 ppm Be, muscovite with 47 ppm and microcline with 8–9 ppm.

The degree of enrichment of beryllium in the later rocks is in agreement with the observation of Beus[2] that the beryllium content of granitic rocks decreases sharply with depth. In the very late stages the granite may contain appreciable amounts of beryllium, Dawson[3] has for example recorded from 23 to 212 ppm Be (average 84 ppm) in thirteen specimens of the Foxdale granite, Isle of Man, whilst Machacek et al.[4] have reported granite specimens with up to 34 ppm Be. Most silicate rocks, however, contain only very small amounts ranging from less than 1 ppm Be in basic rocks to 5 ppm in granites and 7 ppm in syenites.[5–8] There is a suggestion that alkali undersaturated rocks may contain larger amounts up to about 20 ppm Be.[8–10]

The most common beryllium mineral is beryl, of which very large crystals have been found at some localities; the rarer transparent forms have long been prized as gemstones. Other minerals containing beryllium as an essential component include the oxide bromellite, beryllates chrysoberyl and taaffeite, a rare antimonate and carbonate and a variety of silicates, some not yet completely classified.[11] Most of these minerals have formerly presented problems in analysis, due to the difficulties of identifying beryllium in the presence of aluminium, and of obtaining a quantitative separation from it. The composition of beryl is usually given as 3 $BeO.Al_2O_3.6SiO_2$, which would require 14 per cent BeO or nearly 8 per cent Be. Hand-picked mineralogical specimens contain from 12 to 14 per cent BeO, and beryl ore concentrates 10 to 12 per cent. Many pegmatites with beryl contain also a variety of lithium minerals, especially spodumene, amblygonite, zinnwaldite and lepidolite. It is not surprising, therefore, that beryllium has been noted as a frequent im purity in lithium concentrates of these minerals.

Warning

Although beryl, the most abundant ore of beryllium, has long been regarded as inert and non-toxic, due care must always be exercised in preparing and handling any solution or residue containing beryllium. There is no known cure for the pulmonary condition known as beryl-liosis, and even short exposures to minute amounts of beryllium dust have been known to prove fatal. Beryllium containing solutions must not be allowed to come into contact with the skin, as any deposition in an open wound may give rise to a non-healing sore or form of derma-titis that can only be cured by surgery.

Before handling beryllium ores or preparing standard beryllium solutions, the analyst should be familiar with the nature of the hazard and the precautions adopted in those laboratories that regularly handle beryllium containing materials.

The Determination of Beryllium

There is a general lack of precise, accurate and specific methods for the gravimetric determination of beryllium. For many years the principle method was based upon precipitation as hydroxide $Be(OH)_2$ followed by ignition to oxide. This procedure required careful separation of beryllium from both anions and cations that would otherwise inter-fere. The difficulties of ensuring that these separations are complete without loss of beryllium are described by Hillebrand and Lundell.[12]

Beryllium can be precipitated as the ammonium phosphate $Be(NH_4)PO_4$ at a pH of $5 \cdot 2$[13] and ignited to pyrophosphate as a weighing form by a procedure similar to that used for magnesium. This method has been subjected to some criticism, but is given in detail below in the form used at the National Chemical Laboratory.[14] The reagent acetoanilide has been used[15] to precipitate beryllium in the form of a 2:1 complex containing $2 \cdot 49$ per cent of metal. This complex can be dried at 100° and weighed directly. Interference from iron and alumina can be prevented by precipitating the complex at pH 8 in the presence of EDTA.

Many photometric procedures have been devised for beryllium. Reagents in common use include p-nitrobenzeneazo-orcinol,[16] acetyl-acetone[17] and beryllon II.[18] New reagents include beryllon III,[19] chrome azurol S[20] and fast sulphon black F.[21] Morin has been used extensively in fluorimetric methods.[22-25] The procedures given in detail below are based upon beryllon II for beryl ores and silicates rich in beryllium, and upon morin for normal silicate rocks containing a few ppm Be or less.

Gravimetric Determination as Phosphate

Precipitation as ammonium phosphate has been condemned[12] on the grounds that the precipitate does not have the ideal composition indicated by the formula $Be(NH_4)PO_4$. However, acceptable results have been obtained for many years by this procedure, which was the preferred gravimetric method for beryllium at the National Chemical Laboratory.[14] The advantages claimed are the favourable conversion factor ($Be_2P_2O_7:BeO = 1:0 \cdot 2606$), and the absence of interference from phosphate, present in many beryllium ores.

Under the right conditions the phosphate precipitate is granular, and when this is achieved there is a reduced tendency towards the occlusion of impurities. Airoldi,[26] in a review of the method, recommends using the disodium salt of EDTA, but this has been found to be unsatis-factory[14] because the final precipitate may be contaminated with sodium giving a somewhat glassy appearance to the pyrophosphate residue, which then tends to adhere to the platinum dish used for the ignition.

Method

Reagents: *Potassium fluoride*, anhydrous.

Ethylenediaminetetraacetic acid (not the disodium salt), suspend 10 g of the solid reagent in 100 ml of water, and dissolve by adding concentrated ammonia drop by drop.

Diammonium hydrogen phosphate solution, dissolve 20 g of the reagent in 100 ml of water.

Ammonium acetate solution, saturated.

Ammonium acetate wash solution, dissolve 2 g of ammonium acetate in water, add 3 ml of glacial acetic acid and dilute to 1 litre with water. Using a pH meter, adjust the pH of the solution to 5·2 by adding ammonia.

Procedure. Accurately weigh approximately 0·5 g (note 1) of the finely powdered (note 2) beryllium ore into a platinum crucible and fuse with 3 g of anhydrous potassium fluoride. Allow the crucible to cool, add 5 ml of concentrated sulphuric acid and gently evaporate until copious fumes of sulphuric acid are evolved. Allow to cool, lay the crucible on its side in a 250-ml beaker and extract the melt by heating on a steam bath with 50–75 ml of water. Rinse and remove the platinum crucible. At this stage a clear solution should be obtained (note 3). Using a pH meter adjust the pH of the solution to a value of 2 by adding concentrated ammonia drop by drop. Now add 15 ml of the EDTA solution and boil for 2 to 3 minutes to ensure that any chromium is complexed by the EDTA. Allow to cool, add 5 ml of the ammonium phosphate solution and again using the meter, adjust the pH to 5·2 by adding saturated ammonium acetate solution. Boil for 2 minutes.

Stand the beaker on a steam bath only until the white precipitate has become granular and settled. Cool to room temperature, allow to stand for 1 hour and then collect the precipitate on a close-textured filter paper. Wash the precipitate with a total of about 100 ml of the ammonium acetate wash solution, retaining the filtrate and washings (note 4). Dissolve the precipitate in about 100 ml of hot 4 N hydrochloric acid, add 7·5 ml of the EDTA solution, adjust the pH of the solution to 2 by adding ammonia drop by drop, and boil. Cool the solution, add 1 ml of the ammonium phosphate solution and again adjust the pH to 5·2 by adding saturated ammonium acetate solution. Boil for 2 minutes and allow to digest for 2 hours on a steam bath. Allow the precipitate to settle for 4 hours or preferably overnight.

Collect the precipitate on a close-textured filter paper and wash with ammonium acetate wash solution as before. Transfer the paper and precipitate to a weighed platinum crucible, dry, char the paper under an infrared heating lamp and ignite under oxidising conditions in a muffle furnace set at a temperature of 1000°. Allow to cool and weigh as $Be_2P_2O_7$, containing 26·06 per cent BeO or 9·395 per cent Be.

Notes: 1. If the sample material contains less than 5 per cent BeO, use a 1 g sample portion and fuse with 6 g of potassium fluoride.

2. The attack by fusion with alkali fluoride is not likely to be complete unless the sample material is finely ground; grinding to pass a 200-mesh sieve is recommended.

3. Any unattacked minerals can be filtered off at this stage.

4. The filtrates are never entirely free of beryllium. For most purposes this amount can be ignored but if required it can be determined photometrically after destruction of organic matter as follows:

Combine the filtrates and washings and dilute to 250 ml with water. Pipette 50 ml into a Kjeldahl flask, add 1 ml of concentrated sulphuric acid and 25 ml of fuming nitric acid and boil to a small volume. Cool, add 10 ml of fuming nitric acid and 30 ml of concentrated hydrochloric acid and boil down until fumes of sulphuric acid are evolved. Cool and dilute with water to 50 ml in a volumetric flask. Transfer an aliquot of this solution to a separate 50-ml volumetric flask and determine the beryllium with beryllon II as described below.

Photometric Determination with Beryllon II

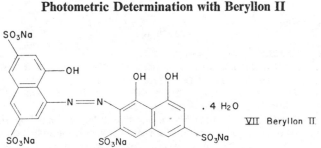

Beryllon II is the trivial name for the tetra-sodium salt of 8-hydroxy-naphthalene-3:6-disulphonic acid-1-azo-2′-naphthalene-1′:8′-dihydroxy-3′:6′-disulphonic acid (VII). The reagent is a deep violet colour in aqueous alkaline solution whilst the beryllium complex is blue. This difference in colour was used by Karanovich[18] as the basis for his determination of beryllium. Only chromium, molybdenum, platinum, thorium and the rare earths are stated to interfere,[14] and then only at ratios of 500:1. These elements are not normally present in this amount in beryllium ores, concentrates or minerals, and chemical separation is therefore not necessary.

In order to increase the selectivity of the reagent, Karanovich added EDTA to complex other metals, and ascorbic acid to reduce iron to the ferrous state in which it does not interfere. Even so interference from iron does occur when it is present in large amounts, and if more than a trace is present it may be necessary to replace the EDTA with N:N-di-2-hydroxyethyl glycine ("Nervanaid F") which gives a clear colourless solution with ferric iron in strongly alkaline solution. An alternative procedure for sample materials containing much iron is to separate the beryllium by a prior precipitation as phosphate using titanium as carrier.

The colour of the beryllium–beryllon II complex fades slowly, and the optical density measurement should not be unduly prolonged. Ideally the sample and standard solutions should be prepared together and measured at the same time—usually after standing for 15 to 30 minutes. Solutions of the beryllium complex do not obey the Beer–Lambert Law[27] (Fig. 22). The precision is very poor when less than about 8 μg of beryllium is present. Reagent solutions have an appreciable absorption at the wavelength used for the beryllium complex[28] (Fig. 23). The reagent solutions must therefore be measured with some care.

Method

Reagents: *Potassium fluoride*, anhydrous.

Ethylenediaminetetraacetic acid, disodium salt solution, dissolve 5 g of the reagent in 100 ml of water.

Ascorbic acid solution, dissolve 0·75 g in 100 ml of water. Prepare freshly every few days.

Beryllon II reagent solution, dissolve 0·1 g of the solid reagent in 100 ml of water. (Samples from some commercial sources have been found[14] to be insufficiently pure for this determination.)

Standard beryllium stock solution, dissolve sufficient beryllium sulphate $BeSO_4.4H_2O$ in 2 N sulphuric acid to give a solution containing approximately 1 mg Be per ml. Standardise by precipitation as phosphate as described above, and then dilute as required to give a solution containing 4 μg Be per ml in 0·5 N sulphuric acid.

Procedure. Accurately weigh 0·1–0·2 g of the finely powdered low-grade beryllium ore or silicate rock into a platinum crucible, add 2–3 g of potassium fluoride and fuse over a burner to give a clear melt. Allow to cool and add 4–6 ml of concentrated sulphuric acid. Transfer the crucible to a hot plate under an infrared heating lamp, heat slowly until all hydrofluoric acid is removed, then more strongly to evaporate sulphuric acid, and finally over a burner to give a clear melt at dull red heat. Allow to cool and then lay the crucible on its side in a 250-ml beaker containing 50–60 ml of water and heat on a hot plate until complete solution is obtained (note 1). Rinse and remove the platinum crucible. Cool the solution, transfer quantitatively to a 100-ml volumetric flask, make up to the mark with water and mix well.

Pipette a suitable aliquot of 10 ml or less (note 2) of this solution into a 100-ml conical flask, and a second aliquot equal in volume to the first into a 50-ml volumetric flask. Titrate the contents of the conical flask with N sodium hydroxide solution using phenol phthalein as indicator and record the titre. Now add to the volumetric flask 2 ml of the ascorbic acid solution, replace the stopper and swirl to mix well. Add 5 ml of the disodium EDTA

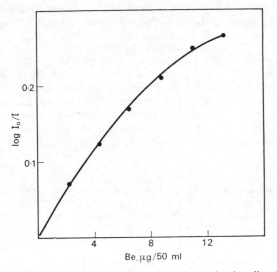

FIG. 22. Calibration graph for beryllium using beryllon II.

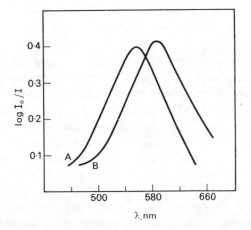

FIG. 23. Absorption spectrum of beryllon II (A, 0·001 per cent in water, pH 12–13) and beryllon II complex with beryllium (B).

solution and again swirl to mix. Now add N sodium hydroxide solution equivalent in volume to the titre plus the volume required to neutralise the sulphuric acid present in any added standard beryllium solution (note 2), together with 5 ml in excess. Replace the stopper and again mix well.

Dilute the contents of the flask to a volume just short of 45 ml (mark the flask at this level) with water, add 5 ml of the beryllon II reagent solution, dilute to volume with water and mix well. Allow this solution to stand for 15 minutes and then measure the optical density in 1-cm cells against the reagent blank solution (note 3) using a spectrophotometer set at a wavelength of 630 nm. Plot a curve relating the optical density of the standard solutions to the beryllium concentration and hence calculate the beryllium content of the sample material.

Notes: 1. Any unattacked mineral grains can be filtered off.

2. This aliquot should contain 8–16 μg Be. It has been noted that the precision of the determination is very poor if less than 8 μg is taken. If the 10-ml aliquot contains less than 8 μg Be, add 2 ml of the standard beryllium solution to the volumetric flask. This standard solution is 0·5 N in sulphuric acid, and it will therefore be necessary to determine the volume of N sodium hydroxide solution required to neutralise the 2-ml aliquot.

3. Prepare also a reagent blank solution by transferring 10 ml of water to a separate 50-ml volumetric flask, adding ascorbic acid and other reagents as described above. The standard solutions are obtained by transferring 2, 3 and 4 ml of the standard beryllium solution to separate 50-ml volumetric flasks, and following the procedure given above. In each case the volume of sodium hydroxide solution required to neutralise the added sulphuric acid must be included.

PRECIPITATION AS PHOSPHATE WITH TITANIUM AS COLLECTOR

This procedure is described by Ponomarev[29] for the determination of beryllium in rocks at the few-ppm level. It involves the decomposition of the rock by evaporation with hydrofluoric acid, fusion of the residue with potassium hydrogen fluoride and evaporation with sulphuric acid to give a sulphate melt. This melt is extracted with water containing hydrochloric acid and the extract neutralised with ammonia. EDTA is added followed by a solution containing 25 mg TiO_2. This titanium is then precipitated as phosphate from the solution by making it just alkaline to methyl red, collecting any beryllium present.

The precipitate is digested with aqueous sodium hydroxide solution, evaporated to small volume and an aliquot of the supernatant liquid taken for the photometric determination of beryllium using beryllon II as described above.

Fluorimetric Determination of Beryllium with Morin

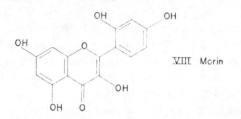

VIII Morin

Morin, said to be a pentahydroxyflavone (VIII), is a colouring matter obtained from old fustic. Recent supplies have been in the form of an iridescent black crystalline material whereas the earlier preferred samples were a pale-buff powder. The iridescent black material is low in active reagent content and contains impurities that mask the fluorescence reaction with beryllium. Purification is difficult, if not impossible. Whenever possible, supplies of the pale buff reagent should be sought. It is insoluble in water but soluble in alcohol and aqueous alkali. In neutral or acidic solution morin reacts with a number of metal ions to give products that fluoresce with varying intensities. Charlot[30] has listed some of the metals that react, giving values for the sensitivities of the reactions. In alkaline solution a strong fluorescence is given only by beryllium, although weak fluorescence is also observed from zinc, scandium, certain of the rare earths, zirconium, thorium, magnesium and calcium. Under these conditions the reagent itself is also weakly fluorescent. The interference from some of these elements can be prevented by adding EDTA or DTPA (diethylene–triaminepentaacetic acid[24] to the solution although these may slightly decrease the intensity of fluorescence of the beryllium complex.

This fluorescence varies linearly with the beryllium content of the solution but fades slowly on standing. Measurements should therefore be made within an hour of preparing the solutions and preferably within 30 minutes. Sill and Willis[31] ascribe this fading to air-oxidation of the morin, and report a slight increase in fluorescence if air is excluded. In a later paper[24] the primary cause of this instability is ascribed to the catalytic oxidation of the reagent by traces of copper present in the distilled water and reagents.

Procedures based upon morin have generally been used to determine beryllium in those rocks containing only small amounts of this metal. At these levels it is important to know how much beryllium is lost in the preliminary processing of the rock material. For this reason Sill and Willis[25] added an active isotope of beryllium—Be-7 to the sample material before decomposition, and determined the radioactivity of the final solution as a measure of the beryllium recovery. Although this step has not been included in the procedure described below this refinement can be added without difficulty (note 1).

In order to separate beryllium from other elements present in silicate rocks, the complex with acetylacetone is formed and extracted into chloroform. Aluminium will similarly extract unless the slightly acid or neutral solution is boiled with EDTA to complete the formation of the Al-EDTA complex. Even so, a few mg of aluminium are usually extracted with the beryllium.

Method

Reagents: *Potassium fluoride*, anhydrous.

Acetylacetone.

EDTA, disodium salt solution, dissolve 10 g of the disodium salt of ethylenediaminetetraacetic acid in water and dilute to 100 ml.

EDTA wash solution, add 10 ml of 20 N sulphuric acid and 10 ml of the 10 per cent EDTA solution to 400 ml of water. Make just alkaline to phenol red indicator and dilute to 500 ml with water.

Morin solution, dissolve 50 mg of the reagent in 100 ml of ethanol. Dilute ten times with water for use as required.

Sodium hydroxide solution, 8·2 M.

Quinine sulphate solution, 0·01 per cent dissolve 50 mg of the reagent in 500 ml of water containing 10 ml of 72 per cent perchloric acid.

Triethanolamine solution, dissolve 5 g of the disodium salt of ethylenediaminetetraacetic acid and 3 ml of triethanolamine in 95 ml of water.

Phenolsulphonphthalein indicator solution, 0·1 per cent solution in water.

Piperidine buffer solution, stir together 10 g of hydrazine sulphate, 5 g of EDTA, 30 ml of water and 50 ml of redistilled piperidine until a clear solution is obtained. Dilute to 500 ml with water.

Alkali stannite solution, dissolve 1·50 g of stannous chloride in 2 ml of water. Add 25 ml of cold water and pour the milky suspension into a cold solution of 2·4 g sodium hydroxide in

10 ml of water. Transfer the solution to a 50-ml volumetric flask and dilute to volume with water. This solution is unstable and should be prepared only immediately before use.

Procedure. Accurately weigh $0 \cdot 5$ g of the finely powdered silicate material into a small platinum dish, moisten with water and add 1 ml of concentrated nitric acid and 10 ml of concentrated hydrofluoric acid. Gently swirl the dish to permit the removal of any liberated carbon dioxide, and then evaporate slowly to dryness on a steam bath or hot plate. When quite dry fuse the residue with the minimum quantity of anhydrous potassium fluoride. Allow to cool, add 1 ml of concentrated sulphuric acid and again evaporate first to fumes then until all the solid material has dissolved and a pyrosulphate melt obtained. Allow to cool, dissolve the melt in 40 ml of 3 N hydrochloric acid and gently boil this solution for 10 minutes to hydrolyse any pyrophosphates formed during the fusion stage. Allow to cool once more.

For samples containing 1 ppm Be or less, the whole of this solution should be taken for the subsequent stages of the analysis. For samples containing larger amounts of beryllium, dilute this solution to volume and use a suitable aliquot diluted to about 40 ml with water.

To this solution contained in a small flask or beaker, add 30 ml of 10 per cent EDTA solution (note 2) and 3–4 drops of phenolsulphonphthalein indicator solution. Now add aqueous ammonia drop by drop until the first trace of the red alkaline form is obtained. Discharge this red colour by adding 1 or 2 drops of dilute sulphuric acid, this should give a pH of about 6. Now gently boil the solution for 5 minutes, cool to room temperature and then transfer to a 125-ml separating funnel.

Add 10 drops ($0 \cdot 25$ ml) of acetylacetone to the solution, stopper and shake vigorously for 10 seconds. Add aqueous ammonia drop by drop until the red colour of the indicator is just produced, and then add 10 ml of chloroform and shake vigorously for a further period of 2 minutes. Run off the organic layer and repeat the extraction with a further 3 drops of acetylacetone and an additional 10 ml of chloroform.

Combine the organic extracts in a clean 100-ml separating funnel, stopper and shake for 1 minute with 20 ml of the EDTA wash solution. Run the organic layer into a 100-ml beaker. Add 2 drops of acetylacetone to the aqueous layer in the funnel, and extract into a further 10 ml of chloroform. Add this additional chloroform layer to the extract in the beaker. Now add to the beaker 3 ml of concentrated perchloric acid and 3 ml of concentrated nitric acid. Cover the beaker with a clock glass and heat gently to evaporate the chloroform, taking care to avoid bumping. The acetylacetone reacts with the nitric acid and the vigorous reaction should be allowed to subside before raising the temperature to expel most of the perchloric acid. The solution should not be allowed to evaporate to complete dryness, or insoluble beryllium oxide will be formed.

Add exactly $3 \cdot 00$ ml of $8 \cdot 2$ M sodium hydroxide solution and 3 drops of quinine sulphate solution to the beaker. Now add concentrated perchloric acid drop by drop until a blue fluorescence just appears when the solution is examined under ultraviolet light. One drop of perchloric acid in excess should be added and the beaker swirled to dissolve any beryllium hydroxide precipitated on the sides of the beaker. Transfer the solution quantitatively to a 25-ml volumetric flask (note 3), ensuring that the solution does not exceed 17 ml in volume. Prepare also a reagent blank solution starting from 5 ml of water, and a beryllium standard containing 0·5 μg Be in 5 ml of water, by adding 3·00 ml of sodium hydroxide solution and 3 drops of quinine sulphate solution and then neutralising with perchloric acid as described above.

Develop fluorescence in the sample solution, the reagent blank and the beryllium standard solution in sequence without delay as follows. To each add 0·5 ml of the triethanolamine solution and N sodium hydroxide solution drop by drop to extinguish the fluorescence due to quinine followed by 2 drops in excess. Now add 0·5 ml of the alkali stannite solution, rinsing down the sides of the flask with a little water. Mix the solution, add 5·00 ml of the piperidine buffer solution and again mix. Dilute the solution to just less than 24 ml, mix again and then introduce 1·00 ml of the morin solution under the surface of the sample solution, preventing contact with the air. Dilute the solution exactly to volume, stopper the flask, mix well and allow to stand in a constant temperature bath for at least 5 minutes before measurement of the fluorescence (note 4).

Measure the intensity of fluorescence using either a permanent standard such as the uranium glass filter described by Sill and Willis,[31] or the fluorescence of a standard solution of beryllium to set the transmittance scale of the instrument. The choice of wavelength used will depend upon the type of instrument and choice of filters available. If a recording spectrofluorimeter can be used, the exciting radiation should be set at about 445 nm, and the fluorescence spectrum recorded over the range 500–550 nm. If only a filter instrument is available, the filters used should be chosen to give a narrow excitation band at about 445 nm with a tungsten lamp, or to isolate the 436 nm radiation of a mercury lamp. Filters with a maximum transmission at about 525 nm should be used to isolate the fluorescence radiation. As can be seen from Fig. 24 both the morin reagent and the beryllium complex have appreciable excitation and fluorescence respectively at these wavelengths.

Notes: 1. If isotope dilution is to be used the volume of beryllium tracer required for addition to the sample material should give about 2×10^4 counts per minute in the final solution used for the fluorescence measurement. The pipette used to transfer the tracer solution to the sample should also be used to transfer an identical volume of the tracer solution to a separate vessel containing inactive beryllium solution for a subsequent comparison with the beryllium activity recovered.

2. The amount of EDTA added should be sufficient to complex all other metallic ions present. If any turbidity appears on neutralising the solution, acidify, add more EDTA and repeat the addition of the aqueous ammonia. Where an aliquot of the rock solution has been taken for the analysis it may be possible to reduce the volume of EDTA required. Not less than 10 ml should be added.

3. By careful working it is possible to obtain an increase in sensitivity by using a

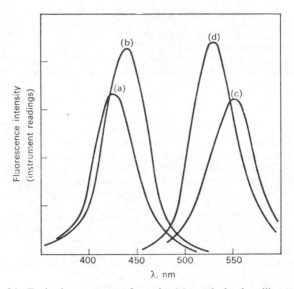

FIG. 24. Excitation spectra of morin (a), and the beryllium–morin complex (b), with the fluorescence spectra of morin (c) and the beryllium–morin complex (d).

10-ml volumetric flask. A more concentrated piperidine buffer solution should then be used.[3]

4. The fluorescence of beryllium-morin solutions is temperature dependent and the blank, standard and sample solutions should all be brought to the same temperature, as close to that of the room as possible. If the sample compartment of the spectrofluorimeter can be thermostatted, both this and the constant temperature bath can be set to room temperature.

The Determination of Beryllium by Atomic Absorption Spectroscopy

Beryllium can be determined by flameless atomic absorption spectroscopy using a high temperature graphite furnace to dry, ash and ignite an aliquot of the rock sample solution. Sighinolfi[32] has given details of a suitable procedure for determining a few ppm beryllium in silicate rocks following decomposition with hydrofluoric and perchloric acids and conversion to sulphates. Advantages claimed include a very much greater sensitivity and the need to use only very small sample volumes. It is clear from this published work that the precision and accuracy of this technique depend very much upon other elements present in the rock matrix. These matrix effects are not yet well understood.

References

1. SOLODOV N. A., *Geokhimiya* (1960) (8), 874.
2. BEUS A. A., *Geokhimiya* (1961) (5), 432.
3. DAWSON J., *Bull. Geol. Surv. Gt. Brit.* (1966) (25), 55.
4. MACHACEK J., SULCEK Z. and VACL J., *Sb. Geol. Ved. Technol. Geochem.* (1966) **7**, 33.
5. WARNER L. A., HOLSER W. T., WILMARTH V. R. and CAMERON E. N., *U.S. Geol. Surv. Prof. Paper* 318, 1959.
6. BEUS A. A., *Geokhimiya* (1956) (5), 75.
7. COATS R. R., BARNETT P. R. and CONKLIN N. M., *Econ. Geol.* (1962) **57**, 963.
8. SHAWE D. R. and BERNOLD S., *U.S. Geol. Surv. Prof. Paper* 501-B, p. 100, 1964.
9. GOLDSCHMIDT V. M., *Geochemistry*, Oxford, 1954, p. 207.
10. SHAWE D. R. and BERNOLD S., *U.S. Geol. Surv. Bull.* 1214-C, 1966.
11. BEUS A. A., *Geochemistry of Beryllium*, Freeman, London, 1966.
12. HILLEBRAND W. F. and LUNDELL G. E. F., *Applied Inorganic Analysis*, Wiley, New York, 2nd ed., 1953.
13. HURE J., KREMER M. and LeBERQUIER F., *Anal. Chim. Acta* (1952) **7**, 57.
14. *The Determination of Beryllium*, National Chemical Laboratory, H.M.S.O., 1963.
15. DAS J. and BANERJEE S., *Zeit. Anal. Chem.* (1961) **184**, 110.
16. POLLOCK J. B., *Analyst* (1956) **81**, 45.
17. ADAM J. A., BOOTH E. and STRICKLAND J. D., *Anal. Chim. Acta* (1952) **6**, 462.
18. KARANOVICH G. G., *Zhur. Anal. Khim.* (1956) **11**, 417.
19. PAKALNS P. and FLYNN W. W., *Analyst* (1965) **90**, 300.
20. PAKALNS P., *Anal. Chim. Acta* (1964) **31**, 576.
21. CABRERA A. M. and WEST T. S., *Analyt. Chem.* (1963) **35**, 311.
22. SANDELL E. B., *Anal. Chim. Acta* (1949) **3**, 89.
23. MAY I. and GRIMALDI F. S., *Analyt. Chem.* (1961) **33**, 1251.
24. SILL C. W., WILLIS C. P. and FLYGARE J. K., *Analyt. Chem.* (1961) **33**, 1671.
25. SILL C. W. and WILLIS C. P., *Geochim. Cosmochim. Acta* (1962) **26**, 1209.
26. AIROLDI R., *Ann. Chim., Roma* (1951) **41**, 478.
27. BELYAVSKAYA T. A. and KOLOSOVA I. F., *Zhur. Anal. Khim.* (1964) **19**, 1162.
28. LUKIN A. M. and ZAVARIKHINA G. B., *Zhur. Anal. Khim.* (1956) **11**, 393.
29. PONOMAREV A. I., *Methods for the Chemical Analysis of Silicate and Carbonate Rocks*, Isdat Akad. Nauk U.S.S.R., 1961, p. 130.
30. CHARLOT G., *Anal. Chim. Acta* (1947) **1**, 233.
31. SILL C. W. and WILLIS C. P., *Analyt. Chem.* (1959) **31**, 598.
32. SIGHINOLFI G. P., *Atom. Absorp. Newsl.* (1972) **11**, 96.

CHAPTER 13

BISMUTH

DISCRETE bismuth minerals do not appear to be normal constituents of silicate rocks, and the presence of bismutite, a basic carbonate, or bismuthinite, Bi_2S_3, can usually be taken to indicate mineralisation in the area. The bismuth content of igneous and other silicate rocks and minerals is quite small, well below the limit of detection of most spectrophotometric and spectrographic methods. For these silicates a variety of new methods using neutron activation and isotope dilution procedures have been developed,[1, 2, 6] enabling reasonable agreement to be reached on the bismuth content of some USGS rock standards, Table 16.

TABLE 16. BISMUTH CONTENT OF SOME ROCK STANDARDS
ppb Bi

Standard	Reference 1	2	3
AGV-1	44	56·4	55·5
BCR-1	35	46·7	49·6
DTS-1	5·9	4·8	5·1
G-1		46	51·9
G-2	63	37·6	41·0
GSP-1	33	36·8	36·7
PCC-1	13	8	5·7
W-1		43·5	51·6

Using a substoichiometric radioisotopic dilution technique Greenland et al.[10] reported the bismuth content of a total of seventy-four rocks from a differentiated tholeiitic dolerite and two calc-alkaline batholiths. All three rock bodies demonstrated an enrichment of bismuth in residual magmas with magmatic differentiation. Bismuth was also enriched, relative to the host rock, in the calcium-rich accessory minerals apatite

151

and sphere, but this Bi–Ca association may be of little significance in the geochemistry of this element. Most of the bismuth may be present as inclusions in a trace mineral phase which, in view of the chalcophilic nature of bismuth, could be a sulphide phase. A value of 50 ppb was suggested for the crustal abundance of bismuth, in good agreement with the figure of 60 ppb of Marowsky and Wedepohl.[11] Somewhat higher bismuth values have been given by some authors, e.g. by Velikii[4] who reported igneous rocks from Southern Fergana (USSR) to contain from 25 ppb to 2·2 ppm and by Brooks et al.[5, 6] who reported values in the range 10–800 ppb.

A variety of "chemical" procedures have been described for determining bismuth in silicate and other rocks, but these are applicable primarily to rock materials that have been enriched in bismuth. These include a photometric method based upon dithizone after extraction of bismuth iodide into isoamyl acetate,[7] an atomic absorption method involving extraction of soluble bismuth into nitric acid and nebulisation into an air–acetylene flame[8] and a polarographic method applied after an extraction with diethylammonium diethyldithiocarbamate.[9]

References

1. SANTOLIQUIDO P. M. and EHMANN W. D., Geochim. Cosmochim. Acta (1972) 36, 897.
2. LAUL J. C., CASE D. R., SCHMIDT-BLEEK F. and LIPSCHUTZ M. E., Geochim. Cosmochim. Acta (1970) 34, 89.
3. GREENLAND L. P. and CAMPBELL E. Y., Anal. Chim. Acta (1972) 60, 159.
4. VELIKII A. S., Forma Nakh. Osob. Raspredel. Vismuta Gidroterm. Mestorozhd. (1969) 52–72 (see C.A. (1970) 73, 122520x).
5. BROOKS R. R., AHRENS L. H. and TAYLOR S. R., Geochim. Cosmochim. Acta (1960) 18, 162.
6. BROOKS R. R. and AHRENS L. H., Geochim. Cosmochim. Acta (1961) 23, 100.
7. MOTTOLA H. A. and SANDELL E. B., Anal. Chim. Acta (1961) 25, 520.
8. WARD F. N. and NAKAGAWA H. M., U.S. Geol. Surv. Prof. Paper 575-D (1967) 239.
9. RUSSEL H., Zeit. Anal. Chem. (1962) 189, 256.
10. GREENLAND L. P. and CAMPBELL E. Y., Geochem. Cosmochim Acta (1973) 37, 283.
11. MAROWSKY G. and WEDEPOHL K. H., Geochim. Cosmochim Acta (1971) 35, 1255.

CHAPTER 14

BORON

Occurrence

Vinogradov[1] has suggested that basic rocks contain less boron than acidic rocks, a conclusion supported by Maurice[2] in an extensive review of the geochemistry of boron. However, the reported values for boron in silicate rocks show an unusually wide variation, and individual specimens with boron contents greatly in excess of the average value for that class, Table 17, are by no means uncommon. Lyakhovich,[3] for example, has reported up to 310 ppm boron in a series of granitic specimens, whilst Bawden,[4] examining silicates from Cornwall, noted up to 0·17 per cent in killas and up to 0·12 per cent in granites.

TABLE 17. THE BORON CONTENT
OF IGNEOUS ROCKS[1]

Rock type	B, ppm
Granites, granitoids	15
Diorites, andesites	15
Basalts, gabbros	5
Dunites	1

In acidic rocks boron occurs predominantly as the mineral tourmaline, and high-boron granites usually contain abundant quantities of this easily recognised mineral. Tourmaline contains about 3 per cent boron, 10–11 per cent B_2O_3. There are a number of rarer minerals containing boron as an essential constituent, but in some silicate rocks much of the boron is present in the common silicate minerals, particularly felspars, pyroxenes and amphiboles.[2] Datolite, a calcium borosilicate occurring in basic rocks such as basalts, contains about 6 per cent boron, 20–22 per cent B_2O_3, giving the approximate upper limit

153

to the concentration range of boron in silicate minerals. The lower limit for common rocks and silicate minerals is less than 1 ppm.

The Determination of Boron

Much of the earlier data on the occurrence of boron is unreliable owing to the lack of accurate and suitable methods of analysis. However, the nuclear properties of boron, in particular its high cross section for the absorption of thermal neutrons, have led to a considerable interest in the determination of small quantities of boron. The older methods of separation, involving distillation as methyl borate, have been re-examined and improved, and new methods based upon pyrohydrolysis, solvent extraction and ion-exchange devised. A large number of colour-forming reagents have been investigated for the photometric determination which, except for those minerals high in boron, have displaced the earlier titrimetric procedures.

THE SEPARATION OF BORON

Distillation as methyl borate is regarded as the classical procedure for the separation of boron, but conflicting reports exist in the literature regarding the adequacy of this method. For example, Strahm[5] (1965) has noted: "Although distillation of methyl borate has been a widely applied method of separation, it is unwieldy and recovery of boron is often low. The fact that many investigators have turned to other methods ... is evident that for general use the method is inadequate and impractical." On the oher hand, Mills[6] (1966) has written: "For siliceous samples no technique was found to be superior to a semi-micro version of the classical Wherry–Chapin methyl borate distillation."[7]

Difficulties can arise from the presence of boron in almost all laboratory glassware, and the distillation apparatus itself should be made from fused silica. Another difficulty that has been noted is the need for maintaining anhydrous or near anhydrous conditions during the esterification and evolution of the borate.

Ion-exchange separation procedures are generally more rapid than distillation methods, and this is true of the methods advocated for removing interfering cations from boron. In addition these procedures are simpler than distillation in regard to both apparatus required and method of working used. Silicate rocks and minerals are brought into solution by an alkaline fusion, and the aqueous extract acidified with hydrochloric acid. The passage of this solution through a column of cation exchange resin removes iron and other interfering ions to give a clear solution that can be used directly for the determination of boron by either titrimetry or spectrophotometry. In a recent version of this procedure Fleet[8] dispenses with the use of a column and adds the resin directly to the acid extract of the fused rock melt. This procedure is described in detail below for application to silicate rocks.

Other procedures used for the separation of boron in silicate materials include extraction into organic solutions (such as a mixture of ethanol and diethyl ether used by Glaze and Finn[9]), and pyrohydrolysis. This latter technique has been used successfully to recover boron from glass,[10] but Mills[6] reports that it was not possible to obtain a quantitative recovery from silicates.

The Determination of Boron in Tourmaline

One of the earliest procedures to be devised for the determination of boron in silicate rocks and minerals was that of titrating liberated boric acid with standard alkali in the presence of a polyhydric alcohol—mannitol being now commonly employed. The boric acid–mannitol complex acts as a strong monobasic acid. When combined with an ion-exchange separation, this procedure can be simply and easily applied to the analysis of tourmaline and other silicate minerals such as axinite, datolite, danburite or dumortierite that contain boron as a major, essential component. The procedure described here has been adapted from that given by Kramer.[11]

Method

Reagents: *Sodium hydroxide solutions*, approximately 5 M, and standard 0·05 M.
Mannitol.
Ion-exchange resin, strongly acid, cationic resin such as Amberlite IR 120(H) or Dowex 50W-X8, in the form of a bed 2 cm in diameter and 25 cm in length. The column should be of borosilicate glass (attack of the glass is negligible) or polypropylene tubing. To prepare the column for use or to regenerate for further use wash the bed with 100 ml of 6 N hydrochloric acid followed by water until the eluate is free from acid.

Procedure. Accurately weigh approximately $0 \cdot 5$ g of the finely powdered tourmaline or other boron mineral into a small platinum crucible, mix with 3 g of anhydrous sodium carbonate and fuse over a Bunsen burner for 30 minutes. Transfer the crucible to a Meker burner and continue the heating for a further 30 minutes. Allow the melt to cool, spreading it around the sides of the crucible in the usual way.

Place the crucible on its side in a 100-ml polythene or polypropylene beaker containing 20 ml of water and add concentrated hydrochloric acid down the sides of the beaker until there is an excess of about 1 ml above the amount required to neutralise the alkali carbonate used for the fusion. Warm the solution and allow to stand until the melt has completely decomposed and all soluble material has passed into solution. Rinse and remove the platinum crucible and lid. At this stage no unattacked mineral grains should be present, and the only residue a few flakes of silica precipitated from the solution. Filter the solution through a 9-cm open-textured paper supported in a polythene funnel into a polythene beaker, and wash the residue with hot water to give a solution volume of about 50 ml. Discard the residue.

Add 5 M sodium hydroxide solution drop by drop until the formation of a precipitate that only just fails to dissolve on warming. Clear this precipitate with a few drops of hydrochloric acid, and transfer the solution to the ion-exchange column, previously washed with water. Allow the eluate to collect at a rate of between 30 and 40 ml per minute in a 400-ml polypropylene beaker, and wash the resin with about 200 ml of water. Add $0 \cdot 5$ ml of concentrated hydrochloric acid and boil the solution for 1 minute but not longer (note 1), to expel any carbon dioxide, and cool the solution to room temperature.

Using a magnetic stirrer and a pH meter, add sodium hydroxide solution, first concentrated then diluted, drop by drop, until the pH of the solution reaches 7. Now add 20 g of solid mannitol and titrate the boric acid with $0 \cdot 05$ M standard sodium hydroxide solution until the pH again reaches 7. Subtract a reagent blank value from the titre before calculating the results. The reaction that occurs may be expressed by the equation

$$H_3BO_3 + NaOH = NaH_2BO_3 + H_2O$$

so that 1 ml of $0 \cdot 05$ M sodium hydroxide solution is equivalent to $1 \cdot 741$ mg B_2O_3 (note 2).

Notes: 1. Prolonged boiling of hydrochloric acid solutions will result in substantial loss of boron.[12]

2. Where boron is included in the summation of the rock or mineral, it is customary to express the results as per cent boric oxide, B_2O_3. Where boron occurs only as a trace constituent, parts per million boron is used.

The Determination of Boron in Silicate Rocks

Of the many reagents described for the photometric determination of boron only curcumin, carmine (carminic acid) and dianthrimide (1,1'-iminodianthraquinone) have been extensively used. Curcumin, the active colour principle of the vegetable product turmeric, has long been used for the detection and determination of small amounts of boron. Considerable difficulties were earlier experienced in obtaining repeatable quantitative results, although Hayes and Metcalfe[13] have now established conditions necessary for a reliable procedure based upon this reagent. Alonso and Sanchez[14] described its application to the determination of boron in geological materials after a preliminary separation using Dowex 50W-X8 anion exchange resin.

The use of dianthrimide for the photometric determination of boron has been carefully examined by Danielsson.[15] This reagent is more sensitive than carmine, but less so than curcumin. It gives a linear calibration in concentrated sulphuric acid solution. The reagent itself has an absorption band with a maximum value below 400 nm, clearly separated from that of the boron complex which has a maximum value at 620 nm. The rate of reaction of boron with dianthrimide is strongly dependent upon temperature. Optical density values are also temperature dependent, although this effect is related to the acid concentration.

Carmine, the name given to a naturally occurring dyestuff from cochineal, is a calcium–aluminium compound of carminic acid, which is a derivative of anthraquinone. Both carmine and carminic acid react with boron in concentrated sulphuric acid solution to give blue coloured complexes, but as carminic acid is deliquescent, carmine is preferred.[16] In the absence of boron the colour of the dye at pH $6 \cdot 2$ is bright red, but in the presence of boron this changes to blue. The wavelength of maximum absorption changes from 520 nm for the reagent to 585 nm for the boron complex.[17] The Beer-Lambert Law is obeyed over the concentration range 0–10 ppm of boron.

The colour development characteristics may vary with the brand of reagent used.[8] Hatcher and Wilcox [17] have reported that the coloured complex with boron can be measured after 45 minutes and then shows no appreciable change at the end of 4 hours. Fleet,[8] however, noted that the absorption reached a maximum after 40 minutes and thereafter decreases. The procedure described by Fleet, using an anion-exchange resin to separate interfering ions from boron, is given below.

Method

Reagents: *Mannitol solution*, dissolve 1 g of reagent in 100 ml of water.
Carmine solution, dissolve 50 mg of the reagent in 100 ml of concentrated sulphuric acid.

Hydrochloric acid, 0·6 N.
Sodium hydroxide solution, 0·1 N.
Standard boron stock solution, dissolve 0·5716 g of recrystallised
    boric acid in water and dilute to 1 litre. This solution contains
    100 μg boron per ml.
Standard boron working solutions, dilute aliquots of the stock
    solution with water to give three new solutions containing 5,
    10 and 20 μg boron per ml respectively.

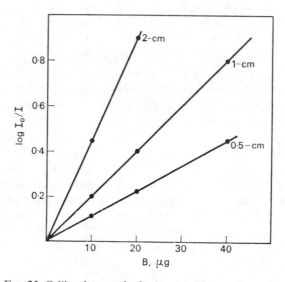

FIG. 25. Calibration graphs for boron with carmine (585 nm).

Procedure. Accurately weigh approximately 0·2 g of the finely powdered
rock sample (or a smaller amount if the sample material contains more than
200 ppm boron) into a 10-ml platinum crucible and add 1·25 g of potassium
carbonate. Mix, and fuse over a Bunsen burner for 1 hour. Allow the crucible to
cool, loosen the melt by warming with a small amount of water, and transfer
the solution and residue to a 50-ml polypropylene beaker. Cover the beaker
and add 2 ml of mannitol solution, followed by 20 ml of the cation exchange
resin and 2 ml of 0·6 N hydrochloric acid. Break up any lumps of residue, mix
with the ion exchange resin and add enough water to give a slurry. Allow to
stand overnight.
    Collect the ion-exchange resin and precipitated silica on a small medium-
textured filter paper, wash well with water and discard it. Collect the filtrate
and washings in a 400-ml polypropylene beaker, add 23 ml of 0·1 N sodium

hydroxide solution and carefully evaporate to dryness on a steam bath. Allow to cool and add by pipette 5 ml of 0·6 N hydrochloric acid. When the residue has dissolved, pour this solution into a centrifuge tube and centrifuge.

Pipette 2 ml of the clear solution into a 50-ml polypropylene beaker, add 2 drops of concentrated hydrochloric acid and *with great care* add 10 ml of concentrated sulphuric acid. Allow the solution to cool, then add 10 ml of the carmine reagent solution. Swirl gently to mix the contents of the beaker and allow to stand for 40 minutes. Measure the optical density of the solution in 1-cm cells with the spectrophotometer set at a wavelength of 585 nm.

For the reference solution, transfer 2 ml of water to a 50-ml polypropylene beaker and add concentrated hydrochloric acid, sulphuric acid and carmine reagent as described for the sample solution. A reagent blank should also be prepared from 1·25 g of potassium carbonate fused without rock material in a separate platinum crucible, and carried through the procedure as described. A series of three standards can be used for calibration. These are obtained by transferring 2 ml aliquots of the three working solutions containing 10, 20 and 40 μg boron respectively to separate beakers and proceeding as described above (Fig. 25).

References

1. VINOGRADOV A., *Geokhimiya* (1962) (7), 555.
2. MAURICE J., *Ann. Agron.* (1966) **17**, 367.
3. LYAKHOVICH V., *Geokhimiya* (1965) (1), 25.
4. BAWDEN M. G., *Proc. Ussher Soc.* (1965) **1**, 11.
5. STRAHM R. D. in KOLTHOFF I. M. and ELVING P. J., *Treatise on Analytical Chemistry*, Interscience, New York, Part 2, Vol. 12, p. 169, 1965.
6. MILLS A. A., *Proc. Soc. Anal. Chem.* (1966) **3**, 160.
7. WHERRY E. T. and CHAPIN W. H., *J. Amer. Chem. Soc.* (1908) **30**, 1687.
8. FLEET M. E., *Analyt. Chem.* (1967) **39**, 253.
9. GLAZE F. W. and FINN A. N., *J. Res. Nat. Bur. Stds.* (1941) A27, 33.
10. WILLIAMS J. P., CAMPBELL E. E. and MAGLIOCCA T. S., *Analyt. Chem.* (1959) **31**, 1560.
11. KRAMER H., *Analyt. Chem.* (1955) **27**, 144.
12. FELDMAN C., *Analyt. Chem.* (1961) **33**, 1916.
13. HAYES M. R. and METCALFE J., *Analyst* (1962) **87**, 956.
14. ALONSO S. J. and SANCHEZ G. A., *An. Quim.* (1972) **68**, 335.
15. DANIELSSON L. *Talanta* (1959) **3**, 138.
16. DANIELSSON L., *Organic Reagents for Metals*, Hopkin & Williams Ltd., Chadwell Heath, Essex. Ed. JOHNSON W. C., Vol. 2, p. 32, 1964.
17. HATCHER J. T. and WILCOX L. V., *Analyt. Chem.* (1950) **22**, 567.

CHAPTER 15

CADMIUM

Occurrence

The cadmium content of normal silicate rocks is below the limit of detection by simple spectrographic or spectrophotometric methods. The average crustal abundance of cadmium has been variously reported to be from $0 \cdot 08$ to $0 \cdot 31$ ppm,[1-4] although a value of about $0 \cdot 2$ ppm seems to be most likely. The cadmium content of shales appears to be of a similar order, but limestones and sandstones contain rather less.

The Determination in Silicate Rocks

To obtain the values listed above it has been necessary to use more sensitive techniques such as neutron activation,[3,4] or polarography.[5-7] Ion-exchange chromatography[1, 8] and solvent extraction[5, 6] procedures can be used to effect a preliminary separation and concentration. The method given below, based upon a procedure of Stanton *et al.*[6] uses a dithizone extraction to concentrate the cadmium, and polarography to complete the determination. A similar method has been outlined by Butler and Thompson.[8] Many of the older pen-recording polarographs do not possess sufficient sensitivity to cadmium for this determination, and a cathode ray-, square wave- or pulse-polarograph is recommended.

Method

Reagents: *Carbon tetrachloride*, high-purity grade.

Sodium citrate buffer solution, dissolve 20 g of trisodium citrate dihydrate and 1 g of hydroxyammonium chloride in 100 ml of water, add ammonia until alkaline to thymol blue (note 1).

Dithizone solution, dissolve 15 mg of analytical grade dithizone in 100 ml carbon tetrachloride. Store in a refrigerator, and prepare freshly at weekly intervals.

Hydrochloric acid solution, $0 \cdot 2$ N.

Hydrazine hydrate solution, dissolve 1·5 g of the reagent in 10 ml of water.

Phosphoric acid, 2 M.

Standard cadmium stock solution, dissolve 0·102 g of cadmium chloride ($CdCl_2.2\frac{1}{2}H_2O$) in water and dilute to 1 litre. This solution contains 50 μg Cd per ml.

Standard cadmium working solution, dilute 1 ml of the stock solution to 100 ml with water. This solution contains 0·5 μg Cd per ml.

Procedure. Weigh approximately 1 g of the finely powdered silicate rock into a platinum crucible, add 20 ml of concentrated hydrofluoric acid and 3 ml of concentrated perchloric acid and evaporate to fumes of perchloric acid in the usual way. Transfer the crucible to a hot plate and evaporate to dryness. Moisten the dry residue with 3 ml of constant boiling hydrochloric acid, add 10–15 ml of water and warm to complete dissolution. Cool, transfer to a 100- or 125-ml separating funnel and add 25 ml of the sodium citrate buffer solution.

Extract the reacting metals with successive 5-ml portions of dithizone solution until there is no further reaction, collecting the separated carbon tetrachloride extracts in a second separating funnel. Before discarding the aqueous phase, check its pH—if less than 8·9 increase to this value by adding ammonia solution, and then extract again with a further 5-ml portion of the dithizone solution. In the second funnel complete the separation of carbon tetrachloride solution from any small amount of accompanying aqueous phase, and transfer the organic layer to a 100-ml beaker. Rinse the separating funnel with a little carbon tetrachloride and add the washings to the beaker.

Evaporate the combined extracts to dryness and then add 2 ml of 20 N sulphuric acid and 1 ml of concentrated perchloric acid. Transfer the beaker to a hot plate and evaporate to dryness. Dissolve the residue by warming with exactly 1 ml of 0·2 N hydrochloric acid, and transfer to a polarographic cell. Add 0·2 ml of hydrazine hydrate solution and 0·2 ml of 2 M phosphoric acid. De-aerate the solution with nitrogen in the usual way and determine the cadmium present in the solution by polarographic reduction, measuring the wave at −0·70 V.

Calibration. Transfer 1 ml of the cadmium chloride solution containing 0·5 μg Cd to the polarographic cell. Add hydrazine solution and phosphoric acid and continue as described for the sample solution above.

Notes: 1. In their description of this method the original authors[6] include also the determination of cobalt, copper, lead, nickel and zinc. For these determinations it is necessary to remove contaminating heavy metals from the buffer solution before it is used. This is done by extraction with dithizone solution.

2. In the original method only one-tenth of the solution was used for the determination of cadmium. To achieve the necessary sensitivity to cadmium in this aliquot it was then necessary to use the high sensitivity obtainable with a square wave polarograph.

References

1. BROOKS R. R. and AHRENS L. H., *Geochim. Cosmochim. Acta* (1961) **23**, 100.
2. SANDELL E. B. and GOLDICH S. S., *J. Geol.* (1943) **51**, 99.
3. VINCENT E. A. and BILEFIELD, L. I., *Geochim. Cosmochim. Acta* (1960) **19**, 63.
4. SCHMITT R. A., SMITH R. H. and OLEHY D. A., *Geochim. Cosmochim. Acta* (1963) **27**, 1077.
5. CARMICHAEL I. and McDONALD A., *Geochim. Cosmochim. Acta* (1961) **22**, 87.
6. STANTON, R. E., McDONALD A. and CARMICHAEL I., *Analyst* (1962) **87**, 134.
7. HUFFMAN C., *U.S. Geol. Surv. Prof. Paper* 450-E, p. 126, 1962.
8. BUTLER J. R. and THOMPSON A. J., *Geochim. Cosmochim. Acta* (1967) **31**, 104.

CHAPTER 16

CALCIUM

Occurrence

Peridotites and dunites, the earliest rocks to crystallise, contain only small amounts of calcium which, with the precipitation of olivine and enstatite, tends to be concentrated in the remaining melt. The succeeding stages of magmatic emplacement involve the crystallisation of much of the monoclinic pyroxenes and the calcium-rich felspars, giving rocks containing a great deal more calcium and culminating in the gravity separation of anorthosites which may contain up to 20 per cent CaO. The residual magma is depleted in calcium and the succeeding rocks contain successively less (Fig. 26).

Silicate minerals containing calcium as an essential constituent include augite and other pyroxenes of the diopside-hedenbergite series, hornblende and amphiboles of the tremolite-actinolite series as well as anorthoclase and the plagioclase felspars. Some of the felspathoids, garnets, epidotes and zeolites contain calcium as a major component as do scapolite, wollastonite ($CaSiO_3$) and sphene ($CaTiSiO_5$). Carbonate minerals include calcite and aragonite $CaCO_3$, dolomite $CaMg(CO_3)_2$ and ankerite with iron and manganese substituting for magnesium. Other non-silicate minerals of calcium include the two sulphates gypsum $CaSO_4.2H_2O$ and anhydrite $CaSO_4$, fluorite CaF_2, perovskite $CaTiO_3$, scheelite $CaWO_4$ and a number of rarer arsenates, phosphates, vanadates and uranates as well as chlorides, carbonates and borates among the evaporite minerals.

The Determination of Calcium as Oxalate

CLASSICAL GRAVIMETRIC PROCEDURE

After the removal of iron, aluminium and other elements of the ammonia group, any calcium present in the filtrate can be precipitated as oxalate, accompanied by much of the small amount of strontium found in most silicate rocks. In the classical procedure for determining calcium, the first oxalate precipitate is redissolved in dilute hydrochloric acid and then re-precipitated from a smaller volume of solution. This gives a precipitate that is almost completely free from magnesium and manganese[1] that can be ignited to oxide in a platinum crucible as described on p. 44.

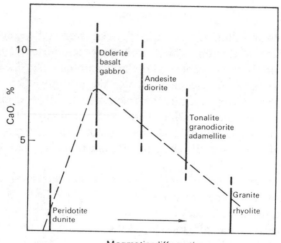

FIG. 26. The calcium content of silicate rocks.

After ignition, calcium oxide tends to increase in weight by absorption of water and carbon dioxide. This does not usually give rise to any serious error, but can be avoided by igniting the precipitate at a temperature of only 500°, converting the oxalate to carbonate, in which form it is weighed.

DIRECT GRAVIMETRIC DETERMINATION

Although it is customary to remove iron, aluminium, titanium and phosphorus before precipitating calcium, the work of Meade[2] has shown that this is not absolutely necessary and that calcium oxalate can be precipitated quantitatively from a weakly acid solution containing citric or other organic acid to prevent the precipitation of the ammonia group elements.

Manganese interferes and difficulties also occur if the rock material is rich in magnesium or titanium. Maynes[3] has, however, shown that manganese, magnesium, iron, aluminium and titanium can all be retained in solution by replacing the citric acid with 8-hydroxyquinoline-5-sulphonic acid.

Method

Reagents: *8-Hydroxyquinoline-5-sulphonic acid solution*, dissolve 10 g of the reagent in a little dilute aqueous ammonia and dilute to 200 ml with water.

Ammonium oxalate.

Ammonium oxalate wash solution, dissolve 1 g of the reagent in 500 ml of water.

Procedure. Accurately weigh approximately 1 g of the finely powdered silicate rock material into a platinum dish, moisten with water and add 1 ml of concentrated nitric acid, 5 ml of concentrated perchloric acid and 10 ml of concentrated hydrofluoric acid. Transfer the dish to a hot plate and evaporate first to fumes of perchloric acid and then to dryness. Allow to cool, rinse down the walls of the dish with a little water, add 4 ml of concentrated hydrofluoric acid and 5 ml of 20 N sulphuric acid and evaporate to fumes of sulphuric acid. Allow to cool, rinse down the walls of the dish with water and again evaporate, this time to dense fumes of sulphuric acid. Allow to cool, add water and digest on a steam bath until all soluble material has passed into solution.

Rinse the solution into a 600-ml beaker with water. If any unattacked residue remains, collect on a small paper, wash with water and fuse in platinum with a little anhydrous sodium carbonate. Extract the melt with water, acidify with dilute sulphuric acid and add to the main rock solution (note 1). Dilute to a volume of about 200 ml.

Add 60 ml of the 8-hydroxyquinoline-5-sulphonic acid solution (note 2) and heat almost to boiling. Add a few drops of methyl red indicator solution and then concentrated ammonia drop by drop until a pure yellow coloured solution is obtained, then add 5 ml of concentrated ammonia in excess. Add 10 g of solid ammonium oxalate and stir until the crystals have dissolved. Cover the beaker with a clock glass, transfer to a steam bath and heat for a period of 2 hours. Allow to cool somewhat and then, using a pH meter, bring the pH to 6·0 by adding dilute hydrochloric acid. Allow the solution to stand overnight.

Collect the precipitated calcium oxalate on a close-textured filter paper and wash with the ammonium oxalate wash solution until the filtrate is completely colourless. Transfer the filter and precipitate to a weighed platinum crucible, dry, ignite in an electric muffle furnace set at 1000° and weigh as calcium oxide, or set at 500° and weigh as calcium carbonate.

Notes: 1. A fine white precipitate of barium sulphate can be filtered off, washed with a little dilute sulphuric acid and discarded.

2. For basic and other rocks rich in iron, aluminium and magnesium the quantity of reagent should be increased to 120 ml.

3. As in the classical procedure, if any strontium is present in the rock it will be largely precipitated as oxalate and will be counted as calcium.

THE SEPARATION OF STRONTIUM

The ignited calcium oxalate precipitates were formerly used for the determination of strontium, and a correction was then applied to obtain the "true" calcium content. However, none of the chemical methods allows a perfect separation and values for calcium are likely to be almost as much in error after the correction as before. The most frequently used method for this separation was based upon the solubility of calcium nitrate in concentrated nitric acid. Strontium nitrate is relatively insoluble and can be collected and weighed on a sintered glass or silica crucible. This procedure has now been displaced by flame photometric and atomic absorption methods that do not require any extensive separation stage.

TITRIMETRIC DETERMINATION WITH PERMANGANATE

After purification, calcium oxalate can be dissolved in dilute sulphuric acid and the liberated oxalic acid titrated with standard potassium permanganate solution. The small amount of strontium present in silicate rocks and collected largely in the oxalate precipitate will also be counted as calcium.

Method

Reagents: *Ammonium oxalate wash solution*, dissolve 1 g of reagent in 500 ml of water and make just alkaline to methyl red indicator.

Potassium permanganate $0 \cdot 1$ N *solution*, standardise by titration with sodium oxalate or arsenious oxide.

Procedure. Precipitate the calcium as oxalate as described above or on p. 44. Collect the precipitate on a close-textured filter paper and wash with the appropriate wash solution as described. Dissolve the calcium oxalate from the filter with a small amount of dilute hydrochloric acid and re-precipitate calcium oxalate by adding $0 \cdot 2$ g of ammonium oxalate and aqueous ammonia until the solution is alkaline, without adding further 8-hydroxyquinoline-5-sulphonic acid. Collect the precipitate on a close-textured filter paper and wash twice with ammonium oxalate wash solution, then with small quantities of cold water until the filtrate is free from oxalate.

Rinse the precipitate into a beaker and dissolve by warming with 100 ml of 3 N sulphuric acid. Filter the hot solution through the paper previously used to collect the purified calcium oxalate and wash thoroughly with water. Heat the solution to a temperature of 60–70° and titrate with standard $0 \cdot 1$ N potassium permanganate solution.

1 ml of $0 \cdot 1$ N potassium permanganate is equivalent to $2 \cdot 80$ mg calcium oxide, giving a titration of about 36 ml for a 1 g portion of silicate rock containing 10 per cent CaO. Thus for anorthosites and similar rocks containing greater amounts of calcium, $0 \cdot 5$ g sample portions should be taken for the determination. For carbonate rocks a 1 g sample weight should be used, and the rock solution diluted to volume in a 250-ml volumetric flask. A 50-ml aliquot can then be taken for the precipitation of calcium and subsequent titration as described above.

Titrimetric Determination with EDTA

Ethylenediaminetetraacetic acid forms complexes with most metals and cannot be used for the titrimetric determination of calcium unless special precautions are taken to avoid interference from trivalent and other divalent elements. In the analysis of silicate rocks this interference is largely from iron, aluminium, manganese and magnesium. Iron and aluminium can be precipitated with ammonia, but traces of aluminium and some part of the manganese can then always be recovered in the filtrate. Small amounts of both calcium and magnesium are usually co-precipitated with the ammonia precipitate, but these can be recovered in a subsequent filtrate following re-precipitation with ammonia. Iron and aluminium can be removed from the rock solution by extraction of the complexes with 8-hydroxyquinoline into chloroform as described by Cluley[5] for the analysis of glass. The interference from iron and aluminium can be considerably reduced by adding triethanolamine. In order to titrate calcium in the presence of magnesium a pH of above 12 is used, at this pH magnesium is precipitated as hydroxide and does not seriously interfere. A detailed procedure for this determination is given on p. 300, where it is combined with the titrimetric determination of calcium plus magnesium in order to obtain values for both elements.

Although suitable for routine analysis this procedure, in common with many others that have been suggested, is subject to certain errors. The end-point of the calcium determination is particularly difficult to determine, especially in the presence of manganese or iron which affect the indicator, even with the addition of triethanolamine as complexing agent. Only small amounts of ammonium salts can be tolerated, as these prevent the complete precipitation of magnesium, which is then titrated with the calcium. Certain indicators cannot be used in the presence of magnesium, although these are undoubtedly the best for pure calcium solutions. In the absence of magnesium, as for example in certain limestones and marbles, acid alizarin black SN (mordant black 25, C.I. 21725),[6, 7] metalphthalein (phthalein complexone) screened with naphthol green B,[8] and methyl thymol blue[9] all give sharp, easily identified end-points. A procedure for this determination is given below.

An alternative approach to the determination of both calcium and magnesium in silicate rocks is that based upon ion-exchange separation from all other interfering elements and from each other. Once this separation has been made there is no difficulty in determining calcium, as a somewhat lower pH of about 10 to 10·5 can be used with erichrome black T as indicator for both calcium and magnesium. This procedure, devised by Abdullah and Riley,[10] takes several days for the complete separation, but most of this time can be used for other determinations. This procedure is given in greater detail on p. 313.

THE DETERMINATION OF CALCIUM IN CARBONATE ROCKS (LOW IN MAGNESIUM)

In this procedure any calcium present in the acid-insoluble fraction is discarded, and only the soluble calcium titrated. Up to about 4 per cent MgO can be tolerated. Acid-soluble iron, aluminium and other metals are not likely to be present in more than trace amounts, and these traces can be complexed by the addition of potassium cyanide and triethanolamine.

Method

Reagents: *EDTA* 0·02 M *standard solution*, dissolve 7·4 g of the disodium salt of EDTA in 1 litre of water and standardise by titration using standard calcium solution.

Triethanolamine solution, dissolve 6·4 g of potassium cyanide in 60 ml of water and mix with 40 ml of triethanolamine.

Hydroxylamine hydrochloride solution, dissolve 10 g of the reagent in 100 ml of water.

Sodium hydroxide solution, dissolve 30 g of the reagent in water and dilute to 100 ml.

Acid alizarin black SN indicator, grind together 0·2 g of the reagent with 10 g of sodium chloride.

Standard calcium solution, dissolve 0·500 g of pure calcium carbonate in the minimum amount of dilute hydrochloric acid, transfer to a 500-ml volumetric flask and dilute to volume with water.

Procedure. Accurately weigh approximately 0·5 g of the finely powdered limestone rock into a 400-ml beaker of the "tall" or "conical" pattern, and moisten with water. Cover the beaker with a clock glass and add dilute perchloric acid down the side of the beaker, until all solid material has dissolved, avoiding an excess. Boil the solution to expel carbon dioxide, allow to cool and dilute with water to volume in a 500-ml volumetric flask. If any residue remains, collect this on a filter paper, wash with water and transfer the combined filtrate and washings to the 500-ml volumetric flask before dilution to volume with water.

Pipette 50 ml of this limestone solution into a 250-ml conical flask, add 5 ml of hydroxylamine hydrochloride solution followed by 5 ml of the triethanolamine solution (N.B. use a measuring cylinder!), 10 ml of the sodium hydroxide solution and enough of the indicator to give a reasonably strong red-to-purple colour to the solution. Titrate with standard EDTA solution until the indicator is pure blue with no trace of a pink colour.

Photometric Determination of Calcium in Silicate Rocks

Very few reagents are known that give colour reactions which are specific or even selective for the calcium ion. One of the most interesting of these few is calcichrome, believed to be cyclo-tris-7-(1-azo-8-naphthalene-3:6-disulphonic acid) (IX), used as an indicator for the titration of calcium with EDTA.[11] This reagent has been used for the photometric determination of calcium,[12] but does not appear to have been applied to this determination in silicate or carbonate rocks, possibly because of interference from magnesium. Murexide (ammonium purpurate) and glyoxal bis(2-hydroxyanil) have also been suggested as photometric reagents for calcium, but also do not appear to have been used for rock analysis. Leonard[13] has, however, used glyoxal bis(2-hydroxyanil) for determining calcium in magnesium carbonate and his method can probably be adapted for use with magnesites.

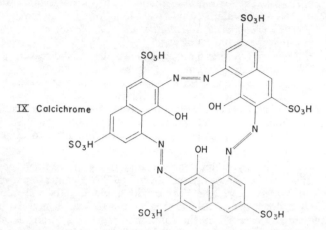

IX Calcichrome

The Determination of Calcium by Flame Photometry

The spectrum obtained when calcium salts are aspirated into a suitable flame is relatively simple, consisting of a resonance line at 422·7 nm and band systems with maxima at wavelengths of 544, 606 and 622 nm. There is further emission in the near infrared, and a doublet at 393/397 nm due to calcium ions present in high temperature flames. Sodium interferes with the determination of calcium by contributing to the background emission at the wavelength of the resonance line, but this can be overcome by using a recording instrument and tracing the emission from about 410 nm to 440 nm.

At high concentration the alkali elements also interfere by reducing the calcium emission, but this type of effect is a more serious problem in the presence of iron, aluminium, sulphate and phosphate. These elements form compounds with calcium, particularly in low-temperature flames. This interference can be completely prevented by separating the calcium by precipitation as oxalate, as in the classical procedure. As a double precipitation of the ammonia group elements is necessary, this method is long and tedious. An alternative rapid procedure is to add an excess of each of the interfering elements to both the rock solution and the calcium standard solutions. The amounts added are such as to give limiting calcium suppression. This technique has been reported by Kramer[14] for the determination of calcium in silicate rocks and minerals.

The Determination of Calcium by Atomic Absorption Spectroscopy

As with flame emission photometry, interference with the determination of calcium arises from the presence of aluminium, iron and other elements that form compounds with calcium in the flame. This interference is very much less than that recorded with emission photometry, and can be reduced still further by using a high-temperature (air–acetylene) flame. Under these conditions, the only serious interference with the determination in silicate rocks is from aluminium. This can be overcome by adding strontium or lanthanum to the solution to serve as a releasing agent. This technique is used also for the determination of magnesium in silicate rocks, and one rock solution can be prepared for both determinations. The determination of both elements by this technique is described in detail on p. 319.

For a review of the mutual interferences between the atomic absorption determination of calcium and some other elements, see Harrison and Ottaway.[15]

References

1. JEFFERY P. G. and WILSON A. D., *Analyst* (1959) **84**, 663.
2. MEADE R. K., *Chem. Eng.* (1895) **1**, 21.
3. MAYNES A. D., *Anal. Chim. Acta* (1965) **32**, 288.
4. RAWSON S. G., *J. Soc. Chem. Ind.* (1897) **16**, 113.
5. CLULEY H. J., *Analyst* (1954) **79**, 567.
6. BELCHER R., CLOSE R. A. and WEST T. S., *Chemist Analyst* (1958) **47**, 2.
7. BELCHER R., CLOSE R. A. and WEST T. S., *Talanta* (1958) **1**, 238.
8. TUCKER B. M., *J. Austr. Inst. Agr. Sci.* (1955) **21**, 100.
9. KORBL J. and PRIBIL R., *Chem. and Ind.* (1957) p. 233.
10. ABDULLAH M. I. and RILEY J. P., *Anal. Chim. Acta* (1965) **33**, 391.
11. CLOSE R. A. and WEST T. S., *Talanta* (1960) **5**, 221.
12. HERRERO-LANCINA M. and WEST T. S., *Analyt. Chem.* (1963) **35**, 2131.
13. LEONARD M. A., *J. Pharm. Pharmacol.* (1962) **14** (suppl.), 63T.
14. KRAMER H., *Anal. Chim. Acta* (1957) **17**, 521.
15. HARRISON A. and OTTAWAY J. M., *Proc. Soc. Analyt. Chem.* (1972) **9**, 205.

CHAPTER 17

CARBON

Occurrence

In view of the importance of the element carbon it is surprising that it is only recently that much interest has been shown in the presence of organic compounds in silicate rocks. Certain sedimentary rocks in particular contain a wide variety of compounds—including saturated alicyclic hydrocarbons, aromatic hydrocarbons, fatty acids, amino acids, porphyrins and organic sulphur compounds. The possibility that both normal and branched-chain hydrocarbons, saturated and unsaturated, could be produced on pyrolysis of igneous rocks was noted by Jeffery and Kipping,[1] who reported compounds in the range C_1 to C_4. The possibility of catalysis in the production of organic compounds was noted by Bear and Thomas[2] in a discussion of the nature of "argillaceous odour", to which they gave the name "petrichor".

Occurrences of graphite, diamond and the various forms of fossil fuel—anthracite, coal, lignite, bitumen and petroleum—have long been of interest economically, but have seldom been the concern of the rock analyst. Of more extensive occurrence and also of considerable industrial importance are the deposits of limestone, other carbonate rocks and the carbonate ore minerals. These materials do not usually present major problems in analysis, as the methods used for the determination of most elements in silicates can be adapted to carbonate rocks. Igneous carbonates, "carbonatites" with abnormal concentrations of certain rare elements, may present unusual problems and require special analytical procedures.

Carbon Dioxide

The carbon dioxide content of some common rocks and minerals is shown in Fig. 27. The maximum theoretical content of over 50 per cent is observed in some samples of magnesite, whilst rocks containing from

35 to 50 per cent exist in many extensive natural formations composed
of chalk, limestone, dolomite and dolomitic limestone. From 5 to 35 per
cent carbon dioxide has been noted in carbonatite and metamorphic
rocks, whilst less than 5 per cent is by no means uncommon in many
silicate rocks. Jeffery and Kipping[3] succeeded in determining parts per
million of carbon dioxide in a number of rocks from which it had not
previously been reported, and it appears likely that most silicate rocks
contain small amounts of this constituent.

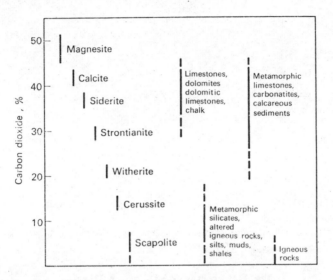

FIG. 27. The carbon dioxide content of rocks and minerals.

The presence of carbonate minerals in silicate rocks is indicated by the
effervescence that is obtained by warming with dilute acid. About 1 g
of the powdered sample, freed from air bubbles by boiling with a little
water, is acidified with a few millilitres of hydrochloric acid, allowed to
stand for a few minutes and after examination, is gently warmed. As
little as 0·1 per cent carbon dioxide can be detected as bubbles of gas.
Calcite is decomposed in the cold, and the liberated carbon dioxide is
easily seen before warming. Other carbonate minerals, particularly
siderite, ankerite and dolomite, will evolve carbon dioxide only on
heating. Care must be taken not to confuse hydrogen with carbon

dioxide. (Hydrogen may be formed by reduction of the hydrochloric acid if the sample contains any tramp iron introduced in the preparation of the material for analysis.) Gaseous hydrogen sulphide may also be liberated from certain sulphide minerals, but this is not likely to be confused with carbon dioxide.

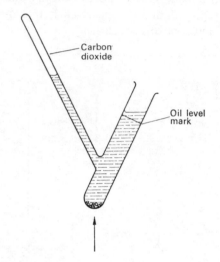

FIG. 28. Shapiro and Brannock's apparatus for the determination of carbon dioxide in silicate rocks.

The Determination of Carbon Dioxide

The simplest procedures for determining carbon dioxide are based upon a measurement of the volume of gas liberated on heating the rock material with mineral acid. These procedures are not the most accurate, and gravimetric procedures are generally to be preferred. A gas chromatographic procedure can be used for those rocks containing only traces of carbon dioxide.

METHOD OF SHAPIRO AND BRANNOCK[4]

Following Fahey,[5] Shapiro and Brannock have devised a special apparatus (Fig. 28) in which the carbon dioxide is liberated and measured. It consists of a boiling tube with a sealed-in side arm. The size and shape represents a compromise between accuracy (requiring a long thin side arm as the measuring tube) and loss of carbon dioxide by solution in the oil (which increases with the length of travel of the gas bubbles). This apparatus should not be used for carbonate rocks, and an upper limit of about 6 per cent carbon dioxide has been suggested. Some samples, particularly those that have been crushed or ground using iron machinery, may contain small amounts of tramp iron. This gives gaseous hydrogen on heating with mineral acid, and an error of as much as 0·2 per cent can be obtained. This error can be avoided by adding mercuric chloride to the sample, converting any tramp iron to ferrous iron without liberation of hydrogen.

Method

Reagents: *Mercuric chloride*, dry powder.
Mercuric chloride solution, saturated aqueous.
Paraffin oil, a heavy fairly viscous oil. That sold in the United Kingdom as "liquid paraffin" is suitable.
Carbon dioxide standard, a silicate rock that has been analysed by a gravimetric method, containing about 2 per cent carbon dioxide is recommended for the calibration of the side arm of the apparatus.

Procedure. Weigh a suitable amount of the finely powdered rock material into the dry tube, taking care that no material adheres to the sides. Add approximately 1 g of the powdered mercuric chloride and 2 ml of the saturated mercuric chloride solution and tap the tube gently to remove all air bubbles. Fill the tube to the mark with oil, and rotate the tube to displace all the air from the side arm and finally bring the tube to a position with the side arm pointing vertically upwards.

Add 2 ml of 6 N hydrochloric acid to the tube and immerse the lower part of the apparatus in an oil bath maintained at a temperature in the range 110–120°. Keep the tube at this temperature for at least 5 minutes and until all the liberated bubbles of gas have collected in the side arm. Remove the tube from the water bath and cool the side arm by running cold water over it for a short while. Using a millimeter scale, measure the length of the side arm occupied by carbon dioxide gas.

Repeat this operation using known weights of the carbon dioxide standard in order to calibrate the side arm.

METHOD USING SCHROTTER'S APPARATUS[6]

This apparatus, shown in Fig. 29, is designed to determine the loss in weight that occurs when carbon dioxide is removed from carbonate minerals. The method is not capable of great accuracy, although with experience, precise and reasonably accurate results can be obtained for carbonate rocks and minerals. It should not be used for samples containing less than about 10 per cent carbon dioxide. The greatest error arises from the failure to dry completely the gases that are evolved on warming.

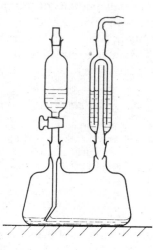

FIG. 29. Schrotter's apparatus for the determination of carbon dioxide.

Method

Apparatus: This consists of a flat-bottomed, 100-ml flask fitted with a tap funnel containing phosphoric acid and a gas bubbler for concentrated sulphuric acid. The use of PTFE sleeves between the ground glass joints is advantageous.

Reagents: *Phosphoric acid*, diluted 1 + 2.

Procedure. Clean, dry and assemble the complete apparatus as shown in Fig. 30. Dismantle the apparatus and weigh the empty flask. Transfer a 1–5 g portion of the ground sample material to the flask, and reweigh to give the weight of sample used. Add about 5 ml of water to the flask, and reassemble the apparatus with 5 ml of phosphoric acid in the tap funnel and 2 ml of concentrated sulphuric acid in the bubbler. Close both the bubbler and the tap funnel with the correct stoppers, and weigh the assembled apparatus.

Remove the stoppers from the bubbler and the tap funnel, and open the tap to allow the phosphoric acid to run slowly into the flask, taking care that the carbon dioxide is evolved slowly, and bubbles gently through the sulphuric acid, interrupting the supply of phosphoric acid, and cooling the flask if the reaction becomes too vigorous. Finally close the tap of the funnel just before the phosphoric acid is completely transferred to the flask.

When the evolution of gas has ceased, gently swirl the flask, place it on a hot plate, and heat the contents to a temperature of about 90°, but do not allow to boil. Allow the apparatus to cool, open the tap and, using a water pump, draw a current of dried air through the apparatus for a period of 2 minutes to displace carbon dioxide remaining in the flask. Replace the stoppers and reweigh the assembled apparatus. Report the percentage loss in weight as the carbon dioxide content of the sample material.

GRAVIMETRIC METHOD USING ABSORPTION TUBES AND ACID DECOMPOSITION

Where greater accuracy is required a gravimetric method should be used. The recommended procedure involves decomposition with diluted phosphoric acid. The gas evolved is dried by passage through an absorption tube packed with anhydrous magnesium perchlorate, and the carbon dioxide is finally absorbed in a weighed tube containing soda–asbestos. A copper phosphate layer or packing is used to retain any hydrogen sulphide that may be formed by the action of the phosphoric acid on certain sulphide minerals that may be present in the sample material.

The apparatus required is shown in Fig. 30; once assembled it should be left permanently in place. If this is not possible, the various items of glassware can be packed into a wooden box for storage.

Borgström[7] has recommended that for certain scapolites where the carbon dioxide content is not readily liberated with hydrochloric acid (or, for that matter, with phosphoric acid), a mixture of hydrochloric and hydrofluoric acids should be used. Although reported in text books of rock analysis, this procedure of Borgström does not appear to have been widely adopted, possibly because of the expense of working with apparatus designed to withstand hydrofluoric acid. The introduction of polypropylene and other similar materials may now provide a means of avoiding these difficulties.

Low reagent blanks are commonly obtained when phosphoric acid is used for the sample decomposition, amounting in most cases to no more than 0·1 mg per hour. This reagent was recommended by Morgan[8] in place of hydrochloric acid, then in use in most laboratories.

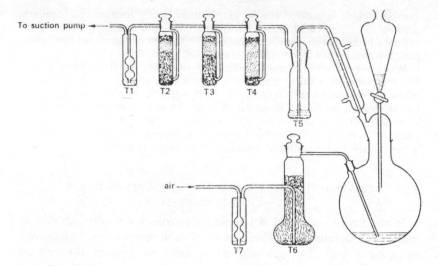

To suction pump →

T1   T2   T3   T4

T5

air →

T7   T6

FIG. 30. Apparatus for the gravimetric determination of carbon dioxide.

EDTA has also been used by the author for this decomposition, but was found to be no improvement on the use of phosphoric acid, and a good deal less effective in liberating carbon dioxide from scapolite and some carbonatite rocks. With EDTA, hydrogen sulphide was liberated from a number of sulphide minerals, but this reagent is less effective than phosphoric acid in promoting partial oxidation of kerogen (insoluble organic matter) to carbon dioxide.[9]

The type of apparatus in use in many laboratories is shown in Fig. 30. One improvement that can sometimes be made, is to use a small electric motor to recycle air through the system, as described by Jeffery and Wilson.[10] This is particularly useful where low "blank" values are important, as for example where the rock material available is limited, and where semi-micro scale working is necessary.

Method

Apparatus: The apparatus (Fig. 30) consists of a 250-ml, round-bottomed flask, tap funnel for the addition of diluted phosphoric acid, a water-cooled reflux condenser and a series of vessels T1 to T7. T2, 3 and 4 are Nesbitt bulbs, T1 and 7 are Arnold bulbs, T5 a 125-ml Drechsel bottle and T6 a Midvale bulb.

The Arnold bulbs contain syrupy phosphoric acid and are used to indicate the rate of flow of gas through the apparatus, the Drechsel bottle contains concentrated sulphuric acid to remove the bulk of the water from the gas, thus prolonging the life of the drying tube T4. The acid level in the Arnold bulbs and the Drechsel bottle should be such as just to cover the end of the delivery tube. T6, used to remove carbon dioxide from the incoming air, is packed with cotton wool to cover the end of the delivery tube and with soda asbestos to the level indicated in the figure. T2, whose increase in weight is used to indicate the weight of carbon dioxide absorbed, is packed with soda–asbestos, covered with a ¾-inch layer of anhydrous magnesium perchlorate. T3, used to remove hydrogen sulphide from the gas stream, is packed with copper phosphate covered with a layer of anhydrous magnesium perchlorate, and T4 contains magnesium perchlorate only. In each Nesbitt tube, the reagent is kept in place with the aid of a plug of cotton wool. The tubes are connected together with butt joints.

The complete apparatus is assembled on a framework of laboratory scaffolding with the aid of clamps and polythene-covered spring clips. The supply of air through the apparatus is obtained from a filter pump or a small electric motor and is regulated with a screw clip. A micro burner is used to heat the flask.

Reagents: *Phosphoric acid*, diluted $1 + 1$, freshly boiled and allowed to cool.
Magnesium perchlorate, anhydrous, 14–22 mesh.
Soda–asbestos, 8–14 mesh.
Cupric phosphate, granular, specially prepared by pressing a stiff paste of the powdered reagent in 1 per cent starch solution through a sheet of perforated metal, and drying in an electric oven at 110°.

Procedure. Set up the apparatus as shown in Fig. 30, with the bulbs packed as described above. With all the taps fully open, start the pump pulling air through the apparatus and observe that the two Arnold bulbs have air at about the same rate of 1 to 2 bubbles per second passing through them. Beginning at T7, close and then open each tap in turn and note that with each tap closed the current of air through the Arnold bulb T1 ceases to flow, indicating that each tap is operating correctly. Finally close all the taps and remove the Nesbitt tube T2 to a balance case and allow to stand for 30 minutes. Carefully wipe the outside of the bulb with a clean cloth and weigh.

Weigh also a sample portion of from 1 to 5 g of the sample material, depending upon the anticipated carbon dioxide content and, using a little water, rinse it into the flask. Transfer approximately 50 ml of diluted phosphoric acid to the tap funnel. Reassemble the apparatus as shown in the figure but with taps T1–7 open once more and with a clip between the suction pump and T1 restricting the flow of air through the apparatus fully closed.

Start the pump and open the screw clip to produce a steady stream of bubbles through the Arnold bulbs. Open the tap of the funnel to admit the phosphoric acid slowly to the flask, temporarily cutting off the current of air if a rapid evolution of carbon dioxide occurs. Close the tap of the tap funnel before all the acid has been added.

When all evolution of carbon dioxide has apparently ceased, bring the solution to the boil and boil gently whilst keeping the air current flowing for a further period of 45 minutes to sweep all the carbon dioxide out of the flask and into the Nesbitt tube. Finally turn off the burner, close all the open taps and open the tap of the tap funnel. Remove the Nesbitt tube T2 to a balance case, allow it to stand as before, wipe the tube and record the increase in weight due to the absorbed carbon dioxide.

GRAVIMETRIC METHOD USING ABSORPTION TUBES
AND THERMAL DECOMPOSITION

The difficulty of liberating all the carbon dioxide from some of the scapolite minerals by boiling with diluted mineral acid can be avoided by using thermal decomposition of the sample. The determination of carbon dioxide may then be combined with that of total water by measuring the increase in weight of separate tubes packed with anhydrous magnesium perchlorate and with soda–asbestos, as described in detail by Riley.[11] He attributed the high blanks that he first obtained for carbon dioxide to the formation of acidic oxides of nitrogen, and subsequently avoided this by replacing the current of air through the apparatus by a current of nitrogen gas from a cylinder supply, and by incorporating a short length of silica tube packed with copper wire and heated to a temperature of 700–750°. Interference from oxides of sulphur was prevented by the inclusion in this short length of silica tube of a layer of silver pumice, and by the use of a bubbler containing a saturated solution of chromium trioxide in phosphoric acid.

The main furnace is heated to a temperature of 1100–1200°, and the samples are inserted into the hot zone of the furnace by means of a stainless-steel push rod. Care must be taken not to push the samples directly into the very hottest part if they contain much water or easily decomposed carbonate minerals such as siderite or magnesite. Failure to observe this precaution can result in mechanical loss of the sample from the containing boat, and to loss of carbon dioxide by incomplete absorption following the too rapid evolution.

Most carbonate minerals readily lose carbon dioxide at 1100°, but strontianite, scapolite and witherite require to be heated to 1200°. The complete recovery of carbon dioxide from the latter may require up to 3 hours heating at this temperature for complete expulsion of all the carbon dioxide.

Method

Apparatus. For details of the construction of the two furnaces used by Riley[11] the original paper should be consulted.

High-temperature furnace (1100–1200°). The furnace tube consists of a 45-cm length of silica tube of internal diameter 1·8 cm, with an insertion device to enable the samples contained in 2-ml alumina boats to be pushed gradually into the heated zone of the furnace. For the majority of rock samples a temperature of 1100° should be used, and only if the sample contains minerals that are difficult to decompose (such as staurolite, cordierite, topaz or the carbonate minerals listed above) should the temperature be raised to 1200°. This is somewhat in excess of the recommended temperature for silica tubing, and with repeated use at this temperature, distortion will gradually occur. Each tube should last for at least three months with regular use at 1100°

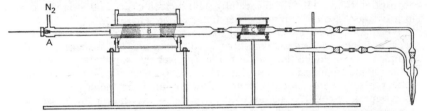

FIG. 31. Apparatus for determining carbon dioxide and water (A— insertion device, B—silica combustion tube, C—silica tube containing copper wire and silver pumice).

Low-temperature furnace (700–750°). The furnace tube consists of a 10-cm length of tubing, 1 cm internal diameter, packed with alternate layers of copper wire and silver pumice, held in place with plugs of asbestos wool. Both ends of the tube are fused to 3-cm lengths of silica tubing of 5 mm external diameter. At least once a week a current of coal gas should be passed through the heated tube to reduce any copper oxide to the metal. The life of the tube is about 3 months of continual use with rocks of low sulphur content.

Gas purification train. Riley has recommended using nitrogen of normal quality (i.e. not "white spot"—oxygen-free quality), from a cylinder fitted with a twin-stage regulator and a needle valve. A bubbler filled with sulphuric acid is used to indicate the flow of gas, regulated to about 3 litres per hour. The nitrogen is purified by passage through tubes containing soda lime, fused calcium chloride and finally anhydrous magnesium perchlorate.

Absorption tubes, those shown in Fig. 31 are those designed by Riley. They are rather smaller than many absorption tubes in common use, and should be replaced with larger tubes for samples containing much carbon dioxide or water.

Bubbler, with samples containing more than 0·5 per cent sulphur, the bubbler, filled with a saturated solution of chromium trioxide in syrupy phosphoric acid, should be interposed between the water absorption tube and the carbon dioxide absorption tube, as shown in the figure. Its side arm contains anhydrous magnesium perchlorate.

Reagents: *Anhydrous magnesium perchlorate*, 14–22 mesh.
Soda–asbestos, 8–14 mesh.
Chromium trioxide.
Copper wire.
Silver pumice, prepared by evaporating 14-mesh pumice with concentrated silver nitrate solution and igniting strongly.
Magnesium oxide, ignited as required.

Procedure. Before use, check that the two furnaces are at their correct temperatures and adjust the flow of nitrogen to about 3 litres per hour. Allow the gas to pass through the apparatus for about 20 minutes, then remove the absorption tubes, wipe them carefully with a clean cloth, and after standing in a balance case for 5 minutes, weigh each tube separately.

Weigh an aliquot of $0 \cdot 5$–$1 \cdot 5$ g of the finely powdered rock material into a previously ignited 2-ml alumina boat lined with a piece of nickel foil. If the sample contains much fluorine or sulphur, cover the sample with a layer of freshly ignited magnesium oxide. Insert the boat into the end of the combustion tube, replace the insertion device and allow 5 minutes for the flow of nitrogen to sweep all introduced air out of the apparatus. Reconnect the absorption tubes and push the sample into the furnace using the push rod. Samples containing a lot of water or easily decomposed carbonate mineral should be pushed only into the cooler region outside the hot zone, and finally be inserted into the hottest part of the furnace tube when the decomposition is nearly complete. After a heating period of 30 to 40 minutes, remove the absorption tubes, wipe them, allow to stand and weigh as before.

Carry out a blank determination in the same way, but without the sample material, both before and after the first determination and after the end of each batch of samples. These reagent blanks should normally amount to no more than about $0 \cdot 1$ mg water per hour and $0 \cdot 2$ mg carbon dioxide per hour respectively. Higher blank values for carbon dioxide usually indicate that the packing in the low-temperature furnace tube is exhausted, which should then be regenerated as described.

The Determination of Carbon Dioxide by Non-aqueous Titration

In the gravimetric procedures described above, the blank values can be of the same order of magnitude as the weight of carbon dioxide absorbed from some samples. To avoid this difficulty Read[12] has described a titrimetric procedure based upon a non-aqueous titration. As with other methods, the carbon dioxide is liberated by heating with diluted orthophosphoric acid. The liberated gas is carried in a stream of nitrogen through absorbents to remove water and hydrogen sulphide and is then absorbed in a 5 per cent solution of monoethanolamine in dimethylformamide containing thymolphthalein as indicator. The carbon dioxide is titrated directly with a solution of tetrabutylammonium hydroxide in toluene.

A comparable method was described by Sen Gupta[13] in which the absorption is into a solution of monoethanolamine in acetone which contains a suitable excess of sodium methoxide together with phenolphthalein as indicator. The excess base is titrated with a standard solution of benzoic acid in methanol.

The Determination of Carbon Dioxide by Gas Chromatography

Even with the more refined techniques using gas-absorption tubes, blank values of the order of $0 \cdot 1$–$0 \cdot 2$ mg per hour can be obtained. These give rise to some uncertainty where the rock material contains $0 \cdot 02$ per cent carbon dioxide or less. For concentrations in this range the procedure described by Jeffery and Kipping[3] using gas chromatography will give a more positive result. The sample material is boiled under reflux with phosphoric acid and the liberated gases are transferred to a chromatograph in a stream of carrier gas, hydrogen being suggested. The separation of carbon dioxide from oxygen and nitrogen takes place on a column packed with silica gel, although activated charcoal could also be used. A thermal conductivity detector is used.

Some silicate rocks were found to contain 50 ppm carbon dioxide, and smaller quantities could undoubtedly be recovered and determined in this way. Jeffery and Kipping[3] suggested that as little as 5 ppm could be detected, but Carpenter[14] has given a lower limit of $0 \cdot 2$ ppm carbon dioxide. The original papers should be consulted for details.

Loss on Ignition

The total weight change that occurs on igniting the finely ground rock material has been used as an indication of the total water content of rocks containing only small amounts of other volatile components such as sulphur, fluorine and carbon dioxide (see p. 270). It has also been used for carbonate rocks as a measure of the carbon dioxide content— all water and organic matter present in the rock being ignored.

A temperature of 1000° is commonly used, this being regarded as sufficient for the complete expulsion of carbon dioxide from limestones, dolomites and siderites, but insufficient for the decomposition of calcium sulphate produced by ignition of anhydrite or gypsum. Strontianite and witherite are unlikely to be completely decomposed at this temperature. Other errors arise from the partial oxidation of siderite and pyrite to ferric oxide.

The results obtained can therefore be regarded only as an approximate measure of the carbon dioxide content, and no substitute for direct determination of total water and carbon dioxide. The loss on ignition should be reported in such a way as to leave no doubt as to the method used.

The Determination of Carbon ("non-carbonate carbon")

Graphitic carbon is sometimes determined by measuring the loss on ignition, although as noted in the previous section, such methods have little to commend them except simplicity and speed. For graphite concentrates the loss on ignition is an important industrial parameter, but it should not be used for silicate rocks as a measure of the carbon content. The only procedures that can be recommended for silicates are those based upon total oxidation of the carbon followed by quantitative determination of the carbon dioxide produced. This may be carried out after decomposing any carbonate minerals present,[15] or at the same time as the decomposition of carbonates. In this latter case a separate carbonate determination must be made and the result calculated as carbon subtracted from the result for total carbon.

Oxidants suggested for the conversion of carbon and carbon compounds include chromic acid, potassium dichromate, sodium peroxide and oxygen. Oxidation with chromic acid[16] is not always complete, as volatile carbon compounds may distil from the acid mixture and give rise to low results. Some improvement can be obtained by using a "closed-circulation" system,[10] but for many samples an alternative procedure should be used. The same objection can be made to the use of potassium dichromate in sulphuric acid solution, which reacts as chromic acid.

Potassium dichromate can be used to determine the reducing capacity of the rock sample.[17, 18] A measured amount is added and the excess titrated with ferrous sulphate solution. In this procedure it is impossible to distinguish between the reduction due to the presence of organic and carbonaceous matter, and that due to sulphide minerals. Some part of the ferrous iron content of the rock material will also be included.

Fusion with sodium peroxide in a closed bomb has been used by the U.S. Geological Survey[17] for determining the total carbon in marine shales. Samples containing more than 30 per cent organic carbon are easily oxidised, but those containing less generally require the addition of a combustion aid such as powdered aluminium or magnesium to provide the heat necessary for ensuring complete oxidation of the carbon. Any sulphides present will be oxidised to sulphates and do not interfere. Only small samples (0·4 g or less) can be taken, and as the sodium peroxide tends to absorb carbon dioxide from the air, a blank correction is necessary. This method is suitable only for those samples containing 0·2 per cent or more of carbon.

Combustion tube methods have generally been preferred for silicate rocks. Where other shorter or approximate methods are in routine use, the combustion tube method is sometimes recommended for control purposes.[16]

Gravimetric Method Using Oxidation with Chromic Acid

Dixon,[16] in his description of this method, noted that the results could be several per cent low, due largely to the formation of carbon monoxide and the distillation of volatile carbon compounds from the oxidising mixture. In addition to using a mercury catalyst, Dixon interposed a second oxidation stage between the reaction vessel and the absorption train. This second vessel served to condense the volatile carbon compounds and to provide the opportunity for further oxidation. In operation this second vessel contained chromic and phosphoric acids, but no mercury catalyst, and was heated to simmering point before the oxidation in the first vessel was initiated.

It is doubtful if the error introduced by omitting the second oxidation, is sufficient to warrant the extra manipulation necessary at least for those samples containing no more than a few per cent carbon.

Method

Apparatus. The apparatus required is shown in Fig. 32. It consists of two round-bottom flasks for the oxidation stages, tap funnel for the addition of the acid reagent, and a carbon dioxide absorption train attached to point A. This train is identical with that shown in Fig. 31 for the same purpose.

Reagents: *Chromic acid reagent*, dissolve 4 g of chromium trioxide in 40 ml of syrupy phosphoric acid.

Chromic acid–mercuric oxide reagent, dissolve 4 g of chromium trioxide and 0·1 g of mercuric oxide in 40 ml of syrupy phosphoric acid.

Procedure. For rock samples containing less than 1 per cent carbon, a 2-g sample portion should be taken for the determination. For rock material

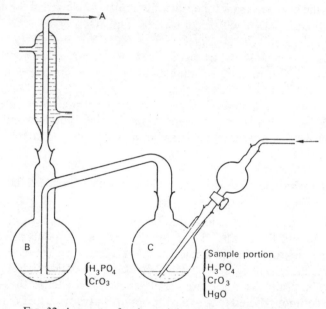

Fig. 32. Apparatus for determining total (organic) carbon.

with several per cent carbon, the sample size should be reduced to 0·5 g. Transfer this sample portion to the reaction flask C, and assemble the apparatus as shown, with phosphoric acid in the tap funnel. Complete the determination of carbon dioxide in the sample, using a few ml of phosphoric acid to decompose the carbonates present, as described on p. 178.

Allow the apparatus to cool, transfer 40 ml of the chromic acid reagent to the second reaction flask B, and 40 ml of the chromic acid–mercuric oxide reagent to the first reaction flask C. Reassemble the apparatus as shown, and draw air through at a rate of about 2 bubbles per second. Heat flask B and when the contents are gently simmering, start to heat flask C, using a small flame in the early stages. The oxidation tends to proceed more rapidly at about 70°, and it may be necessary to remove, temporarily, the source of heat to allow the reaction to moderate.

Allow the oxidation to continue until the colour of the liquid in flask C changes from orange-red to dark green, indicating the complete conversion of chromium(VI) to chromium(III). This may take about 30 minutes, but will depend upon the amount of carbon present and the rate of heating. Allow to cool and continue to draw air through the apparatus in order to collect any carbon dioxide remaining in both reaction flasks. Finally disconnect the absorption train and measure the weight of carbon dioxide absorbed. Hence calculate the carbon content of the sample material.

It should be noted that the dark colour of many carbonaceous shales can be misleading, it does not always indicate a high carbon content. Black shales, often resembling poor coal, may contain no more than 0·5–1 per cent carbon.

Combustion Tube Methods

The apparatus shown in Fig. 31, using a combustion tube for the determination of carbon dioxide, can be used in a somewhat similar way for the determination of total carbon in rocks and minerals. By heating the sample material with a suitable flux, such as potassium dichromate, lead chromate or vanadium pentoxide, complete oxidation takes place and carbon dioxide from both carbonate minerals and non-carbonate carbon is evolved and collected. The carbonate content of the sample must be determined separately. Any sulphides present will be oxidised, and both sulphur dioxide and trioxide formed. These are removed prior to collection of the carbon dioxide in the weighed absorption train.

The determination of non-carbonate carbon in clay materials poses a number of special problems. These were investigated by Ferris and Jepson[19] who recommended a combustion tube method. The clay material is heated at a temperature of 900° in a stream of purified oxygen to convert all carbonaceous material to carbon dioxide, which is absorbed in a solution of monoethanolamine in dimethylformamide. The determination is completed by titration with a solution of tetra-butylammonium hydroxide in a mixture of toluene and methanol with phenolphthalein as indicator—a method similar to that described by Read.[12] Impurities in the gas stream include sulphur dioxide, silicon tetrafluoride and hydrogen fluoride. These interfere with the titration and must be removed.

The Determination of Organic Matter

The separation of organic material from both consolidated and unconsolidated sediments by extraction with organic solvents is an empirical procedure. Such factors as the weight of the sample in relation to the volume of solvent, the grain size to which the sample has been crushed or ground, the length of time during which the sample is extracted and the particular solvent used can all materially affect the results.[20] Extractants that have been used include chloroform, benzene and mixtures of benzene with methanol and acetone. The extracted material, after removal of the solvent, is known as the "bituminous fraction" or more commonly bitumen. It resembles petroleum and can be separated into its component hydrocarbons and other organic compounds, which are then determined by techniques familiar to the organic chemist, for example gas chromatography and infrared spectroscopy.

A further fraction can be separated by extraction of the residue with aqueous sodium hydroxide solution. This fraction is known as the "humic carbon content" or as "humic acids", and may be determined by dichromate oxidation as described by Smeral.[21]

The greater part of the organic content of sedimentary rocks often remains with the mineral fraction after these extractions. This solvent-insoluble fraction, or *kerogen*,[22] can be isolated by the stepwise removal of all the other constituents of the sediment. Carbonates are removed by dissolution in dilute hydrochloric acid, and quartz and silicates by treatment with hydrofluoric acid. Pyrite can be troublesome as it is frequently abundant in carbonaceous sediments. A separation with heavy media is sometimes effective, otherwise a zinc-hydrochloric acid reduction is used. The insoluble organic fraction remaining after this treatment may be contaminated with small amounts of zircon and other similar detrital minerals. The original papers should be studied for details of the analysis procedures adopted.[23, 24, 25] For details of the separation and determination of individual classes of organic compounds (hydrocarbons, fatty acids, amino acids, etc.), the original papers should also be consulted.[26, 27, 28]

References

1. JEFFERY P. G. and KIPPING P. J., *Analyst* (1963) **88**, 266.
2. BEAR I. J. and THOMAS R. G., *Nature* (1964) **201**, 993.

3. JEFFERY P. G. and KIPPING P. J., *Analyst* (1962) **87**, 379.
4. SHAPIRO L. and BRANNOCK W. W., *Analyt. Chem.* (1955) **27**, 1796.
5. FAHEY J. J., *U.S. Geol. Surv. Bull.* 950, p. 139, 1946.
6. SCHROTTER A. R., *Ber. Wien. Acad.* (1871) **63**, 471.
7. BORGSTRÖM L. H., *Zeit. Anal. Chem.* (1914) **53**, 685.
8. MORGAN G. T., *J. Chem. Soc.* (1904) **85**, 1001.
9. WATKINSON J. H., *Analyst* (1959) **84**, 661.
10. JEFFERY P. G. and WILSON A. D., *Analyst* (1960) **85**, 749.
11. RILEY J. P., *Analyst* (1958) **83**, 42.
12. READ J. I., *Analyst* (1972) **97**, 134.
13. SEN GUPTA J. G., *Anal. Chim. Acta* (1970) **51**, 437.
14. CARPENTER F. G., *Analyt. Chem.* (1962) **34**, 66.
15. ELLINGBOE J. L. and WILSON J. E., *Analyt. Chem.* (1964) **36**, 435.
16. DIXON B. E., *Analyst* (1934) **59**, 739.
17. FROST I. C., *U.S. Geol. Surv. Prof. Paper* 424-C, p. B-480, 1961.
18. SCHOLLENBERGER C. J., *Soil Science* (1927) **24**, 65.
19. FERRIS A. P. and JEPSON W. B., *Analyst* (1972) **97**, 940.
20. TOURTELOT H. A. and FROST I. C., *U.S. Geol. Surv. Prof. Paper* 525-D, p. D-73, 1966.
21. SMERAL J., *Pr. Vysk. Ustavu CS. Naft. Dolu, Publ.* (1965) **24**, 27.
22. FORSMAN J. P. and HUNT J. M., *Geochim. Cosmochim. Acta* (1958) **15**, 170.
23. DOUGLAS A. G., EGLINTON G. and MAXWELL J. R., *Geochim. Cosmochim. Acta* (1969) **33**, 579.
24. NISSENBAUM A., BAEDECKER M. J. and KAPLAN I. R., *Geochim. Cosmochim. Acta* (1972) **36**, 709.
25. BROWN F. S., BAEDECKER M. J., NISSENBAUM A. and KAPLAN I. R., *Geochim. Cosmochim. Acta* (1972) **36**, 1185.
26. KVENVOLDEN K. A., PETERSON E., WEHMILLER J. and HARE P. E., *Geochim. Cosmochim. Acta* (1973) **37**, 2215.
27. MATHEWS R. T., IGUAL X. P., JACKSON K. S. and JOHNS R. B., *Geochim. Cosmochim. Acta* (1972) **36**, 885.
28. BELSKY T. and KAPLAN I. R., *Geochim. Cosmochim. Acta* (1970) **34**, 257.

CHAPTER 18

CHLORINE, BROMINE AND IODINE

Occurrence

Only in rare cases does the chlorine content of magmatic rocks reach 1 per cent by weight; values of between 100 and 1000 ppm being commonly reported. Selivanov[1] has given chlorine and bromine values for a number of silicates, the average values being given in Table 18. These indicate chlorine-to-bromine ratios of from 100:1 to 300:1. These bromine figures give no more than an indication of the levels likely to be encountered in igneous silicate rocks. They are somewhat higher than the bromine values of $0 \cdot 25$–$0 \cdot 84$ ppm reported by Filby[2] for a series of geochemical standards and more in line with those of Behne.[3] Goldschmidt[4] has reported iodine values of G. Lunde on rock material that indicated a general magnitude of $0 \cdot 2$–$0 \cdot 3$ ppm I for igneous rocks, in only general agreement with values of $0 \cdot 04$–$0 \cdot 2$ by Crouch.[5]

TABLE 18. THE CHLORINE AND BROMINE CONTENT OF SOME SILICATE ROCKS

Rock type	Cl, ppm	Br, ppm
Granite	330	$1 \cdot 6$
Granodiorite	540	$2 \cdot 57$
Syenite	400	$1 \cdot 07$
Gabbro	200	$2 \cdot 00$
Basalt	260	$2 \cdot 67$

Neither bromine nor iodine form discrete minerals in silicate rocks, although these are well known from evaporite deposits in Chile and elsewhere. There are a small number of very rare ore minerals with bromine and iodine as a major constituent, for example marshite CuI,

190

bromyrite AgBr and embolite Ag(Cl,Br). Chlorine occurs in silicate rocks principally as chlorapatite, scapolite, sodalite and eudialyte, and in metalliferous ore deposits to a small extent as a number of rare minerals such as phosgenite, pyromorphite and vanadinite. The principal occurrence of chlorine minerals is in evaporite deposits as halite NaCl, carnalite $KCl.MgCl_2.6H_2O$ and silvite KCl. Halite has also been reported in some volcanic rocks, and in altered silicates that had been saturated with brine.

The presence of hydrogen chloride in the gases from volcanic vents and fissures, and the occurrence of the simple chlorides of the alkali and other metals—NaCl, KCl, $FeCl_3$, etc.—indicates the abundance of chlorine at the latter stages of magmatic activity. The large size of the chlorine ion ($R = 180$ pm) makes it difficult to find accommodation in silicate lattices and results in this late stage concentration.

Water-soluble Chlorine

The presence of water-soluble chlorine in more than trace amounts is usually indicative of halite, kalsilite, carnalite or other evaporite mineral in the rock material. To determine the water-soluble chlorine, heat a suitable weight of the sample material to boiling with about 100 ml of water. Boil for 30 minutes, filter through a close-textured paper and wash the residue with warm water. Discard the residue. Add 5 ml of concentrated nitric acid to the filtrate and precipitate the chlorine with silver nitrate solution as described below for acid-soluble chlorine.

Acid-soluble Chlorine

Many of the chlorine minerals are completely decomposed by heating with dilute nitric acid. Unless the petrological examination has shown the presence of scapolite and similar minerals that are not decomposed in this way, this can be used in place of the more lengthy procedure described below for the determination of total chlorine. Loss of chlorine may occur if the acid strength is too high or if the heating is prolonged.

Method

Reagents: *Silver nitrate solution*, dissolve 5 g of silver nitrate in 100 ml of water containing 1 ml of concentrated nitric acid.

Procedure. Accurately weigh 2 g or more of the finely powdered rock material into a 150-ml beaker and carefully add 2 ml of concentrated nitric acid and 40 ml of water. Cover the beaker with a clock glass and transfer to a steam bath. When all effervescence has subsided, boil for 2 minutes, but not longer, and allow to cool. Filter the liquid through a close-textured filter paper and wash the residue well with hot water. Combine the filtrate and washings in a 400-ml beaker and allow to cool to room temperature.

Shield the beaker from light either by wrapping it in stout brown paper, or by covering the outside with lacquer, and cover a clock glass in the same way. Now add a slight excess of silver nitrate solution and replace the beaker on the steam bath to coagulate the precipitated silver chloride Transfer the beaker to a dark cupboard and allow to cool overnight. Collect the precipitate on a small, weighed, sintered-glass crucible of medium porosity, and wash with cold 0·16 N nitric acid Finally wash twice with cold water, dry in an electric oven set at a temperature of about 150° and weigh as silver chloride.

Notes: 1. If the amount of acid-soluble chlorine in the rock is too small to permit the silver chloride to coagulate, it may be determined turbidimetrically with a spectrophotometer or by comparing the turbidity with that obtained by adding standard sodium chloride solution to an acid solution of silver nitrate.

2. The sintered-glass crucibles may be cleaned by passing a small quantity of concentrated ammonia solution through them, followed by water to remove silver salts, nitric acid to remove silver stains and finally water again to remove nitric acid.

Total Chlorine

For this determination the sample material is fused with alkali carbonate and the chlorine precipitated as silver chloride from the acidified aqueous extract. The determination may be combined with those of sulphur, barium, zirconium, chromium and vanadium as described by Bennett and Pickup.[6]

The addition of potassium nitrate to the flux serves to increase the fluidity of the melt, and also to oxidise all ferrous iron present in the sample. Some part of the sodium carbonate can be replaced with potassium carbonate to reduce the temperature required for the fusion, and borax glass can be added to assist in the decomposition of refractory oxide minerals.

Method

Reagents: *Sodium carbonate wash solution*, dissolve 10 g of anhydrous reagent in 500 ml of water.

Procedure. Accurately weigh approximately 2 g of the finely powdered rock material into a large platinum crucible and mix with 0·5 g of potassium nitrate and 10 g of anhydrous sodium carbonate. Fuse the contents of the crucible, first over a Bunsen burner and then over a Meker burner for a total of about 1 hour and then allow to cool. Extract the melt into hot water containing 2 drops of ethanol to reduce any manganate formed in the fusion. Allow to cool and collect the residue on an open-textured filter paper and wash well with warm sodium carbonate wash solution. Discard the residue. Combine the filtrate and washings in a 600-ml beaker and dilute to about 400 ml with water.

Add 2 or 3 drops of methyl red indicator solution, 2 N nitric acid until the red form of the indicator is obtained, and then add 1 ml in excess. Stir vigorously to remove the liberated carbon dioxide. Precipitate the chlorine present in the solution and complete the determination as described above for acid-soluble chlorine. Collect, dry and weigh the silver chloride.

Note: 1. Provided that the precipitation is made in a sufficiently large volume, and the solution is made only faintly acid, silica will not be precipitated. If silica does separate, it should be collected with the silver chloride. The latter can then be dissolved in a little hot dilute ammonia solution, filtered from silica and re-precipitated by adding dilute nitric acid.

The Photometric Determination of Chlorine

It is difficult to collect and weigh the small quantity of silver chloride obtained from rocks containing 0·05 per cent or less of chlorine. For these rocks a titrimetric procedure can often be used, but this is imprecise below about 0·01 per cent. A number of spectrophotometric procedures have been described, and these have been adapted for rocks containing small amounts of chlorine.

Kuroda and Sandell[7] have described a procedure using the formation of a colloidal suspension of silver sulphide in aqueous ammoniacal solution. The chlorine present in the silicate rock is recovered by fusion with alkali carbonate, the melt extracted with water and the chlorine present in the filtrate precipitated as silver chloride by the procedure given in the preceding section. The silver chloride is collected on a small sintered-glass filter crucible, washed and dissolved in a small volume of aqueous ammonia. Sodium sulphide is added and the solution diluted to 10 ml for photometric measurement. This procedure extends downwards the concentration range of chlorine that can be accurately determined in silicate rocks, but a lower limit is imposed by the inability to recover very small amounts of precipitated silver chloride.

A photometric procedure not limited in this way is that described by

Bergmann and Sanik,[8] in which mercuric thiocyanate is added to the acidified filtrate from an alkaline fusion. Any chloride ions present displace thiocyanate ions, which are then free to form the intense red-coloured ferric thiocyanate (note 1). A modification of this procedure was described by Huang and Johns,[9] using the sample solution prepared for their determination of fluorine (see p. 242).

Method

Reagents: *Ferric solution*, dissolve 12·0 g of ferric ammonium sulphate, $FeNH_4(SO_4)_2.12H_2O$ in 100 ml of 9 M nitric acid.

Mercuric thiocyanate saturated solution, shake 0·35 g of mercuric thiocyanate $Hg(CNS)_2$ with 100 ml of 95 per cent ethanol and allow to stand overnight to allow the precipitate to settle.

Standard sodium chloride stock solution, dissolve 0·8242 g of dried sodium chloride in 500 ml of water. This solution contains 1 mg chlorine per ml.

Standard sodium chloride working solution, dilute a 5 ml aliquot of the stock solution to 500 ml with water. This solution contains 10 μg chlorine per ml.

Procedure. Decompose the rock material and obtain the chlorine in dilute nitric acid solution as described under fluorine, p. 241. Transfer a 10- or 20-ml aliquot to a 25-ml volumetric flask and add 2 ml of the ferric ammonium sulphate solution and 2 ml of the mercuric thiocyanate solution. Mix the solution, dilute to volume and mix again. Measure the optical density of the solution against water in 1-cm cells with the spectrophotometer set at a wavelength of 460 nm. Measure also the optical density of a reagent blank solution prepared in the same way but omitting the sample solution.

Calibration. Transfer 2- to 10-ml aliquots of the standard sodium chloride solution containing 20 to 100 μg chlorine to separate 25-ml volumetric flasks, dilute each to 20 ml with water. Add 2 ml of ferric ammonium sulphate solution and 2 ml of the mercuric thiocyanate solution, dilute to volume and mix well. Measure the optical density of each solution as described above (note 2).

Notes: 1. Bromine and iodide undergo a similar reaction with mercuric thiocyantate, but it is unlikely that they will be present in silicate rocks in amounts sufficient to interfere.

2. The calibration graph given by this procedure should be checked with each new solution of the ferric ammonium sulphate and mercuric thiocyanate reagent.

The Determination of Chlorine using an Ion-selective Electrode

The introduction of ion-selective electrodes has given rise to new methods of analysis and adaptations of older ones for a number of elements. Chlorine is one of these. In the procedure by Haynes and Clark,[10] a chloride electrode is used to determine the chloride ion *activity* in a method similar to the classical technique involving the precipitation of silver chloride. The equivalence point in the silver nitrate titration is determined directly by measuring the electrode potentials of excess silver with a silver electrode.

The Determination of Bromine

A procedure for the determination of bromine in silicate rocks has been given by Behne.[3] In this a 1–2 g portion of the rock material is fused with sodium hydroxide and the melt extracted with water. Insoluble material is filtered off and the filtrate evaporated to dryness after converting all excess hydroxide to sodium carbonate and bicarbonate with an excess of gaseous carbon dioxide. The dry salts are extracted with ethanol, when any sodium bromide (and iodide) passes into organic solution. The ethanolic solution is filtered from the bulk of the sodium carbonate and the solvent removed by evaporation. The dry residue contains sodium bromide which is converted to bromate by oxidation with hypochlorite:

$$Br^- + 3HOCl = BrO_3^- + 3H^+ + 3Cl^-$$

Potassium iodide is then added:

$$BrO_3^- + 6I^- + 6H^+ = Br^- + 3H_2O + 3I_2$$

liberating six equivalents of iodine for each equivalent of bromine (method of Van der Meulen[11]). The liberated iodine is titrated with a standard solution of sodium thiosulphate in the presence of ammonium molybdate as catalyst.

Sodium hydroxide is used for the decomposition in preference to potassium hydroxide, as sodium bromide is more soluble in ethanol (11·8 g per litre at 30°) than potassium bromide (3·2 g per litre at 30°).

Behne reported that only about 90 per cent of the bromine could be recovered from standard bromide solutions. As the Van der Meulen tritration procedure is quantitative, it must be presumed that the loss of bromine occurs at the extraction stage, although this was not confirmed by Behne. Iodine reacts in the same way as bromine, and the results obtained will include any iodine present in the rock sample. The sensitivity of the method is about 1 μg Br, which is thus unsuitable for those silicates containing less than 1 ppm Br. For rocks containing less than this, a more sensitive procedure must be used—such as that described by Filby,[12] with a sensitivity of 0·001 ppm Br, using a neutron activation technique.

The Determination of Iodine

Crouch[5] has described a procedure for determining iodine in silicate rocks based upon the blue colour given by free iodine with starch. The rock material is decomposed by fusion with sodium hydroxide and the iodine isolated by co-precipitation with silver chloride. Bromine is used to oxidise iodide ions to iodate, which is then reacted with cadmium iodide to give free iodine according to the equation:

$$IO_3^- + 5I^- + 6H^+ = 3I_2 + 3H_2O$$

The limit of detection is of the order of 0·15 μg I, indicating that large-sized samples will be required for most silicate rocks.

A method for iodine has been described by Shneider and Miller[13] using the catalytic action of iodine on the reaction between cerium(IV) and arsenic(III). Silica and manganese must be removed. The sensitivity is given as 0·2 ppm I.

A somewhat simpler method, based upon an extraction of elemental iodine into carbon tetrachloride, was reported by Grimaldi and Schnepfe.[14] The sample is decomposed by sintering with a mixture of sodium carbonate, potassium carbonate and magnesium oxide. Iodide in the aqueous extract is oxidised to iodate with alkaline permanganate and reduced to iodide with stannous sulphate. After adding sodium nitrite and urea, elemental iodine is extracted into carbon tetrachloride solution and the optical density measured at a wavelength of 517 nm. The lower limit of detection is of the order of 1 ppm, making the method suitable only for those rocks that are particularly rich in iodine.

References

1. SELIVANOV L. S., *Compt. Rend., Acad. Sci. U.R.S.S.* (1940) **28**, 809.
2. FILBY R. H., *Geochim. Cosmochim. Acta* (1965) **29**, 49.
3. BEHNE W., *Geochim. Cosmochim. Acta* (1953) **3**, 186.

4. GOLDSCHMIDT V. M., *Geochemistry*, Oxford, 1954, p. 605.
5. CROUCH W. H., Jr., *Analyt. Chem.* (1962) **34**, 1689.
6. BENNETT W. H. and PICKUP R., *Colon. Geol. Min. Res.* (1952) **3**, 171.
7. KURODA P. K. and SANDELL E. B., *Analyt. Chem.* (1950) **22**, 1144.
8. BERGMANN J. G. and SANIK J., Jr., *Analyt. Chem.* (1957) **29**, 241.
9. HUANG W. H. and JOHNS W. D., *Anal. Chim. Acta* (1967) **37**, 508.
10. HAYNES S. J. and CLARK A. H., *Econ. Geol.* (1972) **67**, 378.
11. VAN DER MEULEN J. H., *Chem. Weekblad* (1931) **28**, 82.
12. FILBY R. H., *Anal. Chim. Acta* (1964) **31**, 434.
13. SHNEIDER L. A. and MILLER A. D., *Zhur. Anal. Khim.* (1965) **20**, 92.
14. GRIMALDI F. S. and SCHNEPFE M. M., *Anal. Chim. Acta* (1971) **52**, 181.

CHAPTER 19

CHROMIUM

Occurrence

The occurrence of chromium in silicate rocks was studied by Gold-schmidt[1] who reported that it appeared to a large extent in the earliest magmatic rocks to crystallise, i.e. in the olivine rocks and to a lesser extent in the pyroxenites where the chromium ranges from about 1000 to 5000 ppm Cr. The following table demonstrates the extent to which the latter crystallates contain successively smaller amounts of chromium.

TABLE 19. THE CHROMIUM
CONTENT OF SOME ROCK TYPES

Rock types	Cr, ppm
Peridotite	3400
Gabbro	340
Diorite	68
Granite	2
Nepheline-syenite	0·7

Fröhlich[2] has shown that the chromium content of eruptive rocks varies inversely as the silica content, although the spread of individual values about the mean is considerable. This is illustrated in Fig. 33, for a total of 118 igneous rocks almost all of them from the United Kingdom, for which a chromium content had been reported.[3, 4] Of the 505 complete analyses listed in these two publications, the chromium content was given as "trace", "not detected" or "0·00 per cent" in seventy-nine analyses, including a preponderance of acidic and other rocks with a high silica content. Chromium was not recorded in the remaining 308 analyses, many of which were completed at an early date.

Most of the chromium present in chrome-rich silicate rocks is to be found in the accessory minerals, particularly as chromite and chromi-ferous magnetite. Other chrome minerals occasionally encountered include uvarovite (chrome garnet), fuchsite (chrome mica) and tawma-wite (chrome epidote). As a direct result of the ease with which Cr^{3+} can replace Fe^{3+}, most of the chromium present in primary silicate

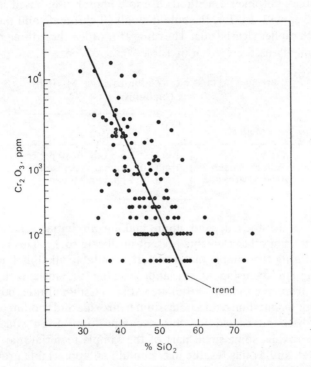

FIG. 33. The chromium content of igneous silicate rocks.

magmas is deposited in the early ferro-magnesian minerals. However, as noted by Ringwood,[5] it is possible for chromium to enter network-forming complexes and occasionally be concentrated in the later stages of magmatic crystallation.

Amongst the sedimentary rocks, high chromium contents have been reported from some bauxites and sedimentary iron ores. Carbonate rocks contain very little chromium.

The Determination of Chromium

No single method is appropriate for all silicate rocks, where the chromium content may range from less than 1 part per million to as much as 10 per cent or more in some chrome-norites. To cover this range three photometric methods are in common use, based upon the colour given with EDTA, the colour of alkali chromate and the colour given with diphenylcarbazide. The ranges for which these three methods are commonly used are given in Table 20.

TABLE 20. USEFUL RANGES OF PHOTOMETRIC METHODS
FOR CHROMIUM

Method	Cr
EDTA	from 0·5 per cent to 10 per cent
Alkali chromate	from 10 ppm to 5 per cent
Diphenylcarbazide	from 0·5 ppm to 200 ppm

All three methods depend upon making an initial separation of chromium from other elements present in silicate rocks. This is usually done by fusing the sample material with a suitable alkaline flux, often containing an oxidising agent, and leaching the melt with water. Sodium carbonate, sodium carbonate with potassium nitrate, potassium chlorate, or magnesium oxide, potassium hydroxide and sodium peroxide have all been advocated for this decomposition. Even when using sodium peroxide, some small part of the sample material may remain unattacked, and as this residue may contain an appreciable proportion of the chromium bearing minerals, it is advisable to re-treat the residue. Even when the sample decomposition is complete, some small part of the chromium may remain with the water-insoluble residue, despite extensive washing of the residue with dilute sodium carbonate solution. This loss of chromium, generally found to be of the order of 1 per cent, was demonstrated[6] with a sample of fuchsite from Nakiloro, Uganda, containing 1·57 per cent total chromium, and a specimen of anthrophyllite from South Harris, Scotland, containing 0·37 per cent total chromium. Both samples were found to be readily decomposed by fusion with sodium carbonate and potassium nitrate, yet the residues after aqueous

leaching of the melts were found to contain chromium equivalent to
0·015 and 0·004 per cent of the sample respectively.

For most purposes, a single fusion and extraction of the melt with
water should suffice. It should, however, be common practice to examine
the residue from the extraction to ensure that all the refractory titanium
minerals have been decomposed.

An earlier procedure for the separation of chromium and vanadium,
together with other similar elements such as molybdenum and tungsten,
involving precipitation with mercurous nitrate now appears to be little
used.

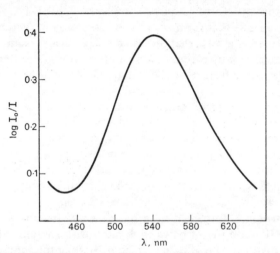

FIG. 34. Absorption spectrum of chromium-EDTA complex (4-cm
cells, 2·5 mg Cr/100 ml).

Determination of Chromium with EDTA

Both ethylenediaminetetraacetic acid (EDTA) and cyclohexanediaminetetraacetic acid (CyDTA) react with chromium(III) to form stable,
violet-coloured complexes that can be used for photometric determination. The complexes formed have similar molar absorbances and have
maximum absorption at the same wavelength of 540 nm (Fig. 34).
CyDTA is a relatively expensive reagent, and there appears to be no
reason for preferring it to the cheaper, more readily available EDTA.

The reaction of chromium(III) with EDTA is of particular value for the determination of chromium when present as a major constituent in such minerals as chromite, but has also been used with advantage for a number of rocks and silicate minerals containing 1 per cent or more of chromium. The lower limit, below which it is necessary to undertake a prior removal of silica, appears to be of the order of $0 \cdot 2$ per cent Cr; below $0 \cdot 5$ per cent the photometric procedure based upon measurement of the yellow colour of alkaline chromate solutions is to be preferred.

All silicate rocks can be decomposed by sinter or fusion with sodium peroxide, followed by extraction of the melt with water. On filtration the alkali chromate is separated from iron and from any cobalt, copper and nickel that also form coloured complexes with EDTA. Aluminium and other elements that form colourless complexes with EDTA do not interfere, provided always that excess complexone is present. Silica does not interfere, except by precipitation from solution, imposing a limit of about $0 \cdot 3$ g to the weight of rock material that can be taken for the analysis.

Iron or zirconium metal crucibles can be used for the fusion; nickel crucibles should be avoided as they may contain appreciable amounts of chromium, some of which passes into solution giving high results, particularly if sodium peroxide has been used for the decomposition. Zirconium crucibles may also contain a little chromium, but the attack of the crucible is very much less, and the chromium contributed to the alkaline filtrate in this way can usually be ignored.

Alkali chromate, formed during the sinter or fusion, is reduced to chromium (III) at a pH of between 4 and 5 with sodium sulphite, and the complex with EDTA is then formed by boiling the solution with an excess of the reagent for a minimum period of 20 minutes. Once formed the complex is very stable, and the solutions obey the Beer–Lambert Law in concentrations of up to at least 12 mg Cr per 100 ml, or 120 ppm (Fig. 35).

Method

Reagents: *Sodium peroxide.*
               Buffer solution, dissolve $10 \cdot 5$ g of sodium acetate trihydrate and 10 ml of glacial acetic acid in water and dilute to 100 ml.
               EDTA solution, dissolve 1 g of the di-sodium salt of ethylenediaminetetraacetic acid in 100 ml of water.

Sodium sulphite solution, dissolve 1 g of hydrated sodium sulphite in 100 ml of water. Prepare freshly each week.

Standard chromium stock solution, dissolve 0·3734 g of dried potassium chromate in water and dilute to 100 ml in a volumetric flask. This solution contains 1 mg Cr per ml.

Standard chromium working solution, dilute 50 ml of the stock solution to 250 ml with water and transfer to a clean dry polythene bottle. This solution contains 200 μg Cr per ml.

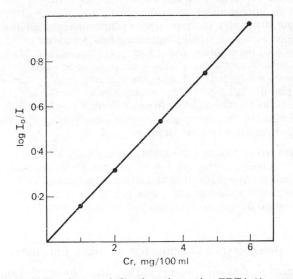

Fig. 35. Calibration graph for chromium using EDTA (4-cm cells, 540 nm).

Procedure. Accurately weigh 0·2–0·3 g of the finely powdered silicate material into a small zirconium crucible, mix with 0·6–1 g of sodium peroxide and sinter at a temperature of 550° in an electric furnace or fuse over a gas burner. Place the crucible on its side in a small beaker, cover with a clockglass, and add sufficient water to cover the melt. Warm gently until the melt has completely disintegrated, then rinse and remove the crucible. Boil the solution for about 10 minutes to decompose the excess peroxide and filter on to a small hardened filter paper. Wash the residue several times with small quantities of a dilute sodium carbonate solution and set the filtrate aside.

Rinse the residue back into the beaker with a jet of water, add more water to bring the total volume to about 50 ml and add concentrated hydrochloric acid warming if necessary to bring all soluble material into solution. If the initial attack of the mineral was complete, no dark-coloured gritty particles should be visible in the acid solution, and this solution can then be discarded. If black grains are discernible, collect on a small filter, wash with water, ignite in the zirconium crucible previously used for the fusion, and fuse with a little sodium peroxide, extracting with water, boiling and filtering as before. Combine the two alkaline filtrates and, if necessary, reduce the volume by evaporation to 75–80 ml.

Using a pH meter, or pH papers, adjust the pH of the solution with hydrochloric acid to a value in the range 4 to 5, add 5 ml of the acetate buffer solution, 5 ml of the EDTA solution and 5 ml of the sodium sulphite solution. Add a few fused alumina granules, cover the beaker with a clock-glass and gently boil for about 30 minutes. In the presence of chromium, the solution gradually becomes violet in colour, usually within the first 5 minutes of boiling. After 30 minutes, cool the solution and dilute to volume with water in a 100-ml volumetric flask. Prepare also a reagent blank solution in the same way, but omitting the sample material. Measure the optical density of the rock solution relative to that of the reagent blank solution in 4-cm cells, with the spectrophotometer set at a wavelength of 540 nm. Calculate the chromium content of the sample material by reference to a calibration graph, or, since this is a straight line, by using a calibration factor.

Calibration. To obtain the calibration graph, pipette aliquots of 5–30 ml of the standard solution containing 1–6 mg of chromium into separate 150-ml beakers. Dilute each to about 50 ml and add acetate buffer, EDTA and sodium sulphite solutions and continue as described in the method above. Plot the relation of optical density to chromium concentration (Fig. 35).

Determination of Chromium as Alkali Chromate

This procedure has the advantage that once the separation as alkali chromate has been made, no further operation other than dilution to volume is required. The maximum light absorption of chromate solutions occurs at a wavelength of 370 nm (Fig. 36) and for a concentration range of up to at least 12 ppm a straight line calibration is obtained (Fig. 37). When a filter photometer or absorptiometer fitted with violet filters is used, the measurement is made at the edge of the absorption band, and a curved calibration may be obtained. The sensitivity is also much reduced.

Similar yellow colours are given by uranium and cerium, but only in rare cases are these elements present in amounts sufficient to interfere. Iron in colloidal solution can also impart a yellow colour to the sample solution, but this is easily removed by digesting the sample solution on a steam bath, standing overnight and filtering in the cold. Platinum metal introduced into the solution from the crucible has also been suggested as a source of interference. This can best be limited by restricting the amount of oxidising agent added to the alkaline melt, and by using as low a temperature as possible for the decomposition. Any platinum that is brought into solution is removed on digesting the melt with water containing a little alcohol. The addition of a little borax to the melt serves to assist in the decomposition of the more refractory chromium-bearing minerals such as magnetite and chromite.

Some filter papers contain a small quantity of organic material that gives rise to yellow-coloured filtrates with the alkaline solutions used. Interference from this source can easily be avoided by washing each paper with dilute alkali carbonate solution before use.

The complete omission of oxidising agent may lead to low recoveries of chromium from some silicate rocks, particularly those that are rich in ferrous iron.[7] Potassium nitrate is commonly added to the flux, but alkali nitrite, formed during the fusion, absorbs strongly at the wavelength used (Fig. 36) and unless all traces of nitrite are destroyed high values of optical density will be obtained. This interference from nitrite can be avoided by measuring the solutions with an absorptiometer fitted with a violet gelatine filter (which has a maximum transmission at about 430 nm) as described by Bennett and Pickup.[7] Such procedures are much less sensitive than methods employing a spectrophotometer set at 370 nm. To avoid the interference from nitrite Rader and Grimaldi[8] used potassium chlorate to complete the oxidation of the rock material. A variation of this procedure is described below.

Many methods for the determination of chromium described a decomposition procedure involving fusion over a gas burner. It has been noted, however, that platinum crucibles are appreciably porous to gases from the burner, and complete oxidation to chromate does not always occur. For this reason, the use of a small electric muffle furnace is recommended. Rather more care is necessary, particularly in the early stages of the attack when carbon dioxide is being liberated, to prevent too rapid a fusion with consequent loss of material and damage to the furnace. A good indication that a satisfactory, completely oxidised, fusion has been obtained is the green colour of the melt when cold (oxidation of some of the manganese to manganate). This is particularly evident with basic and other rocks containing much manganese, but can usually also be seen with granitic and other rocks. Where this green

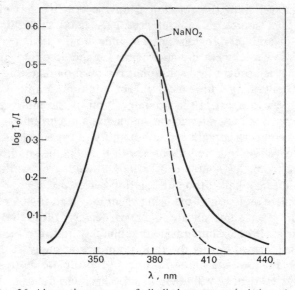

FIG. 36. Absorption spectra of alkali chromate and nitrite solutions.

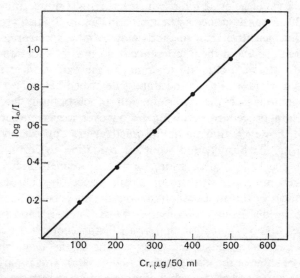

FIG. 37. Calibration graph for chromium as alkali chromate (1-cm cells, 370 nm).

colour is not obtained, a further quantity of potassium chlorate can be added and the fusion continued.

Method

Reagents: *Potassium chlorate.*
Borax glass, powdered.
Sodium carbonate wash solution, dissolve 10 g of the anhydrous reagent in 500 ml of water.
Working standard chromium solution, dilute 20 ml of the stock solution (p. 203) to 1 litre in a volumetric flask and transfer immediately to a clean dry polythene bottle. This solution contains 20 μg Cr per ml.

Procedure. Accurately weigh 1 g of the finely powdered rock material into a platinum crucible and mix with 5 g of anhydrous sodium carbonate and 0·5 g of potassium chlorate. If the rock material contains much chromite or magnetite add also 0·5 g of borax glass. Cover the crucible with a platinum lid and transfer to a cold electric muffle furnace. Gradually raise the temperature to 900–950°, and maintain at this temperature for a full hour. Care must be taken to ensure that mechanical loss of the mixture does not occur in the initial stages of the fusion. Allow the melt to solidify around the sides of the crucible to give a thin layer, tinged green in colour. If this colour is not obtained, continue the fusion with an additional quantity of potassium chlorate.

Extract the melt with water, add a few drops of ethyl alcohol and digest on a steam bath or hot plate to ensure that all soluble material has passed into solution and all manganate decomposed. Crush any lumps of solid material with the flattened end of a glass rod. Collect the residue on a hardened open-textured filter paper, that had previously been washed several times with hot sodium carbonate wash solution, and wash the residue several times with this wash solution. Collect the filtrate and washings in a beaker and again digest on a steam bath or hot plate for at least 1 hour, evaporate to a volume of 40–45 ml, and allow to cool, preferably overnight. When quite cold, filter the solution from any small residue that may have separated, into a 50-ml volumetric flask and dilute to volume with water. Prepare also a reference solution in the same way but omitting the rock sample.

If the residue is not to be used for the determination of barium or zirconium, it should be examined to ensure that the rock portion has been completely decomposed. The residue should dissolve in hydrochloric acid to give a clear solution with no more than a small flocculent precipitate of silica, and no black or gritty particles of unattacked rock material.

Measure the optical density of the solution, relative to the reagent blank solution in 1-cm cells, using the spectrophotometer at a wavelength of 370 nm, and determine the chromium content of the sample by reference to the calibration graph, or by using a calibration factor. If the optical density measurements are made using a filter absorptiometer a straight line graph will not be obtained, and a calibration factor should not be used.

Calibration. Transfer aliquots of 5–30 ml of the standard chromium solution containing 100–600 μg Cr, to separate 50-ml volumetric flasks. Add 5 g of anhydrous sodium carbonate to each flask, sufficient water to dissolve the solid material, and then dilute to volume with water. Measure the optical density of each solution in a 1-cm cell at a wavelength of 370 nm, using a similar solution without added chromium as the reference solution. Plot the relation of optical density to chromium concentration.

Determination of Chromium using sym-Diphenylcarbazide

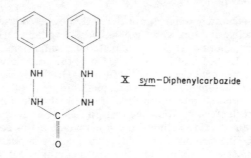

X   sym–Diphenylcarbazide

Chromium(VI) reacts with sym-diphenylcarbazide(X) in dilute acid solution to give a reddish-violet colour that can be used for the photometric determination of chromium. The reaction is of some complexity, involving both an oxidation of the reagent and the formation of a complex molecule containing chromium.[9] The maximum absorption of the coloured solution is at a wavelength of 540 nm (Fig. 38) and the molar absorptivity has been calculated[10] as 31,400. The reaction is therefore one of high sensitivity, and can be used to determine the few parts per million that occur in some silicate rocks. The Beer–Lambert Law is obeyed at the concentrations used (Fig. 39).

Other elements that give colours with the reagent include molybdenum—which forms a similar violet colour, mercury—which gives a blue colour, and iron and vanadium—which both give brown colours. Of these, mercury is most unlikely to be encountered in silicate materials in amounts sufficient to interfere. The reaction of diphenylcarbazide with molybdenum is of a lower order of sensitivity to that with chromium;

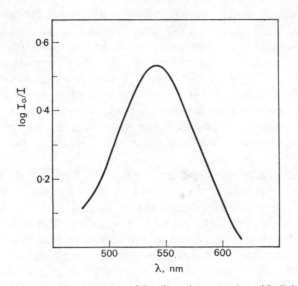

FIG. 38. Absorption spectrum of the chromium complex with diphenylcarbazide (4-cm cells, $6 \cdot 4$ μg Cr/25 ml).

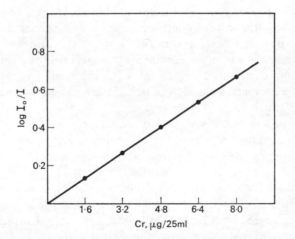

FIG. 39. Calibration graph for chromium using diphenylcarbazide (4-cm
cells, 540 nm).

as most silicate rocks contain considerably less molybdenum than chromium, interference from this source is therefore also unlikely. Iron is usually separated from chromium by alkaline fusion; in extracting the melt with water only traces of iron pass into the filtrate. These can be removed by passing the solution through a cation-exchange column, as described by Fröhlich,[11] but this is not usually necessary as the greater part of these traces of iron precipitates from the solution on standing. Any small amount not collected in this way subsequently reacts with the reagent, but has only a small absorption at 540 nm and can therefore be neglected.

A serious interference is that from vanadium, which is not separated from chromium in any of the procedures used to decompose the sample material. The brown-coloured complex formed with the reagent is unstable—the colour fades rapidly on standing—and nearly correct results can be obtained for chromium by measuring after 10–15 minutes. Partly for this reason and partly because the vanadium complex has a much smaller absorption than the chromium complex at 540 nm, the removal of vanadium is only necessary if its amount grossly exceeds that of the chromium. A procedure for this involving an extraction of the vanadium complex with 8-hydroxyquinoline is described by Sandell.[12]

The violet colour given by chromium with the reagent forms rapidly in $0 \cdot 2$ N sulphuric acid solution; it is unstable at higher acid concentrations, and does not develop completely at lower acidities. The correct acid concentration is often obtained by neutralising the solution and then adding a measured volume of sulphuric acid. Grogan et al.[13] prefer to use a meter to bring the pH of the solution to a value of between $1 \cdot 3$ and $1 \cdot 7$.

The small loss of chromium that occurs when an alkaline fusion is used to decompose the silicate sample material, and to which reference is made above (p. 200), can be avoided by using an acid decomposition. For this Spangenberg et al.[14] describe the use of a potassium bifluoride fusion, although the more usual hydrofluoric-sulphuric acid evaporation can also be used. Iron can be removed by extraction with amyl acetate or other organic solvent from 10 N hydrochloric acid, Chromium III is oxidised to chromate with ceric ammonium sulphate and the excess reagent destroyed with sodium azide, which also reduces any oxidised manganese. The violet-coloured complex with diphenylcarbazide is then formed and measured using a spectrophotometer as described below.

Other difficulties that occur include the tendency of chromium(VI) ions to be adsorbed on the walls of the glass vessels commonly used, noted by Chuecas and Riley,[15] and the loss of chromium that occurs when silica is precipitated from acid solutions. These difficulties may be avoided as described by Fuge,[16] by working in platinum and plastic ware apparatus wherever possible, and by eliminating most of the silica at the extraction stage. This is accomplished by adding magnesium oxide to the sodium carbonate flux used for the decomposition of the silicate rock material, serving to limit the attack of the crucible, and to form the insoluble magnesium silicate which is retained in the insoluble residue. The procedure given below is based upon those described by Fuge[16] and others.

Method

Reagents: *Sodium carbonate–magnesium oxide mixture,* mix intimately 12 g of anhydrous sodium carbonate and 3 g of magnesium oxide. This provides sufficient flux for twenty-five determinations.

Sodium carbonate wash solution, dissolve 5 g of anhydrous sodium carbonate in 250 ml of water.

sym-Diphenylcarbazide solution, dissolve $0·25$ g of a good quality reagent in 100 ml of acetone. This solution deteriorates slowly on standing, and should be replaced when appreciably brown in colour.

Standard chromium working solution, dilute 20 ml of the freshly prepared standard chromium solution containing 20 μg per ml (p. 200) to 1 litre with water and transfer immediately to a clean, dry polythene bottle. This solution contains $0·4$ μg chromium per ml.

Procedure. Accurately weigh approximately $0·1$ g of the finely powdered silicate rock material into a small platinum crucible and mix intimately with approximately $0·6$ g of the sodium carbonate–magnesium oxide mixture. Cover the crucible with a platinum lid and transfer to an electric muffle furnace set at a temperature of about 900°, and maintain at this temperature for 45 minutes. Remove the crucible from the furnace and allow to cool. Add a few millilitres of water, together with 2 drops of ethanol, and heat on a hot plate to disintegrate the melt. Break up any lumps of solid material with the aid of a stout polythene rod, and replace water lost by evaporation as necessary. Digest on the hot plate for 1 hour, then remove the crucible and allow to cool.

Collect the residue on a 9-cm close-textured filter paper supported in a polyethylene funnel, and wash with successive small volumes of the sodium carbonate wash solution. Collect the filtrate and washings in a small polypropylene beaker and dilute to about 20 ml with water. Using a pH meter, bring the pH of the solution to a value in the range 1·3–1·7 by adding 10 N sulphuric acid, swirling to remove as much of the liberated carbon dioxide as possible. Add 1 ml of the diphenylcarbazide solution, transfer to a 25-ml volumetric flask, shake to remove any residual carbon dioxide, and dilute to volume with water.

Prepare also a reagent blank solution in the same way as the sample solution, but omitting the sample material. Allow the coloured solutions to stand for 15 minutes and then measure the optical density in 4-cm cells relative to water using a spectrophotometer set at a wavelength of 540 nm. Measure also the optical density of the reagent blank solution.

Calibration. Transfer aliquots of 0–20 ml of the standard chromium solution containing 0–8 µg Cr to separate 25-ml volumetric flasks. Dilute each solution to about 20 ml with water and add 1 ml of 6 N sulphuric acid and 1 ml of the diphenylcarbazide solution to each. Dilute to volume, allow to stand for 15 minutes and measure the optical density of each solution in 4-cm cells at a wavelength of 540 nm as described above. Plot the relation of optical density to chromium concentration (Fig. 39).

Determination of Chromium by Atomic Absorption Spectroscopy

Both iron and nickel have been found to suppress the chromium absorption in an air–acetylene flame.[17] Interference from these elements can be reduced by using the higher temperature nitrous oxide-acetylene flame, but only at the cost of a reduced sensitivity to chromium. Using this flame, an absorption wavelength of 357·9 nm, and a nitric acid solution of the silicate rock material, Beccaluva and Venturelli[17] reported the chromium content of a range of standard silicate rocks, from 11 ppm Cr in G-2 to 3960 ppm in DTS-1.

References

1. GOLDSCHMIDT V. M., *J. Chem. Soc.* (1937) 655.
2. FRÖHLICH F., *Geochim. Cosmochim. Acta* (1960) **20**, 215.
3. GUPPY E. M., *Chemical Analysis of Igneous Rocks, Metamorphic Rocks and Minerals*, Geol. Surv. Gt. Brit., 1931 (for analyses to 1930).
4. *Ibid.* (analyses 1931–54), Geol. Surv. Gt. Brit., 1956.
5. RINGWOOD A. E., *Geochim. Cosmochim. Acta* (1955) **7**, 242.
6. JEFFERY P. G. and RICHARDSON J. (unpublished work).
7. BENNETT W. H. and PICKUP R., *Colon. Geol. Min. Res.* (1952) **3**, 171.

8. RADER L. F. and GRIMALDI F. S., *U.S. Geol. Surv. Prof. Paper* 391-A, p. A-10, 1961.
9. BOSE M., *Anal. Chim. Acta* (1954) **10**, 201 and 209.
10. ROWLAND G. P. Jr., *Ind. Eng. Chem., Anal. Ed.* (1939) **11**, 442.
11. FRÖHLICH F., *Zeit. Anal. Chem.* (1959) **170**, 383.
12. SANDELL E. B., *Ind. Eng. Chem., Anal. Ed.* (1936) **8**, 336.
13. GROGAN C. H., CAHNMANN H. J. and LETHCO E., *Analyt. Chem.* (1955) **27**, 983.
14. SPANGENBERG J. D., RUSSEL B. G. and STEEL T. W., *Nat. Inst. Metall. Johannesburg S.A.*, Rept. No. 265 (1967).
15. CHUECAS L. and RILEY J. P., *Anal. Chim. Acta* (1966) **35**, 240.
16. FUGE R., *Chem. Geol.* (1967) **2**, 289.
17. BECCALUVA L. and VENTURELLI G., *Atomic Absorpt. Newsl.* (1971) **10**, 50.

CHAPTER 20

COBALT

Occurrence

The occurrence of cobalt in igneous rocks has been summarised by Unksov and Lodochnikova[1] who stressed the relation of both cobalt and nickel to silica content (Fig. 40). This relation of the average values ignores the considerable spread of cobalt results, particularly in granitic rocks. In a study of the geochemistry of cobalt, Carr and Turekian[2] have demonstrated the dependence of the cobalt values upon the magnesium content of granitic rocks, an observation previously reported for American igneous rocks by Sandell and Goldich.[3] For granitic rocks values of less than 1 ppm Co are by no means uncommon, extending to over 10 ppm for those containing more than 1 per cent magnesium. For basaltic rocks cobalt contents averaging about 50 ppm have been reported, again with a wide spread of individual values.

Only rarely are cobalt minerals observed in silicate rocks and then associated with sulphite mineralisation as linnaeite, $(Co,Ni)_3S_4$ or its variety siegenite. Sulphide minerals pyrite, blende, pyrrhotite, pentlandite, arsenopyrite, and other arsenic ores have all been reported as carriers of cobalt in some localities.[4]

The Determination of Cobalt

Little difficulty exists in applying colorimetric methods[5, 6] to the determination of cobalt in basalts and other similar rocks. Care is however required in their application to granitic rocks, particularly those containing less than 1 ppm of cobalt. For these Smales et al.[7] and Turekian and Carr[8] have devised neutron activation procedures based upon irradiation with thermal neutrons, followed by a chemical separation prior to counting.

Spectrographic procedures have been extensively used for determining cobalt in basic rocks,[9] but it is not always possible to detect cobalt in granitic and similar rocks containing little cobalt.

PHOTOMETRIC METHODS FOR THE DETERMINATION OF COBALT

The reaction with ammonium thiocyanate to give a blue-coloured compound that can be extracted from aqueous into iso-amyl alcohol solution has been used as the basis of a photometric method for determining cobalt.[10] This reaction with thiocyanate alone is not of sufficient

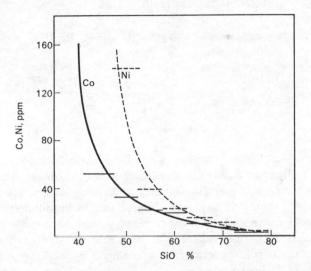

FIG. 40. Cobalt and nickel in silicate rocks.

sensitivity for the determination of cobalt in silicate or carbonate rocks and minerals, but has been extensively used in the examination of sulphide minerals.

Cobalt forms a red-violet coloured complex with dithizone that can be extracted into carbon tetrachloride or chloroform solution for photometric measurement. More frequently, this procedure is used to concentrate the cobalt into a small volume and to separate it from certain other metals. Both Carmichael and McDonald[5] and Rader and Grimaldi[11] have used this technique, combining it with photometric measurement with a separate colour-forming reagent. It has been reported[12] that an excessive concentration of hydroxyammonium chloride in the aqueous solution impedes the extraction of cobalt with dithizone, giving somewhat low recoveries.

Only a few sensitive photometric reagents are known for cobalt. One that has been applied to the analysis of silicate rocks is nitroso-R-salt, and a procedure using this reagent following a dithizone extraction is given below. Other reagents proposed for the photometric determination of cobalt in silicates include CyDTA, 2-nitroso-1-naphthol[13] and thiocyanate with tri-n-butylamine.[14]

Determination of Cobalt with Nitroso-R-salt[11]

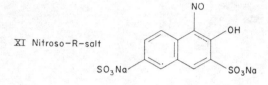

XI Nitroso–R–salt

Nitroso-R-salt (the di-sodium salt of 1-nitroso-2-hydroxynaphthalene-3:6-disulphonic acid) (XI) reacts with cobalt in acid solution to give a stable red-coloured complex, soluble in aqueous solution. The reagent itself is highly coloured and gives an appreciable background absorption. Procedures have been devised for reducing this background absorption by bleaching the excess reagent with nitric acid or bromine, but acceptable results for silicate rocks can be obtained by working at a wavelength where the background absorption is much reduced. See Sandell[15] for a discussion of the most appropriate wavelength to use.

In the procedure described below[11] the cobalt complex of the reagent is developed in a boiling citrate–phosphate–borate buffer solution as described by McNaught.[16] Any nickel or copper extracted from the rock solution by the dithizone extraction will form coloured complexes with the reagent, but these complexes are decomposed by boiling with hot dilute nitric acid.

Method

Reagents: *Citric acid solution*, dissolve 50 g of purified citric acid monohydrate in water and dilute to 100 ml.

Citric acid solution, 0·20 M *solution,* dissolve 4·2 g of the purified reagent (monohydrate) in water and dilute to 100 ml.

Dithizone solution, dissolve 0·05 g of analytical grade diphenyl-thiocarbazone in 100 ml of carbon tetrachloride. Store in a refrigerator.

Nitroso-R-salt solution, dissolve 0·05 g of the reagent in 100 ml of water.

Buffer solution, dissolve 6·2 g of boric acid, 35·6 g of disodium hydrogen phosphate heptahydrate and 20 g of sodium hydroxide in water and dilute to 1 litre.

Standard cobalt stock solution, dissolve 0·807 g of cobaltous chloride hexahydrate in water containing 2 ml of concentrated hydrochloric acid and dilute to volume in a 1 litre volumetric flask. This solution contains 200 μg cobalt per ml.

Standard cobalt working solution, dilute 5 ml of this stock solution to 1 litre with water. This solution contains 1 μg cobalt per ml.

Procedure. Accurately weigh approximately 1 g of the finely powdered silicate rock into a platinum dish (note 1), moisten with water and add 10 ml of hydrofluoric acid, 10 ml of nitric acid and 4 ml of perchloric acid. Cover the dish with a platinum or polypropylene cover and heat on a steam bath for 30 minutes. Rinse and remove the cover, transfer the dish to a hot plate and evaporate to fumes of perchloric acid. Allow to cool, rinse down the sides of the dish with a little water and again evaporate to fumes of perchloric acid. Again rinse down the sides of the dish and evaporate this time almost, but not quite, to dryness. Avoid overheating the moist salts. Add 2 to 4 ml of 6 N hydrochloric acid to the residue followed by 10 ml of water. Digest on a steam bath for a minute or two until all soluble salts have dissolved. If any residue remains, this should be collected and decomposed as described in note 2.

Evaporate the solution if necessary so that it may be diluted to volume in a 25-ml volumetric flask. At this stage some sodium chloride or potassium perchlorate may crystallise out. Allow any precipitated salts to settle, pipette a 10-ml aliquot of the solution into a small beaker and add 5 ml of the concentrated citric acid solution. Using a pH meter, adjust the pH of the solution to 9 by adding aqueous ammonia. Transfer the solution to a 60-ml separating funnel and add 5 ml of dithizone solution, stopper the funnel and shake vigorously to extract the cobalt complex into the carbon tetrachloride solution. Run off the lower organic layer and extract the aqueous phase with further 5-ml portions of the dithizone reagent solution until a green colour remains in the organic phase after shaking for at least 1 minute. Three or four 5-ml portions are usually required for this. Discard the aqueous layer and combine the organic extracts.

Return the organic solution to the separating funnel and add 5 ml of dilute aqueous ammonia (1 + 99). Shake vigorously for 1 minute, allow the layers to separate and transfer the organic solution to a 50-ml beaker. Discard the aqueous fraction. Evaporate the carbon tetrachloride solvent on a water bath.

Add 0·5 ml of 20 N sulphuric acid and between 0·25 ml and 0·5 of perchloric acid to the residue and evaporate on a hot plate until a colourless liquid is obtained, then evaporate the excess sulphuric and perchloric acids, including any drops that have condensed on the sides of the beaker.

Add 1 ml of 6 N hydrochloric acid, and using this acid, wet the inside walls and bottom of the beaker. Rinse down the walls with a very small quantity of water and evaporate the solution to dryness. Complete the drying by removing the last traces of water and acid in an electric oven set at 140°. Add 1·0 ml of 0·2 M citric acid solution to the beaker followed by 1·2 ml of the buffer solution. Stir the solution whilst adding exactly 2 ml of the nitroso-R-salt reagent solution. Boil for 1 minute and then allow to cool.

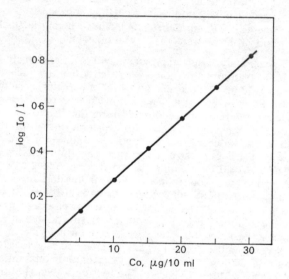

Fig. 41. Calibration graph for cobalt using nitroso-R-salt (1-cm cells, 475 nm).

Add 1·0 ml of nitric acid to the solution and again boil for 1 minute. Cool the solution, transfer to a 10-ml volumetric flask and dilute to volume with water. Measure the optical density of this solution against a reagent blank solution using a spectrophotometer set at a wavelength of 475 nm. The reagent blank solution is prepared by adding 1·0 ml of 0·2 M citric acid solution to 1·2 ml of the buffer solution, adding exactly 2 ml of the nitroso-R-salt reagent and boiling for 1 minute. This solution is then cooled, boiled for a further minute with nitric acid, cooled and diluted to 10 ml in a volumetric flask as described for the sample solution.

Calibration. Transfer 0–30 ml aliquots of the standard cobalt solution containing 0–30 μg cobalt to 50-ml beakers. Evaporate to dryness on a hot plate to remove all traces of acid and proceed as described in the method above beginning with the addition of 1·0 ml of 0·2 M citric acid and 1·2 ml of buffer solution. Plot the relation of optical density of these solutions to their cobalt content (Fig. 41).

Notes: 1. If the sample contains any organic material it should be ignited for a few minutes at a temperature of 700–800°C before moistening with water.

2. Any residue should be carefully collected, washed with a little water, dried, ignited and fused with the minimum quantity of anhydrous sodium carbonate. Dissolve the melt in a little 6 N hydrochloric acid and add to the main rock solution.

Determination of Cobalt with Thiocyanate and Tri-n-Butylamine

This procedure depends upon the formation and extraction of tri-n-butylammonium tetrathiocyanatocobalt(III), described in detail by Stanton *et al.*[14] based upon the earlier work of Zeigler *et al.*[17] Although the extraction of cobalt is complete over the pH range of 1·0–5·2, the addition of sodium hexametaphosphate to mask traces of iron in the aqueous solution, restricts complete extraction to the range 4·5–5·2. Any copper present in the rock solution is reduced by the hydroxyammonium chloride present in the buffer solution, and does not then interfere. No other metal extracted by the dithizone reagent interferes with the cobalt determination unless present in amounts greater than those normally encountered in silicate rocks. In this concentration range solutions of the cobalt complex obey the Beer–Lambert Law, giving a straight-line calibration curve.

Method

Reagents: *Buffer solution*, dissolve 100 g of trisodium citrate and 5 g hydroxyammonium chloride in 400 ml of water, add dilute aqueous ammonia until alkaline to thymol blue. Extract with successive volumes of 0·015 per cent dithizone solution until there is no further reaction, and remove the dithizone from the aqueous phase by successive extractions with chloroform. Finally extract with three 20 ml portions of carbon tetrachloride, adjust the pH to a value between 10·8 and 11·0 with aqueous ammonia, and then dilute to 500 ml with water.

Dithizone solution, dissolve 0·015 g of analytical grade diphenylthiocarbazone in 100 ml of carbon tetrachloride. Store in a refrigerator until required.

Sodium acetate–potassium thiocyanate solution, dissolve 75 g of sodium acetate trihydrate and 2·5 g of hydroxyammonium chloride in water containing 50 ml of 6 N hydrochloric acid and dilute to 250 ml with water. Dissolve also 17·5 g of potassium thiocyanate in water and dilute to 250 ml. When required for use, mix equal volumes of these two solutions: the pH should then be in the range 4·9–5·1.

Sodium hexametaphosphate solution, dissolve 15 g of the reagent in water and dilute to 100 ml.

Tri-n-butylamine solution, mix benzene, pentyl alcohol (note 1) and tri-n-butylamine in the proportions 7:2:1, as required.

Standard cobalt working solution, using 0·2 N hydrochloric acid dilute 25 ml of the stock solution to volume in a 1 litre volumetric flask. This solution contains 5 μg cobalt per ml.

Procedure. Accurately weigh approximately 1 g of the finely powdered silicate rock material (note 2) into a platinum crucible, moisten with water and add 20 ml of hydrofluoric acid and 3 ml of perchloric acid. Cover the crucible, leaving a small gap, and allow the solution to evaporate overnight on a steam bath. Transfer to a hot plate and heat to expel the excess perchloric acid. Warm the residue with 3 ml of 6 N hydrochloric acid and 10–15 ml of water, transfer the solution and residue to a 50-ml beaker and leave on a water bath until all the soluble salts have dissolved and then set aside to cool (note 3).

Add 25 ml of the sodium citrate buffer solution, and transfer to a 150-ml separating funnel. Extract the reacting metals with successive 5-ml portions of dithizone solution until there is no further reaction, collecting the carbon tetrachloride phases in a second separating funnel. Before discarding the aqueous phase, check its pH—if less than 8·9, increase to this value by adding ammonia solution and extract with a further 5-ml portion of dithizone solution. In the second funnel complete the separation of the carbon tetrachloride extracts from any accompanying aqueous phase by transferring the organic layer to a 100-ml beaker. Wash the separating funnel with carbon tetrachloride and collect the washings in the beaker.

Evaporate the combined extracts to dryness on a water bath, and then add 2 ml of sulphuric acid and 1 ml of perchloric acid. Heat the covered beaker on a sand bath until the liquid is colourless, and then remove the cover glass and evaporate to dryness. Dissolve the residue by warming gently with 3 ml of 0·6 N hydrochloric acid and transfer to a 10-ml volumetric flask. Dilute to volume with water and mix well. In addition to the determination of cobalt, this solution can be used to determine cadmium, copper, lead, nickel and zinc.[14]

To determine the cobalt content of the rock solution, transfer 3 ml to a 25-ml stoppered tube or flask and add 5 ml of the sodium acetate–potassium thiocyanate solution, 1 ml of the sodium hexametaphosphate solution and 4 ml of the tri-n-butylamine solution. Stopper the tube or flask and shake vigorously for 1 minute. Allow the phases to separate, remove the benzene layer, filter to remove any water and transfer to a 1-cm spectrophotometer cell. Measure the optical density of the extract relative to pure benzene at a wavelength of 625 nm, and determine the cobalt content by reference to the calibration graph (note 4). Measure also the optical density of a reagent blank solution prepared in the same way as the rock solution, but omitting the sample material.

Calibration. Pipette 0–10 ml aliquots of the working standard cobalt solution containing 0–50 μg of cobalt to separate 50-ml beakers. Evaporate the solutions to dryness, add 3 ml of $0 \cdot 2$ N hydrochloric acid to each and warm to dissolve the residues. Transfer to stoppered tubes or flasks, add 5 ml of sodium acetate–potassium thiocyanate solution, 1 ml of sodium hexametaphosphate solution and 4 ml of the tri-n-butylamine reagent and extract the cobalt complex as described above. Measure the optical densities of the extracts at 625 nm in 1-cm cells and plot the relation of optical density to cobalt concentration in the usual way.

Notes: 1. Zeigler et al.[17] suggest using "pure isoamyl alcohol", but in the application of this reagent to the field determination of cobalt in soils and sediment samples Stanton and McDonald[18] report that both normal and isoamyl alcohol can be used.

2. For basic rock material use only a $0 \cdot 5$-g sample.

3. Any insoluble residue at this stage should be collected, washed with water, ignited, fused with a little sodium carbonate, dissolved in dilute hydrochloric acid and added to the main solution.

4. Somewhat smaller amounts of cobalt can be estimated visually by comparison with a series of standards prepared from $0 \cdot 25$–$4 \cdot 5$ μg cobalt. These and the sample solution should be prepared as described in the method given above, but using only $0 \cdot 5$ ml of the tri-n-butylamine reagent solution.

Ion-exchange Separation and Determination with CyDTA

Anion exchange resins can conveniently be used to separate and concentrate a small number of metallic ions that occur in silicate rocks. Cobalt is one such element that can be separated in this way; it can then be determined by photometric measurement of the violet-coloured cobalt(III) complex with EDTA or CyDTA. The Beer–Lambert Law is obeyed at the concentration range encountered (Fig. 42).

The silicate rock material is decomposed by evaporation with hydrofluoric and nitric acids in the usual way, and the excess nitric acid removed by evaporation to dryness. The metallic nitrates and oxides are converted to chlorides with hydrochloric acid, and the chloride solution transferred to a short column of strongly basic anion exchange resin, such as Dowex 1 × 8, previously washed with 9 M hydrochloric acid. Manganese, titanium, aluminium, calcium, magnesium and the alkali metals are eluted from the column using three column volumes of 9 M hydrochloric acid. Any cobalt present in the sample is then recovered by elution with four column volumes of 5 M hydrochloric acid. (If required, iron, zinc and cadmium can subsequently be recovered from the resin by elution with more dilute hydrochloric acid.)

The eluate containing cobalt is evaporated to dryness and the residue dissolved in an acetic acid–sodium acetate buffer solution (maximum values of optical density were obtained in the pH range 1·5–7[19]). An

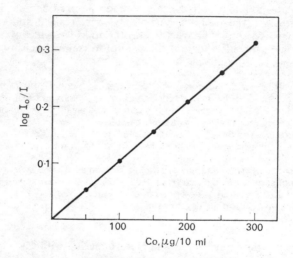

FIG. 42. Calibration graph for cobalt using CyDTA (2-cm cells, 545 nm).

excess of CyDTA is then added followed by hydrogen peroxide to oxidise Co(II) to Co(III), and the solution boiled to complete complex formation. After cooling, the solution can be diluted to 10 ml with water and the optical density measured at 545 nm.

The molar extraction coefficient of the CyDTA-cobalt(III) complex has been given as 305,[19] it is therefore important to use at least 1 g sample portions of the silicate rock, or more if it contains less than 50 ppm Co and to keep the volume of the final solution to the minimum. This procedure is not suitable for acid or other rocks containing less than 10 ppm cobalt. EDTA gives a similar complex with Co(III), but a wavelength of 520 nm has been recommended[20] for measurement.

References

1. UNKSOV V. A. and LODOCHNIKOVA N. V., *Geokhimiya* (1961) No. 9, 732.
2. CARR M. H. and TUREKIAN K. K., *Geochim. Cosmochim. Acta* (1961) **23**, 9.
3. SANDELL E. B. and GOLDICH S. S., *J. Geol.* (1943) **51**, 99 and 167.
4. GOLDSCHMIDT V. M., *Geochemistry*, Oxford, 1954, p. 668.
5. CARMICHAEL I. and MCDONALD A. J., *Geochim. Cosmochim. Acta* (1961) **22**, 87.
6. LODOCHNIKOVA N. V., *Informatsionnyi sbornik VSEGEI* (1956) No. 3.
7. SMALES A. A., MAPPER D. and WOOD A. J., *Analyst* (1957) **82**, 75.
8. TUREKIAN K. K. and CARR M. H., *Geochim. Cosmochim. Acta* (1961) **24**, 1.
9. AHRENS L. H. and TAYLOR S. R., *Spectrochemical Analysis*, Pergamon, Oxford, 1961, p. 227.
10. YOUNG R. S. and HALL A. J., *Ind. Eng. Chem., Anal. Ed.* (1946) **18**, 264.
11. RADER L. F. and GRIMALDI F. S. *U.S. Geol. Surv. Prof. Paper*, 391A, 1961.
12. CARMICHAEL I. and MCDONALD A. J., *Geochim. Cosmochim. Acta* (1961) **25**, 189.
13. CLARK L. J., *Analyt. Chem.* (1958) **30**, 1153.
14. STANTON R. E., MCDONALD A. J. and CARMICHAEL I., *Analyst* (1962) **87**, 134.
15. SANDELL E. B., *Colorimetric Determination of Traces of Metals*, Interscience (3rd ed.), 1959, p. 420.
16. MCNAUGHT K. J., *Analyst* (1942) **67**, 97.
17. ZEIGLER M., GLEMSER O. and PREISLER E., *Mikrochim. Acta* (1956) p. 1526.
18. STANTON R. E. and MCDONALD A. J., *Trans. Inst. Min. Metal.* (1961–2) **71**, 511.
19. JACOBSEN E. and SELMER-OLSEN A. R., *Anal. Chim. Acta* (1961) **25**, 476.
20. GUIFFRE L. and CAPIZZI F. M., *Ann. Chim. Roma* (1961) **51**, 558.

CHAPTER 21

COPPER

Occurrence

The association of copper with basic igneous rocks has been recorded from many localities, where the crystallisation of a sulphide phase includes the deposition of chalcopyrite, $CuFeS_2$. As a result of their study of the strong fractionation of the basic magma at Skaergaard, Wager and Mitchell[1] have shown that copper is present in the parental magma, and could accumulate in the residual magma until a sulphide phase crystallises. At first sight this behaviour is a little anomalous as the cuprous ion could be camouflaged by sodium (Cu^+, $R = 96$ pm; Na^+, $R = 98$ pm), and the cupric ion by ferrous iron (Cu^{2+}, $R = 72$ pm; Fe^{2+}, $R = 74$ pm). The analyses of the Skaergaard minerals show clearly that this does not happen. Ringwood[2] has suggested that this apparent anomalous behaviour arises as a result of the weakness of the copper–oxygen bond, relative to the sodium–oxygen and iron–oxygen bonds. Ahrens,[3] who also commented on the tendency of copper not to enter mafic (iron–magnesium) minerals, associated this behaviour with the strongly covalent nature of the copper–oxygen bond, and the tendency for the copper atom to form coplanar bonds with oxygen.

Sandell and Goldich[4] have shown that the copper content of magmatic rocks from the United States of America decreases with increasing silica content (Table 21). Published determinations since that time (1943)

TABLE 21. THE RELATION OF COPPER TO
SILICA CONTENT

SiO₂ (average), ppm	Cu (average), ppm
48·5	149
62	38
72	16

are in broad agreement with this conclusion, although ultrabasic rocks may resemble the acidic rocks and contain only a few ppm copper. Limestones, sandstones and similar rocks contain very little copper, but shales and argillaceous sediments may contain 40–50 ppm.

The Determination of Copper

The economic importance of copper mineralisation has led to widespread interest in geochemical methods of prospecting for this element. A number of analytical procedures for the determination of copper in soils and silicate rocks have been designed—mostly based upon the need to provide rapid, easy and cheap methods for operation in remote areas. Such procedures invariably determine only that part of the copper present in a readily accessible form; for example, the part that can be extracted into dilute sulphuric acid[5] or into aqueous solution after fusion with potassium pyrosulphate.[6] Although ideal for their purpose, these methods do not give a measure of the total copper content when applied to silicate rocks. For these materials it is preferable to decompose the silicate minerals present, using hydrofluoric acid in combination with either nitric or sulphuric acid in the usual way.[7]

PHOTOMETRIC METHODS

Except in rare cases, the amount of copper present in silicate rocks is insufficient for either gravimetric or volumetric determination, and photometric methods are in general use. Diphenylthiocarbazone (dithizone), diethyldithiocarbamate and diquinolyl are the reagents most commonly used.

TABLE 22.

Reagent	ϵ	λ, nm
Diphenylthiocarbazone (dithizone)	25,000	580*
Diethyldithiocarbamate	13,000	436
Diquinolyl	5500	540

* Charlot[8] also gives $\epsilon = 22,000$ at 430 nm and 35,000 at 620 nm.

The most sensitive of these reagents for copper is dithizone (Table 22), but this reacts also with many other metal ions besides copper. A procedure involving determination with dithizone following a prior extraction of copper, lead and zinc has been described by Rader and Grimaldi[9] and is given in some detail below.

Less sensitive than dithizone are the dithiocarbamate reagents. The diethyl compound is the best known, although the dibenzyl compound has been recommended.[10] These reagents react also with other metallic ions. Although commonly used for copper in many materials, dithiocarbamates have not been extensively used for silicate rocks and minerals, possibly because the third reagent, diquinolyl, is much more specific. Although expensive and considerably less sensitive than the other two, diquinolyl is very much less subject to interference from other metallic ions. It can be used directly for copper in silicate rocks, as described by Riley and Sinhaseni,[7] or after a preliminary isolation of copper, zinc and lead with dithizone, as described by Rader et al.[9]

DETERMINATION BY ATOMIC ABSORPTION SPECTROSCOPY

Atomic absorption spectroscopy provides an alternative technique for the determination of copper in silicate rocks,[11] with a lower limit of about 5 ppm Cu. This method has the advantages of simplicity, rapidity and freedom from interference from other elements present in the rock material. The determination of copper can be combined with that of other elements such as zinc, manganese, magnesium and iron. The procedure is as follows. The rock material is evaporated to dryness with hydrofluoric and perchloric acids, the residue dissolved in dilute hydrochloric acid, almost neutralised with ammonia, diluted to volume, and the flame absorption measured at 325 nm. Further details are given under zinc, p. 499.

Determination of Copper with 2,2′-Diquinolyl

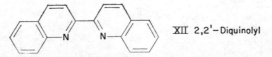

XII 2,2′-Diquinolyl

Diquinolyl(XII), known also by the trivial name "cuproine", is one of a number of specific reagents for copper(I) that contain the "ferron" structure but are sterically inhibited from reacting with iron(II) by the presence of ortho substituents. 2,9-Dimethylphenanthroline and other phenanthroline derivatives, although slightly more sensitive than diquinolyl as reagents for copper, are considerably more expensive. The reaction with diquinolyl was first used for the determination of copper by Breckenridge *et al.*[12] and, because of its specificity, has been extensively used for the examination of a wide range of materials and products. The pink complex formed by the cuprous ion is extracted into a suitable alcoholic solvent at a pH in the range 4·5–7·5[13] Hoste *et al.*[14] has suggested that complete extraction can be obtained in the range 2–9, but Riley and Sinhaseni[7] have found that practically no copper could be extracted at a pH below 2·7, and recommended working in the range 4·2–5·8.

Both normal and isoamyl alcohols have been used for this extraction. They are however appreciably soluble in water; Riley and Sinhaseni[7] noted that the less soluble n-hexanol could also be used. A purification procedure is usually necessary for all suggested solvents. Riley and Sinhaseni[7] have recommended making three extractions of the copper solution, with intervening addition of hydroxyammonium chloride to ensure that all the remaining copper is in the cuprous state. Fading of the colour of the cuprous-diquinolyl extracts has been reported by several workers, this can be prevented by adding a little hydroquinone to the organic extracts.[7]

The cuprous-diquinolyl extracts have a maximum absorption at a wavelength of 540–545 nm (Fig. 43) and the Beer–Lambert Law is obeyed up to at least 10 ppm in the organic extract (Fig. 44). This covers the concentration range normally required for rock analysis.

Method

The procedure given below has been adapted from that described by Riley and Sinhaseni.[7] All solutions used for this determination should be prepared from water that has been distilled from an all-glass or silica still, or has passed through a "demineralising" ion-exchange column. The glass apparatus used for the determination should be thoroughly cleaned by standing overnight with a (1 + 1) mixture of concentrated nitric and sulphuric acids. The acid mixture should then be drained, and the apparatus rinsed several times with distilled water.

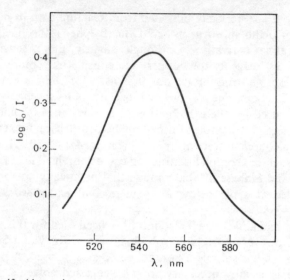

FIG. 43. Absorption spectrum of the copper complex with diquinolyl (2-cm cells, 45 μg Cu/10 ml).

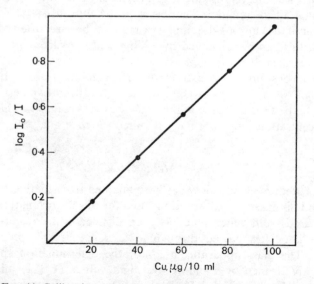

FIG. 44. Calibration graph for copper with diquinolyl (2-cm cells, 540 nm).

Reagents: *Diquinolyl reagent solution,* dissolve 0·03 g of 2,2′-diquinolyl in 100 ml of n-hexanol that has been freshly distilled from solid sodium hydroxide.

Hydroxyammonium chloride solution, dissolve 25 g of analytical grade hydroxyammonium chloride in about 80 ml of water, filter if necessary and dilute to 100 ml with water. If any appreciable amounts of copper are present in the reagent solution, extract with successive 10-ml portions of a 0·01 per cent solution of dithizone in carbon tetrachloride until there is no change in the green colour of the dithizone solution. Then extract the solution with carbon tetrachloride until all colour has been removed from the aqueous solution.

Sodium acetate buffer solution, dissolve 136 g of sodium acetate trihydrate in water and dilute to 1 litre. If the reagent contains more than a trace of copper, purify this solution by extraction with 0·01 per cent dithizone solution in carbon tetrachloride as for the hydroxyammonium chloride solution.

Hydroquinone solution, dissolve 1 g of reagent in 100 ml of redistilled ethanol as required.

Standard copper stock solution, dissolve an accurately weighed 0·1 g portion of pure copper in 3 ml of concentrated nitric acid, add 1 ml of 20 N sulphuric acid and evaporate to fumes of sulphuric acid. Allow to cool, dissolve the residue in distilled water and dilute to 500 ml. This solution contains 200 μg copper per ml.

Standard copper working solution, dilute 5 ml of the stock solution to 250 ml with water. This solution contains 4 μg Cu per ml, and should be used for the calibration of 1-cm spectrophotometer cells. For the calibration of 4-cm cells, dilute 5 ml of the stock solution to 1 litre with water, to give a solution containing 1 μg Cu per ml.

Procedure. Accurately weigh approximately 1 g of the finely powdered rock material into a platinum crucible and add 2 ml of concentrated nitric acid and 15 ml of concentrated hydrofluoric acid. Set the covered crucible aside overnight on a water bath, and then evaporate to dryness. Fuse the dry residue with from 1·5 to 2 g of potassium pyrosulphate at dull red heat for 5 minutes, taking care not to loose any sample material in the early stages of the ignition when excessive effervescence may occur. Dissolve the fused cake by warming on a water bath with 100 ml of water containing 1·5 ml of concentrated hydrochloric acid. When cold, transfer the solution to a 250-ml volumetric flask and dilute to volume with water.

Transfer a 100-ml aliquot containing not more than 80 μg of copper, to a 250-ml separating funnel, and add 2·5 ml of hydroxyammonium chloride solution and 25 ml of sodium acetate buffer solution. Shake with 6 ml of diquinolyl reagent solution for 5 minutes and allow the phases to separate.

Run the lower aqueous layer into another separating funnel, add 2 ml of hydroxyammonium chloride solution and extract again with 2·5 ml of diquinolyl reagent solution. Separate the phases and again extract the aqueous layer with 2 ml of the diquinolyl reagent solution. Combine the three organic extracts in a 10-ml volumetric flask containing 0·5 ml of hydroquinone solution and dilute the solution to volume with n-hexanol. Measure the optical density of the solution in 1- or 4-cm cells with the spectrophotometer set at a wavelength of 540 nm. Measure also the optical density of a reagent blank solution prepared in the same way as the sample solution but omitting the rock material.

Calibration. Use the standard solution containing 4 µg Cu per ml to calibrate the 1-cm cells, and the standard containing 1 µg Cu per ml for the 4-cm cells. Transfer aliquots of 0–25 ml to 250-ml separating funnels, add 1·5 ml of concentrated hydrochloric acid and dilute each solution to 100 ml with water. Add hydroxylamine and buffer solution, extract the copper with diquinolyl reagent solution and measure the optical densities as described above. Plot the relation of optical density to copper concentration for the ranges 0–25 µg Cu (4-cm cells) or 0–100 µg Cu (1-cm cells). A solution containing 25 µg Cu in 10 ml of organic extract should have an optical density of about 0·984 in 4-cm cells, or 0·246 in 1–cm cells.

Note: 1. This method can be used to determine copper in carbonate rocks or minerals as follows. Dissolve 5 g of the carbonate by gradual addition of 30 ml of 4 N nitric acid. If foam tends to rise to the top of the flask it can be broken by the addition of a drop of octyl alcohol. Cautiously evaporate the solution to dryness of a hot plate; if organic matter is present, add 10–15 ml of concentrated nitric acid and repeat the evaporation. Evaporate the residue twice to dryness with 15 ml of concentrated hydrochloric acid to remove the nitric acid. Dissolve the residue in 100 ml of distilled water containing 1·5 ml of concentrated hydrochloric acid and dilute to 250 ml in a volumetric flask. Determine the copper in a 100-ml aliquot of this solution as described above.

Extraction and Determination of Copper with Dithizone

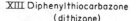

XIII Diphenylthiocarbazone (dithizone)

Diphenylthiocarbazone(XIII), more familiarly known as dithizone, reacts with many metals to give compounds that are red, orange or brown in colour and are soluble in organic solvents. Carbon tetrachloride has commonly been employed as a medium for photometric measurement of these metal compounds, but some workers prefer chloroform for its greater ease of extraction and for the increased solubility of the metallic dithizonates.

The extreme sensitivity of dithizone to copper is of no advantage in the examination of silicate materials. This, coupled with the lack of selectivity and poor stability of the reagent solutions, has led to the preference of many workers for other methods of determinating copper. Rader and Grimaldi,[9] however, describe the use of dithizone for a prior extraction of certain heavy metals including copper, and then complete the determination with either diquinolyl or with dithizone.

The collection of copper, lead and zinc as dithizonates is made by extraction into carbon tetrachloride from a basic citrate solution of pH 9. Lead and zinc are removed from the extract by stripping with dilute acid, leaving the copper dithizonate in the organic, carbon tetrachloride phase. The solution is evaporated to dryness and the copper determined with diquinolyl as described above or with dithizone, following extraction from an aqueous solution at pH 3. Bismuth and other heavy metals that may accompany copper in these extractions are not usually present in silicate rocks in amounts sufficient to interfere.

Method

As with all methods for the determination of heavy metals in trace amounts, the water used for the preparation of solutions should be re-distilled from an all-glass or silica still, or alternatively be passed through an "demineralising" ion exchange column. All items of glass ware must be carefully cleaned before use.

EXTRACTION OF COPPER, LEAD AND ZINC

Reagents: *Dithizone*, dissolve 50 mg of the purified (note 1) reagent in 100 ml of redistilled carbon tetrachloride. Store the solution in a cool, dark cupboard or preferably in a refrigerator.

Carbon tetrachloride, using a pyrex still, re-distil a pure grade of carbon tetrachloride from a little calcium oxide.

Citric acid solution, dissolve 250 g of a pure grade of reagent in about 300 ml of water and dilute to 500 ml.

Procedure. Accurately weigh 0·5–1 g of the finely powdered rock material into a platinum dish. Add 10 ml of concentrated hydrofluoric acid, 5 ml of concentrated nitric acid and 5 ml of perchloric acid. Allow the dish to stand overnight at room temperature, or on a steam bath for about 30 minutes, and then evaporate to fumes of perchloric acid. Cool, rinse down the sides of the dish with a little water and again evaporate to fuming, and this time continue fuming until the volume of acid is reduced to about 1ml. Cool, add 2 ml of 6 N hydrochloric acid and 15 ml of water and then warm to complete the dissolution of all soluble material (note 2).

For acidic and intermediate rocks whose copper contents are not likely to exceed 50 ppm, the whole of this solution should be used for the extraction of heavy metals. The solution obtained from basic rocks should however be transferred to a volumetric flask and a suitable aliquot taken for the succeeding step.

Add 10 ml of the citric acid solution to the sample solution or suitable aliquot of it in a small beaker and, using a pH meter, add ammonia to bring the pH of the solution to 9. If iron or aluminium compounds precipitate add a further quantity of the citric acid solution and again adjust the pH to 9 (note 3). Transfer to a 50-ml separating funnel, add 5 ml of the dithizone solution and shake for 2 minutes. Run off the carbon tetrachloride layer into a second separating funnel. Add a further 5 ml of dithizone solution to the aqueous phase and again shake for 2 minutes and run off the carbon tetrachloride layer as before. Repeat the extractions with further 5-ml portions of dithizone solution until a green-coloured organic layer is obtained. If more than five portions are required, discard the solution and start the extractions again using a smaller aliquot of the sample solution. When the extraction is complete, discard the aqueous layer and combine the organic extracts.

Add 5 ml of dilute aqueous ammonia solution (1 + 99) to the organic solution and shake for 2 minutes. Separate the two layers. Set the organic layer aside, and shake the aqueous layer with 2 ml of dithizone solution. Run off the organic layer and combine this with the carbon tetrachloride extract previously set aside. Discard the aqueous layer. It is important to ensure that at this stage the carbon tetrachloride solution contains no entrained droplets of the aqueous phase.

Add 10 ml of 0·02 N hydrochloric acid to the carbon tetrachloride extract and shake for 2 minutes. If the organic layer remains red, add 1–2 ml of the dithizone solution and shake again. Run off the organic layer into another separating funnel and shake it again with a further 10 ml portion of 0·02 N hydrochloric acid. Any lead and zinc present in the extract will now have been transferred to the aqueous extracts which can be combined and used for the determination of these two elements if required.

Transfer the carbon tetrachloride layer to a small beaker and evaporate to dryness. Add 0·5 ml of concentrated nitric acid and 0·5 ml of concentrated perchloric acid, cover the beaker with a small clock glass and heat on a hot plate until a completely colourless solution is obtained. Cool, rinse down the cover glass and the sides of the beaker and evaporate to complete dryness. Cool, add a few drops of concentrated hydrochloric acid and 10 ml of water, and digest on a steam bath until all soluble material is in solution. Dilute the solution to 50 ml with water in a volumetric flask.

This solution can be used for the determination of copper using diquinolyl as described above, or using dithizone as described below.

DETERMINATION OF COPPER WITH DITHIZONE

Reagents: *Dithizone*, 0.002 per cent w/v, prepare this solution as required by dilution of 0.05 per cent solution in redistilled carbon tetrachloride.

Hydrochloric acid, 0.001 N, prepare by dilution of 0.1 N hydrochloric acid that had been previously standardised by titration with sodium carbonate.

Procedure. Pipette a suitable aliquot of the copper extract containing not more than 5 µg copper into a small beaker and evaporate to dryness on a steam bath. Transfer the beaker to an electric oven set at 120°, and dry thoroughly to complete the expulsion of hydrochloric acid. Allow the beaker to cool, and add by pipette exactly 5 ml of 0.001 N hydrochloric acid and warm to complete the dissolution of all soluble material.

Transfer the solution to a dry separating funnel using a further quantity of exactly 5 ml of 0.001 N hydrochloric acid to rinse the beaker and give a total volume of 10 ml of copper solution. Add, again by pipette, 10 ml of the 0.002 per cent dithizone solution and shake for 2 minutes. If a pure red or red–violet coloured extract is obtained, then the sample aliquot contains too great a quantity of copper and the determination should be repeated using a smaller aliquot.

Filter the carbon tetrachloride solution through a small plug of dry filter paper, rejecting the first 2 ml and then filling a spectrophotometer cell. Measure the optical density relative to the 0.002 per cent dithizone solution. Measure also the optical density of a reagent blank solution prepared at the same time and in the same way as the sample solution.

Calibration. Transfer aliquots of 0–5 ml of the standard solution containing 0–5 µg copper to small beakers and proceed as described above, by evaporating to dryness and drying in an electric oven. It is most important in the calibration as in the determination to ensure that all the acid is expelled before solution in 0.001 N hydrochloric acid, and that the subsequent extraction is made from a volume of exactly 10 ml of solution. Measure the optical densities of the extracts and plot the relation to copper concentration.

Notes: 1. To purify the dithizone proceed as follows. Dissolve about 0.5 g of the reagent in 50 ml of chloroform and filter the solution through a coarse, dry, sintered-glass crucible to remove any insoluble material. Transfer the solution to a 250-ml separating funnel and shake with four successive 50–75-ml portions of diluted (1 + 99) aqueous ammonia solution. Discard the organic layer and combine the ammoniacal extracts. Filter this solution through a small plug of cotton wool to remove any droplets of chloroform.

Add hydrochloric acid drop by drop to the filtrate until it is just acid and the re-agent has precipitated. Add 15 ml of chloroform and shake to extract the dithizone. Separate the organic layer and add further portions of chloroform to the aqueous phase and shake again to complete the extraction.

Combine the chloroform extracts and shake with an equal volume of water. Transfer the chloroform solution to a small beaker and evaporate at a temperature of no more than 50°. Dry the product in a desiccator and store in a cold dark cupboard or preferably in a refrigerator.

2. Any insoluble residue should be collected on a small filter, washed with a little water, dried, ignited and fused with a small quantity of anhydrous sodium carbonate. Disintegrate the melt in a little water, add a slight excess of hydrochloric acid and combine with the main solution.

3. If this determination is to include also that of lead, the steps involving the buffering of the solution to pH 9, and the subsequent extraction of heavy metals at this pH should be carried out as rapidly as possible. Otherwise samples containing significant amounts of calcium and phosphate may give some precipitation of calcium phosphate that may occlude lead and cause low recoveries of this element.

References

1. WAGER L. R. and MITCHELL R. L., *Geochim. Cosmochim. Acta* (1951) **1**, 129.
2. RINGWOOD A. E., *Geochim. Cosmochim. Acta* (1955) **7**, 189.
3. AHRENS L. H., *Physics and Chemistry of the Earth* (1964) **5**, 44.
4. SANDELL E. B. and GOLDICH S. B., *J. Geol.* (1943) **51**, 99 and 167.
5. ALMOND H. and MORRIS H. T., *Econ. Geol.* (1951) **46**, 608.
6. ALMOND H., *U.S. Geol. Surv. Bull.* 1036–A, 1955.
7. RILEY J. P. and SINHASENI P., *Analyst* (1958) **83**, 299.
8. CHARLOT G., *Colorimetric Determination of Elements*, Elsevier, 1964.
9. RADER L. F. and GRIMALDI F. S., *U.S. Geol. Surv. Prof. Paper* 391-A, 1961.
10. MARTENS R. I. and GITHENS R. E. Jr., *Analyt. Chem.* (1952) **24**, 991.
11. BELT C. B. Jr., *Econ. Geol.* (1964) **59**, 240.
12. BRECKENRIDGE J. G., LEWIS R. W. and QUICK L., *Can. J. Res.* (1939) **B17**, 258.
13. GUEST R. J., *Analyt. Chem.* (1953) **25**, 1484.
14. HOSTE J., EECKHOUT J. and GILLIS J., *Anal. Chim. Acta* (1953) **9**, 263.

CHAPTER 22

FLUORINE

Occurrence

An inspection of published analyses suggests that although fluorine is one of the more interesting elements to the petrologist and mineralogist, it is not one that is regularly sought for or determined by the rock analyst. There may have been good reason for this when the only methods of analysis available for fluorine were difficult, tedious, imprecise and insensitive, but there now seems to be no valid reason for excluding fluorine from most so-called "complete analyses".

The fluorine content of magmatic rocks increases with increasing silica content, ranging from 200 to 300 ppm in basic rocks to values of 600 to 1000 ppm in many granites and nepheline syenites. In certain late-stage granites the fluorine content may reach as much as 1 per cent F, whilst pegmatite emplacement may give rise to rocks containing even more fluorine and a number of fluorine minerals—fluorite, amblygonite, topaz, tourmaline and more rarely cryolite, as well as micas and amphiboles rich in fluorine.[1]

The emplacement of carbonatite rocks can also take place under conditions where fluorine is similarly concentrated, giving carbonate rocks rich in fluorine and containing fluorite with fluorine-bearing varieties of mica, amphibole, apatite and pyrochlore. Apatite, mica and amphibole appear to be the hosts for fluorine in normal silicate rocks, as well as tourmaline in the more acidic silicates. Shales contain similar amounts of fluorine to igneous rocks (200–900 ppm), whilst carbonate rocks, excluding carbonatites, contain somewhat less (200–300 ppm).

Fluorine exists as the simple, singly-charged ion F^- in such minerals as fluorite CaF_2, or as a component of complex ions AlF_6^{3-}, BF_4^-, etc., as in cryolite Na_3AlF_6, or avogadrite KBF_4. The close similarity of ionic radius between the fluoride ion ($R = 133$ pm) and the hydroxyl ion

OH^- ($R = 140$ pm), enables a continuous substitution to occur in such minerals as the apatite series:

$$Ca_5 (PO_4)_3OH————Ca_5(PO_4)_3F$$
hydroxyapatite   fluorapatite

The difference in ionic radius between the fluoride ion and those of the remaining halogens (Cl^-, $R = 180$ pm; Br^-, $R = 196$ pm; I^-, $R = 220$ pm) is such as to inhibit substitution of fluorine in chlorine, bromine and iodine minerals and vice versa. Even in the well-known chlorapatite, the chlorine atoms are thought to occupy different atomic positions[2] from those occupied by fluorine in fluorapatite.

The Determination of Fluorine

The oldest reliable method for determining fluorine appears to be that of Berzelius[3] involving the decomposition of the silicate rock by fusion with alkali carbonate, removal of silica and alumina by precipitation with ammonium and zinc carbonates, separation of phosphate and chromate by precipitation with silver nitrate solution, and finally precipitation of the fluorine as calcium fluoride from an acetate solution. Good results could be obtained only by a skilled analyst with the expenditure of much time and patience. However, the method had two considerable advantages: firstly, there was no other reliable method available, and secondly, it was possible to recover silica from the ammonium and zinc carbonate precipitates, giving a method—the only method then available—for the determination of silica in materials containing fluorine. The limit of detection appears to be about $0 \cdot 05$ per cent F.[4]

A considerable improvement was the distillation from sulphuric or perchloric acid solution described by Willard and Winter.[5] Unfortunately the presence of much silica or aluminium hinders the volatilisation of fluorine, so that small amounts of this element are difficult to recover, making the ammonium and zinc carbonate separations essential preliminary stages of the method. The fluorine in the distillate was determined titrimetrically using thorium nitrate as titrant and the pink-coloured zirconium–alizarin lake as indicator. The pink colour is bleached by the fluorine in the distillate, but returns at the end-point of the thorium nitrate titration. The thorium nitrate solution was standardised by titration with known amounts of fluorine. The end-point is poor, necessitating colour-matching and careful lighting when more than a few milligrams of fluorine are present.

Gravimetric and titrimetric procedures based upon precipitation of lead chlorofluoride[6] have found some use in the analysis of samples containing appreciable quantities of fluorite, but are not applicable to normal silicate and carbonate rocks. The useful range is from 1 to about 20 per cent fluorine, for which a titrimetric procedure is given in detail below. A method involving the precipitation of lead bromofluoride has also been described.[7]

One of the earliest colorimetric procedures to be used for fluorine was that of Steiger[8] involving the bleaching action of fluorine on the yellow colour produced by titanium with hydrogen peroxide. The modification of Steiger's method by Merwin[9] gave a procedure more sensitive than Berzelius's gravimetric method, but still "not delicate enough to indicate with certainty the presence of 0·01 per cent of fluorine". It was still necessary to remove silica and aluminium by precipitation with ammonium carbonate.

More recent spectrophotometric methods have been based upon the bleaching action on a number of coloured zirconium and thorium complexes. One such method, devised by Megregian[10] and studied in further detail by Sarma,[11] is given below in the form described by Evans and Sergeant.[12] Other spectrophotometric methods have been based upon the reaction of fluorine with alizarin complexan and either cerous or lanthanum solution.[13, 14] This reaction, the first direct positive colour-forming reaction of the fluoride ion, cannot be used without prior separation from aluminium and iron,[15] although after a preliminary separation with ammonium and zinc carbonates it has been used in the analysis of silicate rocks.[16] The method requires close control of both pH and buffer concentration of the fluoride solution.

When more than about 0·5 per cent of fluorine is present, as in certain pegmatites and mineral concentrates, the technique of pyrohydrolysis can be used with advantage. This is simple and rapid, no prior separations are required, and a relatively simple titrimetric procedure can be used to determine fluorine in the distillate. Details of this procedure are given below.

Also given below is a method for the determination of fluorine in silicate rocks based upon the use of an ion-selective electrode. Elements such as aluminium that form strong complexes with the fluoride ion, tend to interfere with such methods, but in the procedure given, due to Ingram,[17] a high concentration of citrate is used to prevent this interference. These high concentrations tend to delay the equilibration time for the electrode, and due allowance must be made for this.

The Lead Chlorofluoride Method

After a zinc carbonate separation, lead chlorofluoride is precipitated from a solution buffered with sodium acetate, collected, washed and dissolved in dilute nitric acid. The chlorine present is precipitated with an excess of silver nitrate solution, and the amount of the excess determined by titration with standard potassium thiocyanate solution.

Method

Reagents: *Potassium carbonate.*
Lead nitrate.
Sodium acetate trihydrate.
Nitric acid, 1·6 N.
Zinc nitrate solution, dissolve 5 g of zinc oxide in the minimum quantity of 1·6 N nitric acid and dilute to 100 ml with water.
Sodium hydroxide solution, dissolve 10 g in 100 ml of water.
Silver nitrate solution, 0·2 N standard solution.
Potassium thiocyanate solution, approximately 0·2 N, dissolve 20 g of reagent in 1 litre of water.
Sodium chloride solution, dissolve 10 g of the reagent in 100 ml of water.
Lead chlorofluoride wash solution, dissolve 10 g of lead nitrate in 200 ml of water. Mix this with a second solution containing 1 g of sodium fluoride and 2 ml of concentrated hydrochloric acid in 100 ml of water. Wash the precipitate by decantation with 200 ml portions of water five times. Add 1 litre of water to the precipitate, stir occasionally, allow to stand for 1 hour and filter through a close-textured paper before use.
Iron indicator solution, dissolve 14 g of ferric alum in water containing 5 ml of concentrated nitric acid and dilute with water to 100 ml.

Procedure. Weigh an amount of the sample to contain between 0·01 and 0·1 g of fluorine into a clean platinum crucible and add 6 g of potassium carbonate. Fuse the mixture, allow to cool and extract the melt with 100 ml of water in a 250-ml beaker. Boil for 10 minutes and then collect the residue on a close-textured filter paper and the filtrate in a 400-ml beaker. Wash the residue well with hot water before discarding it. Combine the filtrate and washings, heat to boiling and add 20 ml of the zinc nitrate solution. Stir continuously whilst boiling for a period of 1 minute. Collect the precipitate on an open-textured filter paper, and the filtrate in a clean 400-ml beaker. Wash the precipitate well with hot water and discard it.

Combine the filtrate and washings, add a few drops of bromophenol blue indicator solution and 3 ml of the sodium chloride solution and dilute the solution to a volume of about 250 ml. Now add $1 \cdot 6$ N nitric acid until the colour of the indicator just changes to yellow, then restore the blue colour with sodium hydroxide solution added drop by drop. Add 1 ml of concentrated hydrochloric acid followed by 5 g of solid lead nitrate. When the lead salt has completely dissolved, add 5 g of sodium acetate and stir. A white precipitate of lead chlorofluoride indicates the presence of fluorine.

Transfer the beaker to a steam bath and leave for 30 minutes, stirring occasionally, then allow the beaker to stand on the bench for 4 hours or preferably overnight. Collect the precipitate on a fine-textured filter paper and wash it once with cold water, 4 or 5 times with cold saturated lead chlorofluoride wash solution and finally once more with cold water. Place the filter paper with the precipitate into the beaker in which the precipitation was carried out, add 100 ml of $1 \cdot 6$ N nitric acid, stir until the paper is reduced to a pulp and transfer the beaker to a steam bath for 5 minutes.

Add by pipette 25 ml of the standard silver nitrate to the solution, and at the same time a second 25 ml to a separate beaker. Return the beaker with the sample solution to the steam bath for 30 minutes and then allow to cool in the dark. When cold, collect the precipitated silver chloride on an open-textured filter paper and the filtrate in a 500-ml conical flask. Wash the precipitate well with cold water and discard it. Combine the filtrate and washings in the conical flask, add 2 ml of the iron indicator solution and titrate with the ammonium thiocyanate solution. As each drop of solution is added a reddish-brown colour appears that disappears on shaking. Towards the end-point the precipitate flocculates and settles more readily. Finally a single drop of reagent produces a faint brown colour that no longer disappears on shaking. Titrate also the second portion of the silver nitrate solution, acidified with 100 ml of $1 \cdot 6$ N nitric acid and diluted to about 200 ml with water. Calculate the chlorine content of the precipitated chlorofluoride (PbClF), and hence the fluorine content of the sample material.

The Photometric Determination of Fluorine

There are a number of spectrophotometric procedures for fluorine, many of them with a great deal in common. The procedure given here is that described by Evans and Sergeant[12]—the choice has been made largely because a preliminary separation of silica and alumina is not required with the small size of sample portion used. A "Willard and Winter" separation by distillation of hydrofluosilicic acid is used to recover the fluorine from about 0·2 g of rock material. As with other procedures involving distillation, a fluorine recovery of about 95 per cent is achieved. The distillate passes directly to a column of anion exchange resin, which serves to concentrate the fluorine, and from which it is eluted with a small volume of ammonium acetate solution. The final determination is made spectrophotometrically by an eriochrome cyanine R method.

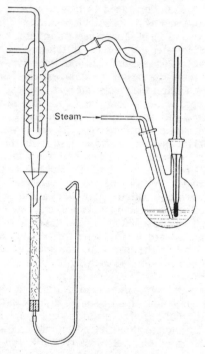

Fig. 45. Apparatus for the separation and recovery of fluorine.

This photometric method is based upon the bleaching action of fluorine on the coloured complex formed between zirconium and eriochrome cyanine R. The conditions for this reaction were studied in some detail by Sarma[11] but have been modified for this particular application. In the range 2–54 μg F per 100 ml of solution, a negative linear calibration is obtained. The reagent blank amounts to about 3 μg F, corresponding to about 15 ppm.

Method

Apparatus. The distillation apparatus and ion-exchange column are shown in Fig. 45. The ion-exchange resin used—De Acidite FF—is converted to the hydroxyl form by treatment with 5 ml of N sodium hydroxide solution, followed by washing free from excess alkali with about 50 ml of water. After use the resin may be regenerated by further treatment in the same way. A column containing about 750 mg of resin is sufficient for about forty determinations before this regeneration becomes necessary.

Reagents: *Sodium acetate solution,* 0·1 M.

Zirconyl chloride solution, dissolve 0·1325 g of zirconyl chloride $ZrOCl_2.8H_2O$ in a little water, add 600 ml of concentrated hydrochloric acid and dilute to 1 litre with water.

Eriochrome cyanine R solution, dissolve 0·8 g of the solid reagent in 1 litre of water. Different batches of this reagent have been found to contain varying amounts of sodium sulphate. This may be removed by dissolving the dye in methanol, filtering and evaporating the solution to dryness under reduced pressure. Store the purified reagent in a desiccator.

Standard fluorine stock solution, dissolve 0·1106 g of sodium fluoride in water and dilute to 500 ml. This solution contains 100 μg F per ml. Store in a polythene bottle.

Standard fluorine working solution, dilute 10 ml of the stock solution to 500 ml with water. This solution contains 2 μg F per ml. Store in a polythene bottle.

Procedure. Accurately weigh 0·2 g of the finely powdered silicate rock into a platinum crucible, add 1 g of anhydrous sodium carbonate and fuse over a Meker burner for 15 to 20 minutes. Allow to cool, add 10 ml of 2 N sulphuric acid slowly and allow the mixture to digest for 30 minutes, then transfer the contents of the crucible to the distilling flask. Rinse the crucible and lid with 45 ml of 18 N sulphuric acid and transfer the acid to the distilling flask. Set up the apparatus as shown in the figure. Pass steam into the flask when the temperature has reached 125°. The distillation temperature should be kept in the range 145–150° with a distillation rate of 6–8 ml per minute. This can be done by appropriate adjustment to the heater and supply of steam. In the original paper a contact thermometer was used to control the heater.

Allow between 300 and 400 ml of distillate to pass through the ion-exchange column, and then elute the absorbed fluorine from the column with 25 ml of 0·1 M sodium acetate solution, collecting the eluate in a 100-ml volumetric flask. Dilute to volume with water.

Pipette an aliquot of this solution containing not more than 50 μg of fluorine into a clean 100-ml volumetric flask and dilute to about 70 ml with water. Add 10 ml of the zirconyl chloride solution, followed by 10 ml of the eriochrome cyanine R solution, dilute to volume with water and mix well. Allow the solution to stand for 30 minutes then measure the optical density in 1-cm cells with a spectrophotometer set at a wavelength of 525 nm. The reference solution is prepared from 70 ml of water, 6 ml of concentrated hydrochloric acid and 10 ml of eriochrome cyanine R solution, followed by dilution to 100 ml as for the sample solution.

Calibration. Transfer 5–25 ml aliquots of the standard fluorine solution containing 10–50 μg F to separate 100-ml volumetric flasks and dilute each solution to 70 ml with water. Add zirconyl chloride and eriochrome cyanine R solutions as described above, dilute to volume with water and measure the optical density against a reference solution as for the sample solution. Plot the relation of optical density to fluorine concentration.

Photometric Determination without Distillation

Huang and Johns[18] also used a spectrophotometric procedure based upon the work of Megregian[10] and Sarma,[11] but by incorporating zinc oxide into the alkaline flux used for sample decomposition, found it possible to dispense entirely with the distillation stage. Iron and phosphate remain entirely with the residue after aqueous extraction. Aluminium is partly extracted, but by making the solution alkaline and allowing the coloured solution to stand for 60 to 90 minutes before photometric measurement, the aluminium content can be tolerated. Sulphates react similarly to fluorine, but the sulphate level in most rocks is too low to give serious interference.

Procedure. Mix 0·5 g of the finely ground rock material, 3·5 g of anhydrous sodium carbonate and 0·6 g of zinc oxide in a large platinum crucible, and fuse in an electric muffle furnace at a temperature of about 900° for 20 to 25 minutes. Allow to cool, add 10 ml of water and 3 drops of ethanol to reduce any manganate formed in the fusion, and heat on a hot plate to disintegrate the melt. Boil for 1 minute and allow to cool. Filter the solution through a close-textured filter paper, collecting the filtrate in a 100-ml polyethylene beaker. Wash the residue five times by decantation with 2-ml portions of hot water and discard it.

Add 4·1 ml of concentrated nitric acid, which should leave the solution faintly acid, and stir to promote the release of most of the carbon dioxide. Transfer the solution to a 50-ml volumetric flask and dilute to volume with water (note 1). Use 5 ml aliquots of this solution for the determination of fluorine. Add 6 drops of 6 M sodium hydroxide solution, mix well and add 3 ml of eriochrome cyanine R solution (note 2) followed by 3 ml of zirconyl chloride solution, dilute to volume with water, mix well and allow to stand for 60 to 90 minutes before photometric measurement.

Notes: 1. This solution can be used also for the determination of chlorine
2. The zirconyl chloride and eriochrome cyanine R solutions should contain twice the weight of reagent as the solutions used by Evans and Sergeant.[12]

Pyrohydrolytic Determination of Fluorine

When fluorine containing minerals are heated in a stream of moist air hydrolysis occurs and the fluorine is evolved—probably as hydrogen fluoride. Complete evolution of fluorine can be achieved by the use of a suitable flux and an accelerator. The flux suggested is a mixture of the oxides of bismuth and vanadium with the addition of tungstic oxide as accelerator. The use of this mixture enables a relatively low temperature of 655–665° to be used. The liberated hydrogen fluoride is collected in sodium hydroxide solution and determined titrimetrically by a cerous-EDTA method.

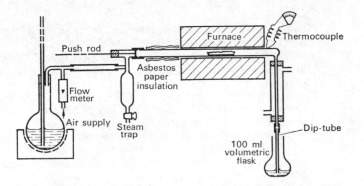

FIG. 46. Apparatus for the recovery of fluorine by pyrohydrolysis.

Method

Apparatus. This is shown in Fig. 46. It consists of a horizontal tube furnace with a chamber 30 cm in length by 4 cm in diameter, together with a reaction tube fabricated from silica with a narrow diameter side tube, water cooled. Silica combustion boats are used, $75 \times 15 \times 12$ mm. One of the commonest causes of incomplete recovery of fluorine is too low a temperature, and the value indicated should be checked from time to time by observing the melting point of silver sulphate (m.p. 652°) in a boat at the hottest part of the furnace.

Reagents: *Vanadium pentoxide.*
            Bismuth trioxide.
            Sodium bismuthate (note 1).
            Tungstic oxide.
            Sodium hydroxide solution, dissolve 50 g of reagent in 1 litre of
                water.

p-Nitrophenol indicator solution, dissolve 0·3 g of the solid indicator in water and dilute to 100 ml.

Chloracetate buffer solution, pH 2·7, dissolve 9·45 g of monochloracetic acid in 50 ml of water. Using a pH meter adjust the pH of the solution to 2·7 by adding sodium hydroxide solution, and then dilute to 100 ml with water.

Xylenol orange indicator solution, dissolve 0·067 g of the solid reagent in a mixture of 25 ml of water with 25 ml of ethanol, dilute to 100 ml with water and filter if necessary. Prepare a fresh solution every few days.

Gelatin, powdered.

Acetate buffer solution, pH 6, dissolve 46 g of ammonium acetate and 18 g of sodium acetate trihydrate in a litre of water. Using a pH meter adjust the pH to a value of 6 by adding glacial acetic acid.

Cerous nitrate 0·05 M *solution*, dissolve 21·7 g of cerous nitrate hexahydrate in water and dilute to 1 litre.

Arsenazo III indicator solution, dissolve 0·02 g of the solid reagent in 100 ml of water containing 2 drops of sodium hydroxide solution.

Pure sodium fluoride, dissolve 2·5 g of "AnalaR" or similar grade sodium carbonate in 25 ml of water in a 50-ml platinum dish. Neutralise by adding carefully "AnalaR" or similar grade hydrofluoric acid (40 per cent) until effervescence ceases, then add 1 ml in excess. Evaporate to dryness on a steam bath, dry in an electric oven at 120°, then ignite to constant weight at 500°.

Ethylenediaminetetraacetic acid, 0·025 M *solution*, dissolve 9·31 g of the disodium salt of EDTA dihydrate in water and dilute to 1 litre. Standardise by weighing out 0·1 g of the pure sodium fluoride, transfer to a 100-ml volumetric flask and dissolve in about 30 ml of water. Add 10 ml of sodium hydroxide solution and follow the procedure given below, beginning with the addition of *p*-nitrophenol indicator.

Procedure. Grind together in an agate mortar 1·5 g of vanadium pentoxide, 0·5 g of bismuth trioxide and 0·2 g of tungstic oxide. Add a weighed portion of the sample material containing from 1 to 30 mg fluorine, mix thoroughly and grind together. Brush the mixture into the silica combustion boat and press down with a spatula.

Check that the furnace temperature is between 655° and 665°. Adjust the air flow to between 100 and 110 litres per hour and the steam rate to condense at about 0·5 ml per minute. Remove the air/steam inlet tube from the reaction tube and wipe the inlet dry. Run 10 ml of sodium hydroxide solution into the volumetric flask and connect to the condenser as shown in Fig. 46. Insert the boat containing the sample into the end of the reaction tube, replace the air/steam inlet and push the boat just inside the furnace. Advance the boat about 4 cm every minute until it is just in the hot zone of the furnace, withdrawing the push rod each time to the cool part of the tube. Allow the pyrohydrolysis to continue for 30 minutes after the furnace has regained its original temperature.

Disconnect the dip-tube and rinse down into the flask using not more than 5 ml of water. Add 1 drop of *p*-nitrophenol indicator solution and neutralise with 5 N nitric acid. Add 0·5 ml of chloracetate buffer and two or three drops of xylenol orange indicator solution. If the colour is red, titrate just to yellow with EDTA solution to complex any traces of metal compounds that may have distilled or been carried over. Add 5–10 mg of powdered gelatine and exactly 20 ml of cerous nitrate solution and warm to about 40° in a water bath for 5 minutes. Add 10 ml of approximately 0·1 N nitric acid, dilute to volume with water and filter through a *dry* open-textured filter paper. Pipette 50 ml of the filtrate into a 250-ml conical flask, add 10 ml of acetate buffer solution and 1–2 ml of arsenazo III indicator solution. Titrate to a red end-point with the EDTA solution.

Carry out a blank determination, beginning with the addition of 10 ml of sodium hydroxide to the volumetric flask.

Notes: 1. In the presence of appreciable amounts of lead or barium it may be necessary to increase the time of pyrohydrolysis to 40 minutes to complete the evolution of fluorine. In these cases, and also for sulphide minerals, 0·3 g of the bismuth trioxide should be replaced with an equal weight of sodium bismuthate. This prevents the tendency for sulphur compounds to distil.

2. 1 ml of 0·05 M cerous nitrate solution is equivalent to 2·85 mg of fluorine. As only one-half of the hydrolysate is titrated, each ml of 0·025 M EDTA titre is equivalent to 2·85 mg fluorine.

The Determination of Fluorine using an Ion-selective Electrode

This procedure is essentially that of Ingram,[17] in which the fluoride ion is determined without separation of aluminium in a citrate buffered solution.

Method

Apparatus: *Ion-selective electrode*, a fluoride-selective electrode, such as the Orion Model No. 94-09.

Reference electrode, a saturated calomel electrode.

pH meter, an expanded-scale pH meter is required for measuring the solution potentials.

Reagents: *Sodium citrate solution*, dissolve 59 g of sodium citrate dihydrate and 20 g of potassium nitrate in water and dilute to 1 litre.

Sodium fluoride standard solution, dissolve 1·105 g of pure sodium fluoride in water and dilute to 500 ml in a volumetric flask. This solution contains 1 mgF per ml, or 1000 ppm.

Procedure. Accurately weigh approximately $0 \cdot 1$ g of the finely powdered silicate rock material into a 10 ml platinum crucible and mix with $0 \cdot 5$ g of anhydrous sodium carbonate and $0 \cdot 1$ g of zinc oxide. Ignite at a temperature of 900° for 30 minutes and allow to cool. Place the crucible in a 50-ml beaker, add 30 ml of water, cover and allow to digest overnight on a steam or water bath.

Rinse and remove the crucible, break up any lumps of solid material with the aid of a glass rod flattened at the end and allow the solution to cool to room temperature. Filter into a 100-ml volumetric flask using a close-textured filter paper such as a Whatman No. 42, wash the paper and residue several times with small quantities of a $0 \cdot 1$ per cent solution of sodium carbonate and discard the residue.

Add to the filtrate slowly and with care, 2 ml of 6 N hydrochloric acid, shaking vigorously and often to remove carbon dioxide. Dilute to volume with water, and shake to mix well. Pipette 10 ml of this solution into a 50-ml beaker and measure the potential of the solution after the addition of 10 ml of the sodium citrate solution and standing for the appropriate length of time (notes 1 and 2).

Notes: 1. Ingram recommended the use of two electrodes, one for solutions containing $0 \cdot 02$ to $0 \cdot 4$ ppm F, and the other for determining the range of the fluoride ion concentration as well as for measuring in the range $0 \cdot 4$ to 4 ppm F. If the fluoride ion concentration exceeds 4 ppm, the solution should be diluted to bring it into this range.

2. In the range $0 \cdot 02$ to $0 \cdot 4$ ppm F, measure after 5 minutes; in the range $0 \cdot 4$ to 4 ppm F after 10 minutes. The standards appropriate to each range should also be read after these time intervals.

3. Standards are prepared from the standard sodium fluoride stock solution to cover the ranges $0 \cdot 02$ to $0 \cdot 4$ and $0 \cdot 4$ to 4 ppm F, each to contain similar quantities of sodium carbonate, hydrochloric acid and sodium citrate solution to the rock solution prepared for measurement.

References

1. RANKAMA K. and SAHAMA TH. G., *Geochemistry*, Univ. Chicago Press, 1949, p. 759.
2. GOLDSCHMIDT V. M., *Geochemistry*, Oxford, 1954, p. 572.
3. BERZELIUS J. J., reported in HILLEBRAND W. F. (ref. 4).
4. HILLEBRAND W. F., The Analysis of Silicate and Carbonate Rocks, *U.S. Geol. Surv. Bull.* 700, p. 225, 1919.
5. WILLARD H. H. and WINTER O. B., *Ind. Eng. Chem., Anal. Ed.* (1933) **5**, 7.
6. HOFFMAN J. I. and LUNDELL G. E. F., *Bur. Stds. J. Res.* (1929) **3**, 581.
7. CHEBURKOVA E. E., *Zavod. Lab.* (1950) **16**, 1009.
8. STEIGER G., *J. Amer. Chem. Soc.* (1908) **30**, 219.
9. MERWIN H. E., *Amer. J. Sci.*, 4th Ser. (1908) **28**, 119.
10. MEGREGIAN S., *Analyt. Chem.* (1954) **26**, 1161.
11. SARMA P. L., *Analyt. Chem.* (1964) **36**, 1684.

12. EVANS W. H. and SERGEANT G. A., *Analyst* (1967) **92**, 690.
13. BELCHER R., LEONARD M. A. and WEST T. S., *J. Chem. Soc.* (1959) 2577.
14. BELCHER R., LEONARD M. A. and WEST T. S., *Talanta* (1959) **2**, 92.
15. JEFFERY P. G. and WILLIAMS D., *Analyst* (1961) **86**, 590.
16. JEFFERY P. G., *Geochim. Cosmochim. Acta* (1962) **26**, 1355.
17. INGRAM B. L., *Analyt. Chem.* (1970) **42**, 1825.
18. HUANG W. H. and JOHNS W. D., *Anal. Chim. Acta* (1967) **37**, 508.

CHAPTER 23

GALLIUM

Occurrence

Gallium occurs to the extent of a few parts per million in most silicate rocks and minerals, where, as a result of similar ionic radii and ionisation potentials, it tends to be camouflaged by aluminium (Ga^{3+}, $R = 62$ pm; Al^{3+}, $R = 51$ pm, Ahrens[1]). Using a fluorimetric method Sandell[2] estimated the abundance of gallium in the upper lithosphere to be 15 ppm, a value in good agreement with the spectrographic results of Goldschmidt and Peters[3] and the spectrophotometric determinations of Burton, Culkin and Riley.[4]

Olivine and pyroxene rocks, deposited early in the magmatic sequence, are poor in both aluminium and gallium, both Brunfelt et al.[5] and De Laeter[6] reporting less than 1 ppm Ga in PCC-1 (peridotite) and DTS-1 (dunite). Both aluminium and gallium occur to a greater extent in later crystallates—granite pegmatites may contain up to about 60 ppm. Specimens of muscovite, phlogopite and lepidolite examined by Burton et al.[4] were found to contain more than 100 ppm.

Carbonate rocks and minerals, evaporites and sulphide minerals contain only small amounts of gallium, although Burton et al.[4] reported one unusual sample of sphalerite, containing 111 ppm Ga.

The Determination of Gallium

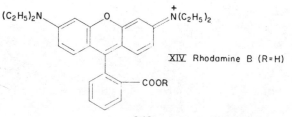

XIV. Rhodamine B (R=H)

248

The most frequently used reagent for the determination of gallium is rhodamine B (XIV, $R =$ H). Kuznetzov and Bol'shakova[7] have,

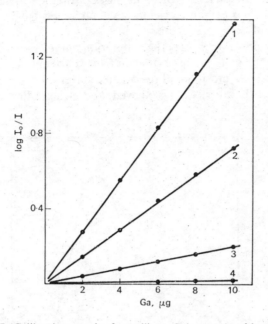

FIG. 47. Calibration graphs for gallium. Solvents: 1—chlorobenzene plus carbon tetrachloride and acetone, 2—benzene plus acetone, 3—chlorobenzene plus carbon tetrachloride, 4—benzene.

however, noted the unfavourable effect of the carboxyl group upon the solubility of the rhodamine complex and recommended the use of the butyl ester ($R =$ n-butyl) under the name butylrhodamine. This latter reagent was used by Skrebkova[8] for the determination of gallium in unroasted lead dust, bauxites, copper–zinc ores and quartz–topaz greisens.

In the procedures based upon rhodamine B, the red-coloured compound formed with the chlorogallate anion is extracted into organic solution prior to measurement of the optical density. Onishi and Sandell[9] have shown that the most favourable extraction into benzene is achieved from a 6 N hydrochloric acid solution containing 0·036 per cent rhodamine B. Under these conditions the extraction coefficient

Ga$_{benzene}$/Ga$_{water}$ has a value of 0·57 at about 25°. This extraction can be increased by adding acetone to the solution as described by Kuznetsov and Tananaev,[10] by replacing the benzene with a mixture of chlorobenzene and carbon tetrachloride (Culkin and Riley[11]), or by a combination of both these methods.

These four procedures gave the calibration curves shown in Fig. 47. In each case the gallium was extracted from a volume of 25 ml of 6 N hydrochloric acid, into 10 ml of the organic solvent. At these concentrations the Beer–Lambert Law is obeyed, and straight-line calibrations were obtained.

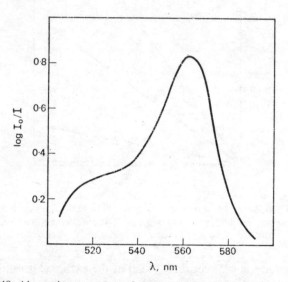

FIG. 48. Absorption spectrum of rhodamine B chlorogallate (6 μg Ga, with chlorobenzene plus carbon tetrachloride and acetone).

The composition of the rhodamine B chlorogallate has been investigated by Culkin and Riley[12] and shown to be (RH) GaCl$_4$, where R = rhodamine B. The absorption spectrum of the chlorogallate in chlorobenzene–carbon tetrachloride solution is shown in Fig. 48. The maximum absorption is at a wavelength of 562 nm.

Under the conditions used to extract the chlorogallate, coloured organic extracts are also given by rhodamine B with antimony(III), gold(III), thallium(III) and iron(III). Kuznetsov and Tananaev[10] reduced all these elements to lower valency states with titanous chloride before extracting gallium. Onishi and Sandell[7] separated gallium from these interfering elements by a prior extraction into di-isopropyl ether. This separation was used also by Burton et al.[4] following the extraction of germanium with carbon tetrachloride, in a procedure for the determination of these two elements in the same sample portion. The procedure below describes the separation with di-isopropyl ether, followed by the extraction of the rhodamine B chlorogallate into chlorobenzene–carbon tetrachloride in the presence of acetone.

Method

Reagents: *Titanous chloride solution*, approximately 15 per cent.

Di-isopropyl ether, freshly distilled from solid sodium hydroxide, b.p. 68°. As with diethyl ether, peroxide formation can give unstable, explosive mixtures.

Rhodamine B, 0·5 g dissolved in 100 ml of water.

Acetone.

Chlorobenzene–carbon tetrachloride solvent, add 125 ml of carbon tetrachloride to 375 ml of monochlorobenzene.

Standard gallium stock solution, dissolve 0·5 g pure gallium in 50 ml of 6 N hydrochloric acid. Transfer to a 1-litre volumetric flask, add 60 ml of concentrated hydrochloric acid and dilute to volume with water. This solution contains 0·5 mg Ga per ml in N hydrochloric acid.

Standard gallium working solution, prepare a dilute solution containing 1 μg Ga ml by serial dilution of the stock solution with N hydrochloric acid.

Procedure. Accurately weigh about 0·2 g of the finely powdered rock material into a small platinum crucible and add 2 ml of 20 N sulphuric acid, 0·5 ml of concentrated nitric acid, and 5 ml of hydrofluoric acid. Heat the crucible on a hot plate or sand bath until copious fumes of sulphuric acid are evolved. Allow to cool, rinse down the sides of the crucible with a little water and again evaporate, this time to dryness. Dissolve the residue in approximately 25 ml of 6 N hydrochloric acid and transfer the solution to a 100-ml separation funnel.

Add a 15 per cent titanous chloride solution drop by drop until the yellow colour of the ferric ion is completely discharged, and then add a few drops in excess. Add 10 ml of di-isopropyl ether and shake for 20 to 30 seconds. Run the aqueous phase into a second separating funnel and shake again with 5 ml of di-isopropyl ether. Repeat the extraction with a further 5 ml portion of ether, discard the aqueous layer and combine the organic extracts in a 50-ml beaker.

Evaporate the ether in a fume cupboard and add 15 ml of 6 N hydrochloric acid to the dry residue. Warm gently and then transfer the solution to a 100-ml flat-bottomed flask. Rinse the beaker with a little 6 N hydrochloric acid, add the washings to the flask and dilute to a volume of 25 ml, also with 6 N hydrochloric acid. Add titanous chloride solution drop by drop until a faint violet colour persists, indicating that an excess is present, and allow the solution to stand for 10 minutes. Add 2 ml of the aqueous rhodamine B solution followed by 5 ml of acetone, mix well and allow to stand for a further period of one hour. Add by pipette 10 ml of the chlorobenzene–carbon tetrachloride solvent and shake vigorously to extract the chlorogallate into the organic phase.

Allow the phases to separate. Transfer the lower, organic layer to a 1-cm spectrophotometer cell and measure the optical density at a wavelength of 562 nm. Measure also the optical density of a reagent blank solution prepared in the same way as the rock solution but omitting the sample material.

Calibration. Transfer 0–10 ml aliquots of the standard gallium solution containing 0–10 μg gallium to separate 100-ml flat-bottomed flasks and dilute each to 25 ml with 6 N hydrochloric acid. Extract the gallium as chlorogallate using the procedure described above and measure the optical densities in the same way. Plot the relation of optical density to gallium concentration (Fig. 48.)

Note: 1. Where reagent grade solvents were used, the chlorogallate extracts were found to fade appreciably. This was avoided by allowing the solvent mixture to stand over concentrated sulphuric acid for a few days, separating the organic layer, shaking with aqueous sodium hydroxide, and finally shaking with water.

Determination of Gallium by Atomic Absorption

This technique is not sufficiently sensitive for the determination of the small quantities of gallium that occur in most silicate and other rocks unless a large sample is taken, and a preliminary concentration stage introduced. Such a procedure was described by Lypka and Chow[13] for the determination of gallium in ores.

Four- to five-gram portions of the ore samples were evaporated with sulphuric and hydrofluoric acids and the excess sulphuric acid removed by fuming. The residues were dissolved by heating with 7 M hydrochloric acid and the gallium extracted into purified isopropylether after titanium(III) chloride reduction. After evaporation of the organic solvent, the residues were dissolved in 7 M hydrochloric acid and nebulised into a lean air–acetylene flame for measurement.

References

1. AHRENS L. H., *Geochim. Cosmochim. Acta* (1952) **2**, 155.
2. SANDELL E. B., *Amer. J. Sci.* (1949) **247**, 40.

3. GOLDSCHMIDT V. M. and PETERS C., *Nachr. Ges. Wiss., Göttingen, Maths. Phys. Kl.* III (1931) **31**, 141.
4. BURTON J. D., CULKIN F. and RILEY J. P., *Geochim. Cosmochim. Acta* (1959) **16**, 151.
5. BRUNFELT A. O., JOHANSEN O. and STEINNES E., *Anal. Chim. Acta* (1967) **37**, 172.
6. DE LAETER, J. R., *Geochim. Cosmochim. Acta* (1972) **36**, 735.
7. KUZNETZOV V. I. and BOL'SHAKOVA L. I., *Zhur. Anal. Khim.* (1960) **15**, 523.
8. SKREBKOVA L. M., *Zhur. Anal. Khim.* (1961) **16**, 422.
9. ONISHI H. and SANDELL E. B., *Anal. Chim. Acta* (1955) **13**, 159.
10. KUZNETSOV V. K. and TANANAEV N. A., *Ivz. Vyssh. Uchebn. Zavedenii, Khim. i Khim. Tekhnol.* (1959) **2**, 840.
11. CULKIN F. and RILEY J. P., *Analyst* (1958) **83**, 208.
12. CULKIN F. and RILEY J. P., *Anal. Chim. Acta* (1961) **24**, 413.
13. LYPKA G. N. and CHOW A., *Anal. Chim. Acta* (1972) **60**, 65.

CHAPTER 24

GERMANIUM

Occurrence

There is very little difference in the germanium content of the various types of igneous rock, most silicates containing 1–2 ppm. Onishi[1] has reported the following averages for silicate rocks mostly from Japan.

TABLE 23. THE GERMANIUM CONTENT
OF SOME IGNEOUS ROCKS

Rock type	Ge, ppm
Granitic rocks	1·3
Intermediate rocks	1·5
Basalts and diabases	1·3
Gabbros	1·2
Ultramafics	1·0

Similar although slightly higher results were reported by Burton, Culkin and Riley[2] and slightly lower results for igneous rocks mostly from the United States of America by El Wardani.[3] The germanium content of shales and similar sedimentary rocks is comparable to that of igneous silicate rocks.[4] Carbonate rocks contain very little germanium. Sulphide minerals also contain about 1 ppm germanium, although Burton *et al.*[2] have reported 116 ppm from one specimen of sphalerite from Derbyshire, and Brewer, Cox and Morris[5] up to 670 ppm in sphalerites from a number of localities, mostly in Great Britain. The concentration of germanium in coal of particular seams has been noted by a number of workers, values of up to 1 per cent germanium being reported in the ash.

There are a number of very rare minerals containing germanium, such as germanite a copper–germanium–arsenic sulphide and argyrodite said to be Ag$_8$GeS$_6$, neither of which is likely to be met with by the rock analyst in an examination of silicate rocks. Many silicate minerals have been reported as carrying up to about 100 ppm germanium, possibly as Ge^{4+} ($R = 44$ pm) substituting for silicon Si^{4+} ($R = 39$ pm).

The Determination of Germanium

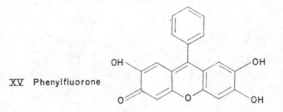

XV Phenylfluorone

Interest in the properties of germanium has resulted in a considerable number of papers concerned with its determination. A number of reagents have been described for spectrophotometric methods, of which the most popular is phenylfluorone(XV) (9-phenyl-2,3,7-trihydroxy-6-fluorone or 2,3,7-trihydroxy-9-phenylxanthen-6-one). This reagent, introduced by Gillis, Hoste and Claeys[6] as a spot test reagent for germanium, has been studied by Cluley,[7, 8] for the determination of germanium in the presence of a wide variety of other elements. The complex formed by germanium with phenylfluorone has a maximum optical density at a wavelength of 504 nm (Fig. 49), and the Beer–Lambert Law is obeyed at the concentrations encountered in silicate rocks (Fig. 50).

Although a recent paper[9] suggests that prior separation of the germanium is not necessary, most authors have recommended either distilling or extracting germanium from hydrochloric acid as a means of separation from other elements. The distillation serves also to concentrate the germanium in a small volume, free from all elements except some of the sulphur, arsenic, antimony and tin present in the original rock. In the amounts in which they are normally present in silicate rocks, these elements do not interfere.

The extraction procedure, described by Schneider and Sandell,[10] is also usually applied from a hydrochloric acid solution. For the amounts of germanium normally encountered, a single extraction with an equal volume of carbon tetrachloride was regarded as sufficient. More recently, however, Burton *et al.*[2] suggest that only 85–90 per cent of the germanium is extracted in a single operation, and recommend three successive extractions. The addition of EDTA has been recommended[11] as a means of preventing the extraction of other elements into the carbon tetrachloride solution.

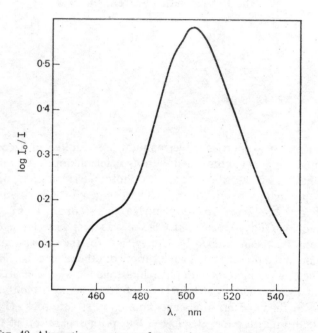

FIG. 49. Absorption spectrum of germanium–phenylfluorone complex (1-cm cells, 5 μg Ge/10 ml).

Silicate rocks and minerals can be decomposed by evaporation with a mixture of sulphuric, nitric and hydrofluoric acids. After the removal of excess sulphuric acid the residue can be dissolved in hydrochloric acid. Hybbinette and Sandell[12] have shown that this method of attack can be used for samples containing even as much as 0·05 per cent chlorine, without loss of germanium by volatilisation of germanium tetrachloride. This is presumably due to the formation of the involatile fluogermanic acid.

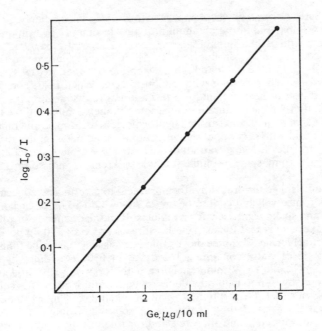

FIG. 50. Calibration graph for germanium using phenylfluorone (1-cm cells, 504 nm).

Method

Reagents: *Carbon tetrachloride.*

Ethylenediaminetetraacetic acid solution, dissolve 1 g of the disodium salt in water and dilute to 200 ml.

Gum arabic solution, dissolve 0·1 g in about 80 ml of boiling water, filter and dilute to 100 ml with water.

Phenylfluorone solution, dissolve 0·05 g of the solid reagent in 75 ml of ethanol and 5 ml of 5 N sulphuric acid by warming on a steam bath. When cold, dilute to 100 ml with water.

Standard germanium stock solution, weigh 0·036 g of pure germanium oxide into a small beaker, add 0·5 g of sodium hydroxide and 20 ml of water. When solution is complete transfer to a 250-ml volumetric flask and dilute to volume with water. This solution contains 100 μg Ge per ml.

Standard germanium working solution, dilute 10 ml of the stock solution to 1 litre with water as required. This solution contains 1 μg Ge per ml.

Procedure. Accurately weigh approximately 0·5 g of the finely powdered silicate rock material into a platinum dish and decompose by evaporation to fumes with 3 ml of 20 N sulphuric acid, 0·5 ml of concentrated nitric acid and 5 ml of concentrated hydrofluoric acid. Allow to cool, rinse down the sides of the dish with a little water and again evaporate to fumes. In each case avoid prolonged or strong fuming with sulphuric acid. When cold, add 5 ml of water and heat to near boiling for a few minutes to disintegrate the residue. Cool and transfer the mixture to a separating funnel, completing the transfer with the aid of 5 ml of concentrated hydrochloric acid. Stopper the funnel and allow to stand with occasional shaking until all solid material has dissolved, or for 30 minutes if it does not all dissolve. If the sample material is rich in calcium or barium, some insoluble sulphate will remain undissolved, but this does not interfere.

Add sufficient concentrated hydrochloric acid to give an acid concentration of 9 molar, mix well and cool to below 25°. Now add 10 ml of carbon tetrachloride and shake vigorously for 2 minutes. Run the organic layer off into a small beaker and repeat the extraction with two successive 10 ml portions of carbon tetrachloride. Combine the organic extracts in a separating funnel, add *exactly* 5 ml of water containing 1–2 drops of 0·01 M sodium hydroxide solution and shake for 2 minutes. Discard the organic phase and transfer the aqueous solution to a 10-ml volumetric flask. Add 1 ml of 10 N sulphuric acid, 0·4 ml of EDTA solution, 2·4 ml of ethanol, 0·4 ml of gum arabic solution and 0·6 ml of the phenylfluorone solution. Dilute the solution to volume with water, mix and allow to stand for 1 hour if the temperature is 25° or below. Measure the optical density of the solution relative to water in 1-cm cells, with the spectrophotometer set at a wavelength of 504 nm. If the temperature is above 25°, place the flask in a water bath at about 20° for 30 minutes, then allow the flask to stand at room temperature for a further 30 minutes before measurement. Measure also the optical density of a reagent blank solution prepared in the same way as the sample solution, but omitting the rock material.

Calibration. Transfer 1–5 ml aliquots of the standard germanium solution containing 1–5 μg Ge to a series of 10-ml volumetric flasks. Add to each 10 N sulphuric acid, EDTA solution, ethanol, gum arabic and phenylfluorone solution as described above. Dilute each solution to volume with water, allow to stand and measure the optical density at 504 nm in the same way as the sample solution. Plot the relation of optical density to germanium concentration.

References

1. ONISHI H., *Bull. Chem. Soc. Japan* (1956) **29**, 686.
2. BURTON J. D., CULKIN F. and RILEY J. P., *Geochim. Cosmochim. Acta* (1959) **16**, 151.
3. EL WARDANI S. A., *Geochim. Cosmochim. Acta* (1957) **13**, 5.

4. HEIDE F. and KOERNER D., *Chem. Erde* (1963) **23**, 104.
5. BREWER F. M., COX J. D. and MORRIS D. F. C., *Geochim. Cosmochim. Acta* (1955) **8**, 131.
6. GILLIS J., HOSTE J. and CLAEYS A., *Anal. Chim. Acta* (1947) **1**, 302.
7. CLULEY H. J., *Analyst* (1951) **76**, 523.
8. CLULEY H. J., *Ibid.* (1951) **76**, 530.
9. DEKHTRIKYAN S. A., *Izv. Akad. Nauk Armyan. S.S.R., Nauka Zemle* (1966) **19**, 97.
10. SCHNEIDER W. A. and SANDELL E. B., *Mikrochim. Acta* (1954) 263.
11. BURTON J. D. and RILEY J. P., *Mikrochim. Acta* (1959) 586.
12. HYBBINETTE A. G. and SANDELL E. B., *Ind. Eng. Chem., Anal. Ed.* (1942) **14**, 715.

CHAPTER 25

HYDROGEN

Occurrence

Hydrogen occurs in a number of roles in silicate rocks, including free or elemental hydrogen, organically bound hydrogen, adsorbed moisture, water of crystallisation, "combined water", and as hydroxyl groups present within the lattice structure of a number of minerals. It is not always possible to differentiate between some of these forms, particularly where some of the silicate minerals have been altered and are represented by newer, more hydrous species.

Free Hydrogen

Although silicate rocks release considerable volumes of hydrogen on ignition, it is by no means certain that this gas is always present in a free or elemental form. Jeffery and Kipping[1] have noted that for the small number of igneous rocks that they examined, the amounts of hydrogen liberated were only roughly reproducible, and were in each case less than the amounts that could have been produced by a reduction of the water present (Table 24).

Tokhtuyev and Frantsuzova[2] have demonstrated that free or elemental hydrogen occurs in metamorphic rocks, by recovering the gases evolved during the powdering of specimens under vacuum, without heating—a technique developed earlier by Elinson.[3] The presence of hydrogen can possibly be ascribed to the formation of magnetite in the process of metamorphism:

$$3FeO + H_2O = Fe_3O_4 + H_2$$

although this is considered to be unlikely, as hydrogen can be recovered from silicates containing little or no magnetite. Shorokhov[4] has also reported the wide occurrence of free hydrogen in sedimentary rocks.

260

TABLE 24. HYDROGEN GAS PRODUCED ON IGNITION OF SOME IGNEOUS ROCKS

Sample material		Hydrogen content		
		found		calculated*
		% by vol.	% by wt.	% by wt.
R 117	granite, Shetland	55·0	0·0019	0·005
		65·0	0·0022	
R 138	granite, Cornwall	720	0·025	0·030
		700	0·024	
G-1	granite, Westerly, R.I.	122	0·0042	0·014
		125	0·0043	
		127	0·0045	
W-1	diabase, Centerville, Va.	260	0·0087	0·122
		210	0·0068	

* Based on FeO content, assuming hydrogen to be produced by the reaction:

$$2FeO + H_2O = Fe_2O_3 + H_2$$

The methods commonly used to determine the free hydrogen content of silicate rocks rely upon either heating the samples, or upon comminution under vacuum. The gaseous mixtures obtained by both these processes contain gases other than hydrogen, and some form of gas analyser is therefore required. This can take the form of a traditional constant pressure or constant volume apparatus ("Haldane" or "Bone and Wheeler" apparatus), an apparatus for the conversion of hydrogen to water with subsequent absorption and weighing, or a more instrumental approach such as that based upon gas chromatography.[5]

Organic Hydrogen

Organic material is an essential component of many silicate rocks, particularly those of sedimentary origin where up to 5 per cent "carbonaceous matter" has commonly been reported. Organic solvents can often be used to extract a small proportion of this, usually 5–15 per cent, which in appearance and composition is related to petroleum and can be analysed as such. A further portion of organic matter known as "humic acids", can be extracted with aqueous alkali. The remaining part known as insoluble organic material or "kerogen" can be recovered by acid digestion (hydrochloric and hydrofluoric to remove carbonates and silicates) as suggested by Forsman and Hunt,[6] and described more fully on p. 188. The insoluble organic material is converted to carbon dioxide and water by combustion in a stream of oxygen, and is finally absorbed and weighed.

Water Evolved at 105°

"Hygroscopic water", "moisture", "water evolved at 105°", "non-essential water" are all terms used to describe the loss in weight experienced by a powdered rock specimen on heating to constant weight at 105°. The results of this determination are used to give a measure of the "free" as distinct from the "combined" water. In this respect it is an unsatisfactory measure in that it includes a certain amount of the essential water of some easily decomposed minerals (such as zeolites) whilst at the same time not including some of the non-essential water of many other minerals. This determination is, however, always included in the list of thirteen constituents accepted as the minimum required for a full analysis of any silicate rock.

It is the custom in some laboratories to remove this "moisture" or "hygroscopic water", by drying the ground rock powder at a temperature in the region 105–110°, and to report the analysis as on, for example, "material dried at 105°". This procedure should not be used where the presence of much chlorite would indicate that the dried material is likely to be extremely hygroscopic. The day-to-day variation of the weight of such samples is often considerable, even for samples that have not previously been dried. This is shown in Fig. 51; the weight changes of a chloritised basalt were as much as 0·73 per cent of the initial weight of the material. In contrast, a granite, ground and exposed to the atmosphere at the same time as the altered basalt, changed weight by only 0·03 per cent during the same period. Where such chloritised rocks are being examined it is advisable to weigh out at one time all the portions required for "hydroscopic water", total water and the major constituents.

The water content of the silicate rock sample will vary with the extent of grinding of the sample. Hillebrand,[7] for example, shows that for a number of rocks excessive grinding results in an increase of the water content of the sample, and that in the presence of certain minerals, gypsum for example, the reverse may occur, and water be lost by over-grinding.

The simplest way of determining "water evolved at 105°", is to measure the loss in weight of about 2 g of the powdered sample when heated in an electric oven for 2 to 3 hours or even longer, until a constant weight is obtained. The former practice of using toluene baths now appears to be obsolete. Certain rocks containing much ferrous iron may not reach constant weight, due to the slow continual oxidation

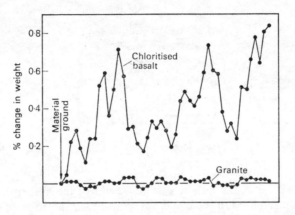

FIG. 51. Weight changes of two ground silicate rocks on exposure to air.

that takes place at this temperature. Many rocks are hygroscopic when completely dry; for this reason a squat weighing bottle with a closely fitting lid, as in Fig. 52, is commonly used for the determination.

The difficulties associated with the weighing of a very hygroscopic material and with the presence of much easily-oxidised ferrous iron can be avoided by making a direct determination of the water evolved. An apparatus for this, described by Jeffery and Wilson,[8] is shown in Fig. 53. It consists of a closed-circulation system comprising a small electric pump, a heating chamber maintained at a temperature of 104–105° by means of boiling isobutyl alcohol (a thermostatically controlled heating block set at 105° may be more convenient), two absorption tubes and a bubbler containing orthophosphoric acid to indicate the rate of flow of air through the apparatus. Before inserting the sample, tube A is replaced by a short piece of glass tubing, and the air in the system is dried by operating the pump for about 30 minutes. Tube A is weighed and replaced, the sample boat inserted, and the circulation of air continued until all the water evolved from the sample has passed into the absorption tube. In practice the apparatus is kept running for about 2 hours, which is sufficient to drive over all the water, including any that may condense on the cooler parts of the heating chamber.

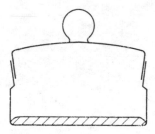

FIG. 52. Apparatus for determining loss in weight by heating at 105°.

The blank value, i.e. the increase in weight of the absorption tube A when the determination is carried out without the insertion of a sample, is less than $0 \cdot 1$ mg. Tube A, which is used in place of a more conventional U-tube, contains a paper spiral that prevents clogging of the magnesium perchlorate when large amounts of water are absorbed. The open end of the tube, fitted with a B7 ground glass joint, is closed with a cap during weighing. When fully packed, the tube and cap together weigh 20–25 g.

Method of Dean and Stark. Shales and certain altered igneous rocks that contain several per cent of moisture, often lose part of this water content in the process of sample preparation. For these materials and for clays that cannot be prepared for chemical analysis without drying, an approximate moisture content can be obtained by the method of Dean and Stark,[9] in which the roughly crushed or broken material, 20–50 g in weight, is boiled with toluene under reflux apparatus (Fig. 55). Moisture evolved at 105° collects in the side arm, which is calibrated directly in millilitres.

Total Water

The methods used for determining total water rely upon the direct collection and weighing of all the water than can be expelled from the sample by heating to a high temperature. One of the earliest methods was that of Penfield[10] in which the sample, contained in a bulb blown at the end of a hard-glass tube, was heated at a temperature just below fusion point of the glass. The water released condensed on the cooler parts of the tube, from which the heated part could be drawn away. The water content of the sample was then obtained by weighing the cooler part before and after drying. The method suffers from the difficulty of

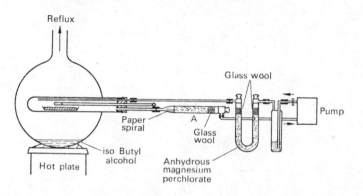

FIG. 53. Apparatus for the direct determination of water evolved by heating at 105°.

ensuring that all of the water does condense, particularly under conditions of low humidity, and also of ensuring that all the water present in the sample had been liberated at the particular temperature used.

The Penfield method does have the advantage of being both simple and rapid, and for this reason a number of modifications to it have been described, notably those of Courville,[11] Shapiro and Brannock[12] and Harvey.[13]

Courville–Penfield method.[11] This method is close to the original Penfield method, in that after heating the sample, the containing bulb is removed and the remaining part of the glass tube is weighed before and

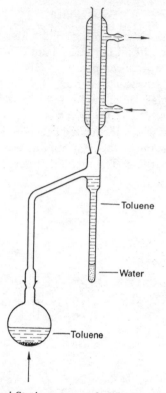

FIG. 54. Dean and Stark apparatus for the determination of water.

after drying. In order to increase the efficiency of liberation of water from the sample, it is mixed with lead oxide (litharge) before transfer to the ignition tube. The mixture fuses on heating and the tube can then be gently rotated. Additional precautions are necessary when certain volatiles, such as fluorine, are present in the sample.

A similar method for use on a micro scale was described earlier by Sandell,[14] 20–30 mg of sample were used with litharge if fluorine or other volatile matter was present, or ignited lime if sulphur trioxide or hydrogen chloride were likely to be evolved on heating.

Shapiro and Brannock–Penfield method.[12] This method has been used as part of a scheme for the rapid, complete analysis of silicate rocks. The sample is heated with sodium tungstate in a pyrex glass boiling tube, and the water evolved is collected in a weighed strip of filter paper. Not all the water is retained by the filter paper, and an empirical correction is necessary; this is in the form of an addition of 10 per cent of the recorded weight of water if less than 20 mg are obtained, or 2 mg if more than 20 mg are obtained.

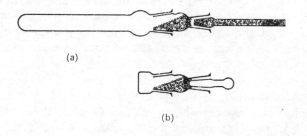

(a)

(b)

FIG. 55. Wilson's apparatus for determining the total water content of silicate rocks, (a) assembled for use, (b) capped and plugged for weighing.

Harvey–Penfield method.[13] In this method the sample is contained in a silica tube, and the water that is evolved on heating is absorbed in anhydrous calcium chloride contained in a tube attached to the heating chamber. Improvements to this early procedure of Harvey were described by Wilson,[15] who used anhydrous magnesium perchlorate and the apparatus shown in Fig. 55, in which the ground glass joints were lubricated with PTFE sleeves. These sleeves permit the use of higher working temperatures. A heating period of 1 hour was used, although 10 minutes were stated to be sufficient in some cases. Before and after the ignition, the absorption tube is sealed with a cap and plug for weighing.

The Penfield method, and the modifications described, suffer from the disadvantage that many minerals are not completely decomposed at the temperature used for the ignition, and that further quantities of water can be obtained by heating these minerals to a higher temperature. Minerals that behave in this way include talc, topaz, staurolite, cordierite and epidote. Other disadvantages of the original method and some of the modifications are that the collection of the water evolved is not always complete, and that other volatile constituents (sulphur and fluorine have been noted) are sometimes collected and weighed as water. The alternative to the Penfield method is to collect the liberated water in an absorption tube that can be weighed before and after the ignition. The sample material, contained in an alumina or platinum boat, can be heated to a temperature of 1000° in a silica tube furnace heated by either gas burners or electricity, and the water evolved can be collected using magnesium perchlorate or other suitable absorbent. This method, described in detail by Groves,[16] can incorporate provision for retaining gases and vapours such as fluorine, sulphur and oxides of sulphur. Using a platinum boat and a flux, such as sodium tungstate, even refractory minerals such as staurolite can be decomposed in a reasonable time. Sodium carbonate and metafluoborate have also been used as fluxes in this way. Blank values are always rather high when fluxes are used, and are by no means negligible with simple ignition.

In an effort to reduce these blank values Jeffery and Wilson[8] used the "closed-circulation system" as a modification to both the ignition method as described by Groves, and a simple fusion using a mixture of equal weights of sodium tungstate and borax glass. The blank value for either determination should be no more than $0 \cdot 1$ mg if the air is recycled within the apparatus as described. If the air is drawn through the apparatus as described by Groves,[16] a blank value of several milligrams will probably be obtained when the apparatus is first set up, but this should fall to a constant value of about $0 \cdot 5$–1 mg after several days operation.

Procedure

Apparatus. This consists of a silica tube containing the sample in an alumina or unglazed porcelain boat that had previously been ignited to 1000°. A current of air from the small circulating electric pump is passed into an absorption tube containing magnesium perchlorate and soda asbestos, over the sample, through a basic lead chromate packing and then to a weighed absorption tube containing anhydrous magnesium perchlorate. The circulation is completed through a bubbler containing orthophosphoric acid, which indicates the rate of flow of air through the apparatus, and finally back to the pump. A packing of basic lead chromate, retained in place by copper spirals, and kept at a temperature of 300–400°, serves to retain oxides of sulphur.[17] Copper wire and silver pumice at a temperature of 700–750° have also been used for this purpose. The wide end of the silica tube is closed with a silicone rubber bung.

The absorption tube is replaced with a short length of glass tubing, and the air within the apparatus dried by operating the pump for 1 hour. During this time the packing heater is allowed to reach its operating temperature. The tube is then weighed and replaced. The weighed sample, contained in an inert boat is inserted into the furnace, which is then allowed to reach its maximum temperature, and is maintained at this temperature for 1 hour. The tube A is then removed from the apparatus, allowed to stand in a balance case for 45 minutes and is then reweighed to give the weight of total water released.

The chief disadvantage of the methods similar to that described by Groves is that no more than three determinations can be completed in any one day, as it is necessary to allow the furnace tube to cool before the next sample can be inserted. In order to increase this number Riley[18] has described an apparatus (Fig. 31), in which the sample is pushed gradually into the hot zone of a furnace maintained at a sufficiently high temperature to release all the water within about 20 minutes. This enables twelve determinations to be made in a single working day. For most rocks a temperature of 1100° was found to be sufficient, but for refractory minerals, topaz, epidote, staurolite, etc., 1200° was required. At this higher temperature it was found possible to determine also any carbon dioxide present in the material by including an absorption tube packed with soda–asbestos in the absorption train. Interference from sulphur compounds was prevented by including a layer of silver pumice and a bubbler containing a saturated solution of chromium trioxide in orthophosphoric acid in the train. A current of nitrogen gas, obtained from a gas cylinder is used, and blank values of $0 \cdot 1$ mg per hour of water and $0 \cdot 2$ mg per hour of carbon dioxide were reported. Details of the procedure using this apparatus for the determination of carbon dioxide and water are given on p. 180.

Gooch–Sergeant method. The problems associated with the releasing of water from refractory materials do not arise when using a technique based upon fusion with anhydrous sodium carbonate, due originally to Gooch.[19] In this method, the use of an alkaline flux ensures the retention of both sulphur and fluorine. Sergeant[20] has reported the adaptation in which the rock sample is heated with the flux in a platinum or other refractory metal crucible supported inside a silica vessel using a high-frequency induction furnace. Nitrogen is used as carrier gas and sodium chlorate is added to avoid possible loss of water by interaction with any ferrous iron present in the silicate rock.

"Loss on Ignition"

On heating a ground rock sample to a temperature of 1000° or more a number of changes occur. Hydrogen and other gases are evolved, adsorbed water and water of crystallisation are removed together with some or all of the combined water of the minerals present. Carbonate minerals are more or less completely decomposed with evolution of carbon dioxide, some but by no means all of the alkalis present are vaporised, loss of fluorine and of sulphur may occur, and some of the ferrous iron and sulphur will be oxidised. The arithmetic total of the weight changes involved is described as "loss on ignition". In rare cases this does give a measure of the total water content of the material, but it should not be regarded as a substitute for a water determination. Its measure is of little importance in rock analysis.

References

1. JEFFERY P. G. and KIPPING P. J., *Analyst* (1963) **88**, 266.
2. TOKHTUYEV G. V. and FRANTSUZOVA T. A., *Geokhimiya* (1963) 961.
3. ELINSON M. M., *Izv. Acad. Nauk. S.S.S.R., otd. teknich, nauk.* No. 2, 1949.
4. SHOROKHOVA N. R., *Trudy Soyzn. Geologporsk. Kontura Gluagaza pri Sov. Min. U.S.S.R.* 1960, 64.
5. JEFFERY P. G. and KIPPING P. J., *Gas Analysis by Gas Chromatography*, Pergamon, Oxford, 1964.
6. FORSMAN J. P. and HUNT J. M., *Geochim. Cosmochim. Acta* (1958) **15**, 170.
7. HILLEBRAND W. F., *U.S. Geol. Surv. Bull.* 700, 1919.
8. JEFFERY P. G. and WILSON A. D., *Analyst* (1960) **85**, 749.
9. As in, e.g., B.S.S. 756:1952.
10. PENFIELD S. L., *Am. J. Sci.*, 3rd Ser. (1894) **48**, 31.
11. COURVILLE S., *Canad. Mineral.* (1962) **7**, 326.
12. SHAPIRO L. and BRANNOCK W. W., *Analyt. Chem.* (1955) **27**, 560.
13. HARVEY C. O., *Bull. Geol. Surv. Gt. Brit.* (1939) (1), 8.
14. SANDELL E. B., *Mikrochim. Acta* (1951) **38**, 487.
15. WILSON A. D., *Analyst* (1962) **87**, 598.
16. GROVES A. W., *Silicate Analysis*, Allen & Unwin, 2nd ed., 1951, p. 95.
17. ROODE R. DE, *Amer. Chem. J.* (1890) **12**, 226.
18. RILEY J. P., *Analyst* (1958) **83**, 42.
19. GOOCH F. A., *Amer. Chem. J.* (1880) **2**, 247.
20. SERGEANT G. A., *Rept. Govt. Chem. Lond.* (1971) p. 111.

CHAPTER 26

INDIUM

Occurrence

Very few analysts have reported indium in igneous rocks, and the abundance values quoted for silicate rocks by workers using material from various localities are not in good agreement, Table 25.

TABLE 25. THE INDIUM CONTENT OF SOME SILICATE ROCKS
In, g/ton (ppm)

Granitic	Intermediate	Basaltic	Gabbroic	Ultramafic	Ref.
0·05			*		1
0·12					2
2·0			0·3		3
0·26		0·22	0·015	0·013	4
0·026 (G-1)		0·064 (W-1)			5
(0·09)†			0·054–0·18		6
0·07			0·12	0·33	7
		mafic rocks			
0·17	0·05	0·07		0·04	8

* "Too low to measure."   † "Acid granophyre or microgranite, 75% SiO_2."

The technique of neutron activation[5] has proved to be particularly valuable for indium, for which the required degree of sensitivity is difficult to obtain by either emission spectrography or spectrophotometry.

A spectrophotometric procedure based upon the use of large samples was used by Rozbianskaya, and described briefly by Ivanov.[8] A 5-g portion of the silicate material was dissolved in a mixture of sulphuric and hydrofluoric acids, and the insoluble material treated by fusion with potassium bisulphate. Iron, titanium and other elements including

271

indium were precipitated with ammonia, and after solution in 5 N
hydrobromic acid the indium was recovered by extraction into ether. The
indium was returned to the aqueous phase by back-extraction into 6 N
hydrochloric acid containing hydrogen peroxide. The complex with
rhodamine C was then formed in 2 N hydrobromic acid containing
ascorbic acid, and extracted into a 3:1 mixture of benzene and ether.
The determinations were completed by a visual comparison with
similar extracts prepared from a standard indium solution. For indium
contents exceeding 2 μg per ml, the solutions were viewed in transmitted
light, for contents less than 2 μg per ml the solutions were examined
under ultraviolet light and the fluorescence compared.

References

1. BOROVIK S. A., PROKOPENKO N. M. and POKROVSKAYA T. L., *Doklady Akad. Nauk S.S.S.R.* (1939) **25**, 620.
2. VINOGRADOV A. P., *Geokhimiya* (1956) (1), 6.
3. PREUSE F., *Zeit. angew. Mineral.* (1940) **3**, 8.
4. SHAW D. M., *Geochim. Cosmochim. Acta* (1952) **2**, 185.
5. SMALES A. A., SMIT J. VAN R. and IRVING H. M., *Analyst* (1957) **82**, 539.
6. WAGER L. R., SMIT J. VAN R. and IRVING H. M., *Geochim. Cosmochim. Acta* (1958) **13**, 81.
7. BROOKS R. R. and AHRENS L. H., *Geochim. Cosmochim. Acta* (1961) **23**, 145.
8. IVANOV V. V., *Geokhimiya* (1963) (12), 1101.

CHAPTER 27

IRON

Occurrence

Iron is one of the commonest of elements, occurring to an extent of about 5 per cent by weight of the Earth's crust, and fourth in the list of abundances after oxygen, silicon and aluminium. Native iron, although rare, has been reported from a number of localities. Other iron-containing minerals (Fig. 56) are widely distributed. Silicate minerals vary considerably in iron content, and this variation is reflected in the iron content of silicate rocks. Hortonolite and other similar basic rocks may contain 35–40 per cent iron (calculated as FeO), whilst many acidic rocks contain no more than 1 per cent total ferric and ferrous oxides.

In the analysis of silicate rocks, both ferrous and ferric iron are reported, usually expressed as oxides. Ferric iron is frequently associated with aluminium, and ferrous iron with magnesium. This latter association is observed in the formation of minerals with composition and properties intermediate between the pure iron and pure magnesian end members. The simplest series formed in this way is that of the olivine orthosilicates, whose end members are

$$Mg_2SiO_4 \quad \text{and} \quad Fe_2SiO_4$$
forsterite $\qquad$ fayalite

Common olivine is usually represented as $(Mg,Fe)_2SiO_4$. Other series of rock-forming minerals where this type of continuous substitution of one element for another occurs, include the pyroxenes where iron- and magnesium-substitution gives the enstatite–hypersthene series, and calcium-, iron- and magnesium-substitution gives the diopside–hedenbergite series. Substitution also occurs extensively in the amphibole and mica mineral series. As Mg^{2+} has a somewhat smaller ionic radius than Fe^{2+} (Mg^{2+}, $R = 66$ pm; Fe^{2+}, $R = 74$ pm), the first minerals to crystallise from a rock magma will be the magnesium-rich variety; iron will appear to an increasing extent in the later crystallates.

273

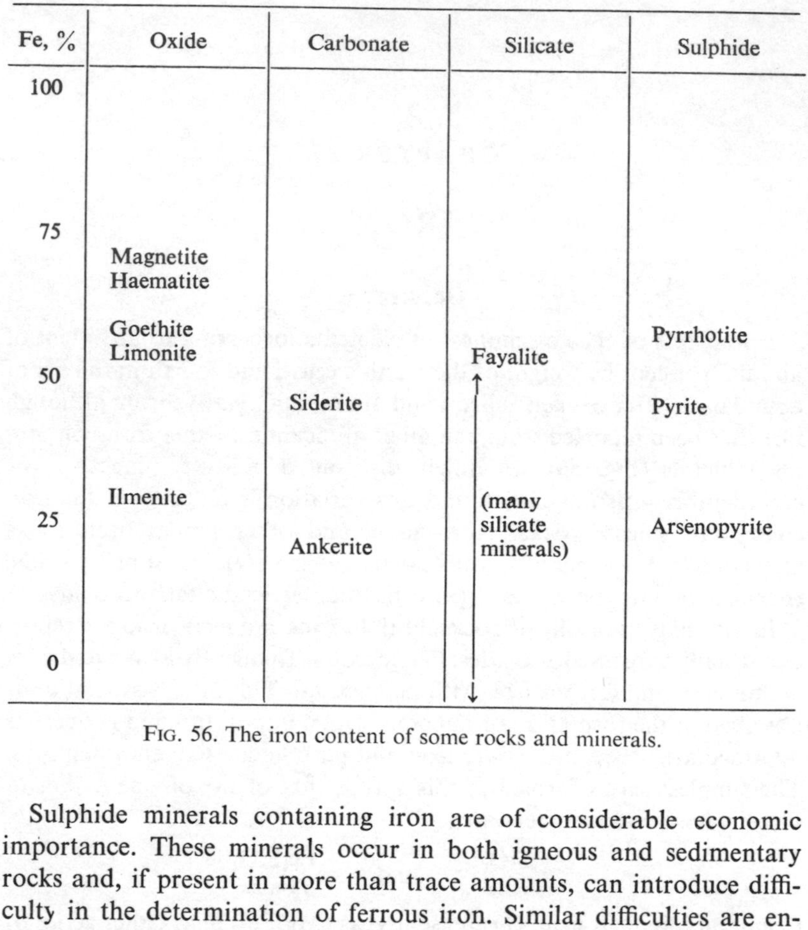

Fe, %	Oxide	Carbonate	Silicate	Sulphide
100				
75	Magnetite Haematite			
50	Goethite Limonite	Siderite	Fayalite ↑	Pyrrhotite Pyrite
25	Ilmenite	Ankerite	(many silicate minerals)	Arsenopyrite
0			↓	

Fig. 56. The iron content of some rocks and minerals.

Sulphide minerals containing iron are of considerable economic importance. These minerals occur in both igneous and sedimentary rocks and, if present in more than trace amounts, can introduce difficulty in the determination of ferrous iron. Similar difficulties are encountered in the presence of carbonaceous material—particularly common in mudstones, marls and shales and in the presence of native or tramp iron.

Metallic Iron

Native iron occurs in both massive form and as disseminated grains, as in the basalt from the Giant's Causeway, Ireland. The method of analysis given below has been adapted from that described by Easton and Lovering[1] for the metallic phase of chondritic meteorites. It can be used for silicate rocks provided that the sample material has been ground sufficiently fine to liberate all the metallic phase—usually to pass a 200-mesh sieve.

The procedure is based upon an extraction of the metal by heating the sample with an aqueous solution of mercuric and ammonium chlorides as proposed by Friedheim.[2] In the scheme of analysis suggested by Easton and Lovering, an anion-exchange resin is used to separate iron, cobalt and nickel ions from each other and from the excess mercuric salt. This method of separation has been retained for silicate rocks, but the elution of nickel and cobalt has been omitted.

Method

Apparatus. *Anion-exchange column*, with a resin bed 2 cm diameter and 12 cm in length, packed with Dowex 1 × 4, 100–150 mesh (or other similar strongly basic anion exchange) resin. Before use, wash this column with 100 ml of 6 N hydrochloric acid, 150 ml of 0·5 N hydrochloric acid and again with 100 ml of 6 N hydrochloric acid.

Reagents: *Mercuric chloride.*
Ammonium chloride.
Hydroxyammonium chloride solution, dissolve 10 g of the reagent in 100 ml of water.
1,10-Phenanthroline solution, dissolve 0·2 g of the solid reagent in 200 ml of water.
Sodium citrate solution, dissolve 10 g of the dihydrate in 100 ml of water.
Standard iron stock solution, dissolve 0·100 g of pure iron wire in 10 ml of 6 N hydrochloric acid and dilute to 500 ml with water. This solution contains 0·2 mg Fe per ml.
Standard iron working solution, dilute 25 ml of the stock iron solution to 250 ml with water. This solution contains 20 μg Fe per ml.

Procedure. Accurately weigh about 0·5 g of the finely powdered (200-mesh) sample material into a stoppered 100-ml volumetric flask, add 0·6 g of mercuric chloride, 0·6 g of ammonium chloride and about 70 ml of warm water. Swirl the flask to ensure thorough mixing, wrap in foil to exclude light and then stand on a heated water bath for 4 to 5 days. Collect the residue on an open-textured filter paper and wash several times with water to ensure the complete recovery of all soluble salts.

Evaporate the combined filtrate and washings to dryness on a steam bath, and dissolve the residue in the minimum quantity of 6 N hydrochloric acid.

Transfer this solution to the anion exchange column and elute with three column volumes of 6 N acid. Discard the eluate. Elute the iron from the column with four column volumes of 0·5 M hydrochloric acid. The excess mercuric chloride is retained on the anion exchange resin, which can then be discarded. Transfer the solution containing the eluted iron to a 250-ml volumetric flask and dilute to volume with water.

Transfer an aliquot of this solution, containing not more than 500 μg of iron, to a 100-ml volumetric flask, dilute to a volume of about 25 ml, add 5 ml of hydroxyammonium chloride solution and allow to stand for a period of 5 minutes. Now add 10 ml of 1,10-phenanthroline solution and 10 ml of sodium citrate solution and dilute to volume with water. Mix well, allow to stand for 2 hours and measure the optical density in a 1-cm cell with a spectrophotometer set at a wavelength of 508 nm. Use the calibration graph to obtain the concentration of iron in the solution and hence calculate the metallic iron content of the sample.

Calibration. Transfer aliquots of 0–25 ml of the standard iron solution containing 0–500 μg of iron, to separate 100-ml volumetric flasks and dilute each solution to 25 ml with water. Add hydroxyammonium chloride, 1,10-phenanthroline and sodium citrate solutions and measure the optical densities as described above. Plot the relation of optical density to concentration of metallic iron. A straight-line graph should be obtained.

Sulphate Iron

The minerals pyrite and marcasite, both FeS_2, are of common occurrence. Where deposits containing these minerals include a "zone of oxidation", the mineral copperas or melanterite, $FeSO_4.7H_2O$ may be found. It is an unstable mineral, losing water and oxidising on exposure to dry air to form a basic ferric sulphate. The total sulphate present in the friable, extensively altered rocks containing copperas and in the mineral concentrates obtained from them, can readily be extracted into water containing a few drops of diluted sulphuric acid, and the iron determined by titration with potassium dichromate solution (p. 297). Carbonate minerals do not survive in the acid environment under which copperas is produced.

Carbonate Iron

Siderite or chalybite, $FeCO_3$, ankerite (a mixed carbonate of iron, calcium and magnesium), and other carbonate minerals containing iron, dissolve only slowly in dilute hydrochloric acid at room temperature but more readily on heating. The iron that passes into solution can be determined by titration with potassium dichromate solution (p. 297). Other iron minerals present may be partially or completely decomposed, and some iron may be leached from certain silicate minerals during this dissolution. A mineralogical examination of the rock specimen may indicate whether or not the iron result obtained using an acid dissolution procedure is an acceptable measure of carbonate iron.

Sulphide Iron

The difficulty of determining ferrous iron in the presence of pyrite and pyrrhotite is discussed in greater detail in the following section. The same difficulties exist when attempts are made to determine the iron present in combination with sulphur, namely that some of the sulphide will be oxidised during the grinding of the rock and that sulphur and hydrogen sulphide liberated in the course of attacking sulphide minerals will affect the valency state of iron derived from other minerals. Once the iron is obtained in solution there is no means of differentiating between the iron derived from sulphide and that obtained from other minerals. In addition, sulphide iron occurs in a number of minerals that vary considerably in composition and in resistance to decomposition.

Once again a mineralogical examination may assist in the interpretation of results or indicate a suitable method of analysis. Many rocks contain only small amounts of pyrite as the predominant sulphur-containing mineral. For these an approximate value of the sulphide iron can be obtained by determining the nitric acid-soluble sulphur content of the rock.

A direct determination of sulphide iron in rocks containing pyrite based upon the resistance to decomposition with hydrofluoric and sulphuric acids has been described by Trusov.[3] Any residual ferrous or ferric iron (as may occur in the minerals chromite or magnetite) will be counted as sulphide iron and give rise to positive errors. These may be offset to some extent by the negative error arising from the slight oxidation of the mineral pyrite. An adaptation of this method is given below. It is not applicable to rocks containing the mineral pyrrhotite which is soluble in the acid solution used.

Procedure. Decompose a $0 \cdot 5$ or 1 g portion of the not too finely ground rock sample by evaporation with 10 ml of 20 N sulphuric acid and 10 ml of hydrofluoric acid in a platinum or PTFE dish. After cooling pour the mixture into 80 ml of saturated boric acid solution. Collect the insoluble residue containing any pyrite originally present in the rock, on an open-textured filter paper, and wash well with water until the washings no longer give a colour reaction with ammonium thiocyanate solution. Discard the filtrate and washings.

Transfer the filter and residue to a silica crucible and ignite at a temperature of about 800°. Fuse the residue with the minimum quantity of potassium pyrosulphate and dissolve the melt in dilute sulphuric acid. A clear solution should be obtained, although occasionally a small crystalline precipitate of insoluble sulphates can be observed at the bottom of the beaker. Cool the solution, transfer to a volumetric flask and dilute to volume with water. Determine the iron present in this solution by one of the procedures given below and report the sulphide iron as FeS_2.

Ferrous Iron

The ease with which ferrous iron is converted into the ferric state by atmospheric oxidation is well known and, during the course of the preparation of the rock sample and at every subsequent stage of the determination, care must be taken to ensure that such oxidation is kept to the minimum. Hillebrand has shown that partial oxidation of the ferrous iron in the rock may occur during the grinding of the sample, and for that reason recommended that samples rich in ferrous iron should be ground only until they passed a 70-mesh sieve. The introduction of mechanical mortars has increased the tendency to over-grind rock samples, and frequent sievings are necessary if serious oxidation is to be avoided. In special cases the grinding may be completed under absolute alcohol.

The presence of an appreciable amount of "acid decomposable sulphide" invalidates the ferrous iron determination, although the approximate figures can often be quoted. Pyrite is not appreciably attacked by the mixture of hydrofluoric acid and sulphuric acid used to decompose silicate rocks, but other sulphides such as pyrrhotite are more extensively decomposed, liberating hydrogen sulphide which results in the reduction of some of the ferric iron of the rock. It is possible to correct the ferrous and ferric iron figures of the rock for the pyrrhotite content but such corrections are seldom valid as some sulphur is usually lost and the attack of the mineral is often not complete.

Other constituents of silicate rocks that affect the ferrous iron determination are manganese dioxide which occurs in some sedimentary rocks, vanadium which is oxidised to V(V) during the titration (a correction can be applied if the V(III) content of the rock is known), and organic matter. Graphite itself does not appear to have any effect upon the ferrous iron determination, but other forms of organic matter may completely invalidate the determination.

The method commonly used for the determination of ferrous iron is known in some laboratories as the "Pratt method". Although the form in use differs somewhat from that originally described by Pratt,[4] the principle of the method is unchanged. A "Pratt crucible" is used for the

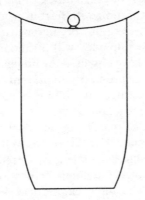

FIG. 57. Platinum crucible used for the determination of ferrous iron ("Pratt crucible").

determination, a platinum or platinum–iridium alloy crucible of approximately 80-ml capacity as shown in Fig. 57.

The rock sample is decomposed in this crucible using a mixture of hydrofluoric and sulphuric acids at the boiling point. The crucible and contents are plunged into boric acid solution and the ferrous iron liberated during the acid decomposition titrated with an oxidising agent. Originally potassium permanganate was used, but in the presence of hydrofluoric acid a very poor and transient end-point is obtained. A more satisfactory end point is obtained with potassium dichromate using barium diphenylamine sulphonate as indicator.

Improvements to the Pratt method have included a number of devices to exclude air from the sample during the decomposition—such as the copper apparatus described by Harris,[5] and the lead box described by Treadwell,[6] in which atmospheres of carbon dioxide are maintained.

In a modification by Lo-Sun Jen,[7] the platinum crucible ("Pratt crucible") is replaced by a 400-ml PTFE beaker with a well-fitting lid. The boric and phosphoric acids are added to the beaker after the decomposition of the rock material is complete.

An alternative procedure for the titrimetric determination of ferrous iron in silicate minerals and rocks is that due to Wilson[8, 9] who described the decomposition of the rock material at room temperature in the presence of an excess of vanadate solution. The excess of oxidising agent is determined by titration with ferrous iron. Vanadium as V(III) in the rock does not interfere but organic matter, manganese dioxide and "acid-decomposable sulphide" will invalidate the determination.

Reichen and Fahey[10] have also used the presence of an added oxidant —in this case potassium dichromate—to prevent the atmospheric oxidation of ferrous iron liberated by sulphuric and hydrofluoric acids. However, there is some reaction between the dichromate and hydrofluoric acid, the extent of which appears to be proportional to the amount of excess dichromate. For this reason Wilson's procedure based upon the use of ammonium vanadate is preferred.

When refractory minerals containing ferrous iron—such as chromite or staurolite—are present in the rock, the acid decomposition must be replaced by a fusion procedure such as that described by Rowledge,[11] involving fusion with sodium fluoride and boric oxide in a sealed glass tube. Modifications to this have been described by Vincent,[12] Groves[13] and Meyrowitz.[14]

Colorimetric procedures for the determination of ferrous iron in small sample weights have been described by Shapiro[15] and Wilson[9] based upon 1,10-phenanthroline and 2,2'-dipyridyl respectively. Wilson's method involves the use of highly dangerous beryllium solutions to complex fluoride ions; the procedure described below is based upon that of Shapiro, who used boric acid for this purpose.

The difficulties of the Pratt method were ventilated by French and Adams[16] who recommended the use of wide-mouthed, screw-capped polypropylene bottles and a small domestic pressure cooker for the sample decomposition. These authors also recommended the use of a mixture of hot sulphuric and hydrofluoric acids. Although this may be ideal for some rocks, it produces a too-violent reaction with others and should therefore be used with caution. Similarly the practice of using beryllium solutions to complex fluoride ions is hazardous. Such operations should be carried out only with strict observance of the code of practice relating to the handling, care and disposal of beryllium-containing solutions.

THE "PRATT METHOD"

Reagents: *Boric acid, saturated solution.*

Barium diphenylamine sulphonate solution, dissolve 0·15 g of reagent in 50 ml of water.

Potassium dichromate standard solution, dissolve 3·268 g of pure dry potassium dichromate in 2 litres of water. For most purposes this solution can be considered as standard N/30, but for accurate work it should be standardised by titration against an iron solution prepared from pure iron wire.

Procedure. Accurately weigh 0·5 g of the finely powdered rock material into a Pratt platinum crucible, moisten with a little recently boiled distilled water and add 10 ml of 20 N sulphuric acid followed by freshly boiled distilled water until the crucible is approximately half full. Cover the crucible with its platinum lid. Support the crucible on a triangle over a low flame protected from draughts and rapidly bring the contents of the crucible to the boiling point. Displace the lid slightly and add 10 ml of concentrated hydrofluoric acid, replace the lid and rapidly bring to the boil. Continue to boil for 7–10 minutes. During this boiling a constant jet of steam should issue from beneath the lid of the crucible, ensuring the complete exclusion of atmospheric oxygen. If the sample is appreciably decomposed by sulphuric acid alone, then the crucible should be filled with carbon dioxide from a small generator before the addition of the acid.

Remove the flame and plunge the crucible and lid below the surface of a boric acid solution prepared by adding 30 ml of saturated boric acid solution to 150 ml of cold, recently boiled distilled water. Rinse the crucible and lid and remove them. Add 10 ml of syrupy phosphoric acid and 5 drops of barium diphenylamine sulphonate solution, and titrate immediately the ferrous iron with potassium dichromate solution. After completing the titration, allow the beaker to stand and then decant off the aqueous solution. If any gritty or dark-coloured particles remain, indicating an incomplete attack of the rock powder, they should be collected and subjected again to the procedure described. If an appreciable quantity of the rock material remains unattacked after prolonged boiling, the rock powder should be ground more finely before repeating the decomposition.

The indicator does not change colour in the absence of ferrous iron, and a zero "blank" determination will not therefore give an end-point. A small quantity of a standard ferrous ammonium sulphate solution should therefore be added to the "blank" solution before titrating, and the appropriate correction applied to the volume of potassium dichromate solution used.

THE "WILSON METHOD"

Apparatus. A polyethylene vessel of 75–100 ml capacity, with tightly fitting lid, as shown in Fig. 58 (the vessel may also be made of polypropylene or polycarbonate).

Reagents: *Ammonium vanadate solution,* dissolve 5 g of ammonium metavanadate and 2 g of sodium hydroxide in 100 ml of water.

Procedure. Accurately weigh approximately 0·5 g of the finely powdered rock material into the polyethylene vessel, add by pipette 2 ml of the vanadate solution and with a measuring cylinder 10 ml of concentrated hydrofluoric acid. Close the vessel with its tightly fitting cap and set it aside. Allow the mixture to stand until the decomposition of the rock powder is complete, as indicated by the absence of gritty particles. These should not be confused with the white fluoride precipitate that usually separates on standing. For most rocks the decomposition will be complete after standing overnight or at the most over two nights. In rare cases a more prolonged digestion may be required.

When the decomposition is complete, add 30 ml of 10 N sulphuric acid to the vessel and rinse the contents into a 800-ml beaker containing 250 ml of saturated boric acid solution. Add 5 drops of barium diphenylamine sulphonate solution, stir until the fluoride precipitate has largely dissolved and titrate the solution with a standard ferrous solution. Titrate also a "reagent blank"

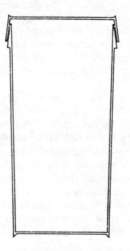

Fig. 58. Polyethylene vessel used for the determination of ferrous iron by Wilson's method.

similarly prepared but omitting the rock sample, and also a 2 ml aliquot of the ammonium vanadate solution. This quantity of ammonium vanadate

(0·1 g) is sufficient for rocks containing up to 12 per cent ferrous iron and should be increased or decreased if the rock material contains much or little ferrous iron.

PHOTOMETRIC DETERMINATION OF FERROUS IRON

In the titrimetric procedure devised by Wilson and described above, the ferrous iron liberated on dissolving the silicate rock in hydrofluoric acid is immediately removed from the solution by reaction with ammonium vanadate. An alternative method of removing the free ferrous iron is by combining with a suitable chelating reagent. 1,10-Phenanthroline, 2,2′-dipyridyl or any of a number of similar reagents can be used for this, with the added advantage that the extent of ferrous chelation (and hence of ferrous iron present in the rock sample) can be observed by measuring the optical density of the coloured solution. The procedure described below is based upon that of Shapiro,[15] who prefers this procedure for rocks containing oxidisable material such as organic material or sulphide minerals for which reliable results are not otherwise obtained.

On decomposing most rocks a cloudy solution is obtained due to the presence of insoluble fluorides and sulphates. The extent to which this cloudiness obscures the light path of the spectrophotometer is determined by measuring the optical density at a second, higher wavelength. The value obtained is subtracted from the optical density recorded at the lower wavelength.

Under the steaming conditions used the ferrous-phenanthroline complex is not completely stable, the colour decreasing by about 1 per cent for 3 minutes of standing. Too short a steaming period may result in incomplete attack of the sample, too long a period in a greatly reduced sensitivity. A 30-minute period was selected as a reasonable compromise. The error introduced by the bleaching action is limited by using a silicate rock of known ferrous iron content to calibrate the procedure.

Method

Apparatus. Polythene bottles, about 25-ml capacity (about 1 oz).

Reagents: *1, 10-Phenanthroline*, powdered solid reagent.
Boric acid, dissolve 5 g of the reagent in 100 ml of hot water and allow to cool.

Sodium citrate solution, dissolve 50 g of the dihydrate in 500 ml of water and filter if necessary.

Reference sample of silicate rock, choose a silicate rock of known ferrous iron content, similar in composition to the samples being examined.

Procedure. Accurately weigh about 10 mg of the finely powdered silicate material into a dry polyethylene bottle and a similar amount of the reference sample into a second bottle. Use a third bottle for the reference blank solution. Add approximately 20 mg of 1,10-phenanthroline to each followed by 3 ml of 4 N sulphuric acid and 0·5 ml of concentrated hydrofluoric acid. Transfer the bottles to a steam bath in a fixed order, so that the first one on will be the first off, and leave for 30 minutes.

While the bottles are on the steam bath, transfer 5 ml of the boric acid solution to each of a series of 100-ml volumetric flasks. Using the same order as previously, remove the bottles from the steam bath and, as rapidly as possible, add 20 ml of sodium citrate solution to each bottle. As far as possible each bottle should be left on the steam bath for the same length of time. Transfer the contents of each bottle to one of the 100-ml volumetric flasks, using water to rinse the bottles. Dilute each solution to volume with water and mix well.

Using a spectrophotometer set at wavelengths of 555 nm and 640 nm measure the optical densities of the solutions using the blank as the reference solution. For each solution, subtract the optical density at 640 nm from that obtained at 555 nm, and determine the ferrous iron content of the sample by reference to the optical density difference obtained for the silicate rock of known ferrous iron content.

THE DETERMINATION OF FERROUS IRON IN CARBONACEOUS SHALES

The importance of knowing the ferrous iron content of carbonaceous shales has led Nicholls[17] to propose a method of determination in which up to 4 per cent of carbon can be tolerated. The shale is decomposed by heating with hydrofluoric and sulphuric acids, and the excess fluoride complexed with boric acid in the usual way. The ferrous iron is allowed to react with iodine monochloride, and the liberated iodine titrated with standard potassium iodate solution. This method of determination was proposed by Heisig[18] in 1928, and was used at a later date by Hey[19] for the determination of ferrous iron in silicate rocks. The procedure is described in detail below. As with other methods for the determination of ferrous iron following sample decomposition with hydrofluoric and sulphuric acids, some interference may be encountered from the presence of sulphide minerals.

Method

Reagents: *Carbon tetrachloride.*

Iodine monochloride solution, dissolve 10 g of potassium iodide and 6·44 g of potassium iodate in 150 ml of 6 N hydrochloric acid.

Potassium iodate standard solution, 0·1 N (M/40) dissolve 5·350 g of dried pure potassium iodate in water and dilute to volume in a 1 litre flask.

Procedure. Accurately weigh about 0·5 g of the finely powdered shale sample into a platinum crucible, moisten with water and add 10 ml of 20 N sulphuric acid and 5 ml of concentrated hydrofluoric acid. Cover the vessel with a suitable lid and simmer gently for a period of 5 minutes, or until the decomposition is judged to be complete. Pour the contents of the vessel into a 100-ml beaker containing 2 g of solid boric acid and rinse the platinum vessel and lid with cold, boiled distilled water and add these washings to the beaker to give a solution approximately 50 ml in volume.

Pour this solution into a 200–250-ml bottle containing 75 ml of concentrated hydrochloric acid and 6 ml of the prepared iodine monochloride solution. Rinse the beaker, adding the washings to the bottle and dilute the solution to about 150 ml. Now add 10 ml of carbon tetrachloride, stopper the bottle and shake for 20 seconds. At this stage the carbon tetrachloride layer should be coloured deep purple with the liberated iodine extracted from the aqueous phase. The colour may be partially obscured by the presence of carbonaceous matter which tends to gather at the interface. Titrate the solution by adding standard potassium iodate solution from a burette, with intermittent shaking of the bottle, until the purple colour has almost disappeared from the organic layer. Now add a further 10 ml of carbon tetrachloride to form a zone free from carbonaceous material at the bottom of the bottle in which the end-point (the complete disappearance of the purple colour) can easily be detected. Towards the end of the titration shake the contents of the bottle after the addition of each drop of potassium iodate solution.

Ferric Iron

The direct determination of ferric iron is seldom necessary in silicate rock analysis, an adequate measure usually being obtained by deducting the ferrous iron from the separately determined total iron—both expressed as Fe_2O_3. As noted in the previous section, the presence of sulphide minerals or carbonaceous material may, by the reduction of trivalent iron, give rise to appreciable errors in the ferrous iron content which in turn are reflected in the calculated value for the ferric iron. An assumption is sometimes made that where sulphide minerals are present they are insoluble in the hydrofluoric-sulphuric acid mixture used for the determination of ferrous iron, and therefore do not affect this determination. This assumption is largely but not completely true for pyrite, FeS_2, the sulphide mineral occurring most frequently in silicate rocks. Where the rock sample contains more than a trace of pyrite, the sulphide iron, calculated as Fe_2O_3, is added to the ferrous iron (also calculated as Fe_2O_3) before the calculation of ferric iron. Pyrrhotite and certain other sulphide minerals are appreciably soluble in the acid mixture used for the ferrous iron determination.

When required, a direct determination of ferric iron can be made by a direct titration of the rock solution without the addition of nitric acid or other oxidising agent, using titanous chloride or sulphate solution as titrant. This method should in principle, provide a more accurate value for ferric iron than that derived by difference, especially when the ratio of ferric to ferrous iron is low.*

Total Iron

There is a wide selection of procedures for the determination of the total iron content of silicate and carbonate rocks. Photometric methods are of greatest use for those sample materials that contain only small amounts of iron, but as these procedures are some of the most precise of their kind, they have been widely adopted for the determination of iron even when present as a major component. Mercy and Saunders[20] have, however, shown that titrimetric procedures are more precise than photometric ones, although the differences they obtained were quite small. In a comparison of two titrimetric and three photometric procedures, the titrimetric methods were found to be consistent with each other, but there were some differences in the results from the three photometric methods, although all five methods gave approximately the same average values.

A number of separation procedures are available for the recovery of iron from solution. The simplest is precipitation with aqueous ammonia, followed by filtration and dissolution of the residue in dilute hydrochloric acid. Titanium, vanadium, chromium, phosphorus, most of the aluminium and some of the manganese will accompany the iron. A double precipitation is necessary to remove all the calcium and magnesium from the iron. Ion-exchange procedures have been developed with both cation and anion exchange resins. The anion-exchange procedure is particularly useful for the separation of iron from aluminium, titanium, manganese and other metallic elements.

*See Note added in proof, page 507.

Solvent extraction procedures have rarely been applied to the determination of iron in silicate rocks, but Kiss[21] in a scheme for the determination of a number of major components utilises a well-known separation based upon the extraction of the chloroferrate ion from seven molar hydrochloric acid solution with methyl isobutyl ketone. The iron in the organic extract is recovered by shaking the extract with water.

PHOTOMETRIC REAGENTS FOR IRON

XVI Tiron (disodium salt of
     catechol – 3:5 – disulphonic acid)

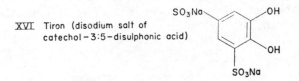

There is no lack of variety in the colour-forming reagents suggested for the determination of iron in silicate or carbonate rocks. Tiron (disodium salt of catechol-3,5-disulphonic acid (XVI)) has been proposed for the determination of both iron and titanium in the same solution.[22] The iron complex is violet coloured with an absorption maximum at a wavelength of 560 nm. This solution can be decolorised by reduction of the iron using sodium dithionite, leaving the yellow-coloured titanium complex in solution. This has an absorption maximum at 430 nm and does not absorb at a wavelength of 560 nm, used for the iron determination. There is disagreement as to the best pH to use for this determination.[23, 24] This, together with the recorded instability of the reducing agent leading to the formation of collodial sulphur, suggests that little advantage is to be gained in the use of tiron for the determination of iron, particularly as superior reagents are available for titanium.

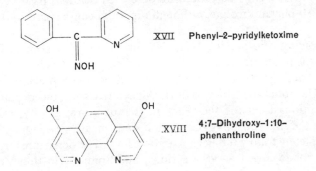

XVII   Phenyl–2–pyridylketoxime

XVIII   4:7–Dihydroxy–1:10–
        phenanthroline

Both phenyl-2-pyridylketoxime (XVII)[25, 26] and 4,7-dihydroxy-1,10-phenanthroline[27] (Snyder's reagent XVIII) react with ferrous iron in alkaline solution to give coloured products that have been used for the determination of iron in silicate materials. The latter reagent is extremely expensive and appears to offer little if any advantage over the former. Unpublished work[28] has shown that any time gained by working with silicate rocks in alkaline solution was more than offset by the time spent in removing the iron alloyed with the platinum of the crucibles used. A considerable loss of iron can occur in this way, although the amounts are rather variable.

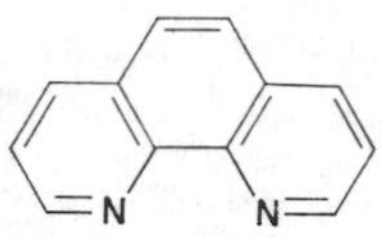

XIX 1:10–Phenanthroline

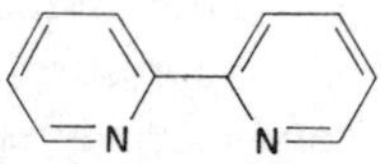

XX 2:2'–Dipyridyl

1,10-Phenanthroline (XIX) and 2,2'-dipyridyl (XX) are two of the most widely used reagents for iron. Both reagents readily form red-coloured complexes with ferrous iron that are stable over wide ranges of salt concentration and temperature. Precise control of pH is not necessary and interference from other elements is negligible. Both reagents are completely stable in the solid state. The ferrous complexes form rapidly, are completely stable in aqueous solution, and such solutions obey the Beer–Lambert Law. The procedure described below is based upon the use of 1,10-phenanthroline, although dipyridyl can be used in the same way.

Ammonium thiocyanate, NH_4CNS, has in the past been extensively used for the photometric determination of iron. It suffers from a number of disadvantages, particularly as compared with the previous two reagents. The optical densities of ferric thiocyanate solutions depend upon the conditions used for the reaction (temperature, acidity, excess of reagent), the solutions may suffer from some measure of fading, and do not completely follow the Beer–Lambert Law. The departures from this law are not regarded as very serious, and ammonium thiocyanate is still used in a number of laboratories. Other reagents that have been used for the photometric determination of iron in silicate materials include salicylic acid, EDTA and hydrogen peroxide, acetylacetone and sulphosalicylic acid.

PHOTOMETRIC DETERMINATION OF IRON IN SILICATE ROCKS

Some workers have recommended using a mixture of perchloric and hydrofluoric acids for the decomposition of silicate rocks and minerals and for the volatilisation of silica. This use of perchloric acid is ideal where the sample contains only easily decomposable minerals such as felspars, but the somewhat higher temperatures that can be obtained with sulphuric acid are preferred for general silicate rock analysis where a variety of accessory minerals are likely to be present. Even using sulphuric acid, a small unattacked residue is often obtained consisting of such minerals as tourmaline, zircon, ilmenite and rutile, together with barium precipitated as sulphate. Any iron present in this residue is recovered following fusion with sodium carbonate, and is added to the main rock solution. If it is required, the barium content of the rock sample can be determined in this insoluble residue.

The iron present in a suitable aliquot of the rock solution is reduced to the ferrous state using hydroxyammonium chloride, and the red colour given with 1,10-phenanthroline is formed in the solution buffered with ammonium tartrate. The absorption maximum occurs at a wavelength of 508 nm (Fig. 59) and a calibration graph is shown in Fig. 60.

The reagents used for this determination all contain small amounts of iron and it is therefore particularly important that the reagent blank solution should be carefully prepared. Particular attention should be paid to the cleanliness of the apparatus used, especially the platinum crucibles and the spectrophotometer cells. For silicate materials containing only very small amounts of iron (e.g. vein quartz or quartzite) ferrous ammonium sulphate can be used to prepare the standard iron solution used for the calibration, but for rocks containing more than about 1 per cent of iron, pure metallic iron should be used.

Method

Reagents: *Tartaric acid solution*, dissolve 10 g of the solid reagent in 100 ml of water.

p-Nitrophenol indicator solution, dissolve 1 g of the solid reagent in 100 ml of water.

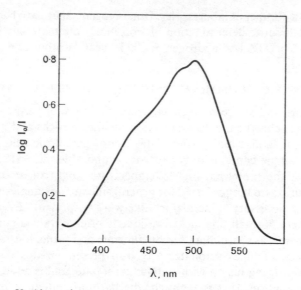

FIG. 59. Absorption spectrum of the iron complex with 1,10-phenanthro-line (1-cm cells, $0\cdot6$ μg Fe/100 ml).

Hydroxyammonium chloride solution, dissolve 10 g of the solid reagent in 100 ml of water.

1,10-Phenanthroline solution, dissolve $0\cdot1$ g of the reagent in 100 ml of water.

Standard iron stock solution, accurately weigh $0\cdot112$ g of pure iron wire into a small beaker, add 20 ml of water and 5 ml of 20 N sulphuric acid. Warm gently until the metal has completely dissolved and then dilute to 1 litre with water. This solution contains 160 μg Fe_2O_3 per ml.

Standard iron working solution, pipette 25 ml of the stock solution into a 100-ml volumetric flask and dilute to volume with water. This solution contains 40 μg Fe_2O_3 per ml.

Procedure. Accurately weigh approximately 1 g of the finely ground rock material into a clean platinum crucible, and add $0\cdot5$ ml of 20 N sulphuric acid (note 1), a few drops of concentrated nitric acid and 10 ml of hydrofluoric acid. Transfer the crucible to a hot plate and evaporate just to fumes of sulphuric acid Remove the crucible, allow to cool, add a further 5 ml of hydrofluoric acid and again evaporate to fumes of sulphuric acid. Allow to cool once again, rinse down the sides of the crucible with a little water, add a further 2 ml of water, break up any solid cake that has formed using a small platinum rod and again evaporate on the hot plate, this time to copious fumes of sulphuric acid. Allow the crucible to cool.

Rinse down the sides of the crucible and add a further quantity of water

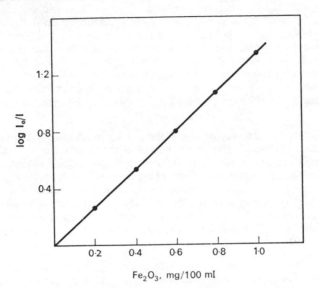

Fe$_2$O$_3$, mg/100 ml

F$_{IG}$. 60. Calibration graph for iron using 1,10-phenanthroline (1-cm cells, 508 nm).

to about half-fill the crucible. Allow to stand on a hot plate for a few minutes for the solid residue to disintegrate and pass partly into solution. Rinse the solution and any remaining residue into a 250-ml beaker, set the crucible aside, and heat the solution until all soluble material has dissolved. Collect any remaining insoluble material on a small, close-textured filter paper, wash well with very dilute sulphuric acid and transfer the paper and residue to the platinum crucible previously used. Collect the filtrate and washings in a 250-ml volumetric flask and set it aside.

Dry and ignite the paper, fuse any remaining residue with a small quantity of anhydrous sodium carbonate for a period of at least 30 minutes and then allow to cool. Extract the melt with water and rinse the solution and residue into a small beaker. Warm the crucible with a little dilute sulphuric acid, and rinse this acid solution into the beaker. At this stage no unattacked particles of the original silicate rock should remain, and the only precipitate should be barium sulphate which will form if the sample material contains an appreciable amount of barium. Collect this precipitate on a small close-textured filter paper, wash with very dilute sulphuric acid and discard it. Add the sulphuric acid solution and the washings of the barium sulphate precipitate to the solution contained in the 250-ml volumetric flask and dilute to volume with water.

Pipette a suitable aliquot of this solution into a 100-ml volumetric flask, add 10 ml of tartaric acid solution and a drop of *p*-nitrophenol indicator solution, followed by concentrated aqueous ammonia until the solution becomes pure yellow in colour. Add dilute hydrochloric acid drop-by-drop until this colour is just discharged. Cool the solution to room temperature, add 2 ml of hydroxyammonium chloride solution and 10 ml of 1,10-phenanthroline solution and dilute to volume with water. Mix well, allow to stand for an hour and measure the optical density at a wavelength of 509 nm, relative to a reagent blank solution, prepared in a similar way but without the sample material.

Calibration. For the preparation of the calibration graph for 1 cm spectrophotometer cells, pipette aliquots of 0–25 ml of the standard solution containing 0–1 mg Fe into separate 100-ml volumetric flasks and add to each 10 ml of tartaric acid and 1 drop of *p*-nitrophenol indicator solutions. Adjust the pH by adding aqueous ammonia and hydrochloric acid as described above. finally adding hydroxyammonium chloride and 1,10-phenanthroline solutions and diluting to volume with water. After standing, measure the optical densities of these solutions at 508 nm and plot the values obtained against the concentration of iron (Fig. 60).

Note: 1. This amount of sulphuric acid is sufficient for a high silica material (> 96 per cent SiO_2), and should be increased to as much as 10 ml for rocks containing much smaller amounts of silica and proportionally greater amounts of iron aluminium and other elements.

THE TITRIMETRIC DETERMINATION OF TOTAL IRON

Titration procedures for the determination of total iron are based upon the conversion of all the iron present into one valency state, which is then converted in the course of the titration into the other. The titration of ferric to ferrous iron has never been popular and the most widely used procedures are those based upon the titration of ferrous to ferric.

The titration of ferrous iron. Only a small number of oxidants have been successfully used for this determination, and either potassium dichromate or ceric sulphate is recommended. Potassium permanganate may oxidise chloride ion and should be avoided for use with hydrochloric acid solutions.

A variety of reducing agents have been suggested for the reduction of ferric iron, although some of them are far from ideal. Sulphur dioxide and hydrogen sulphide for example, although effective reducing agents, must be added in excess, and the excess is then difficult to remove. Prolonged boiling is recommended for this, but some trace of sulphur compounds usually remains. Titanous sulphate and chloride have been recommended, but these solutions are unstable and the solutions cannot be stored for long periods. They offer no advantage over the more conventional reductant–stannous chloride solution. A great excess of stannous chloride should be avoided, otherwise the mercuric chloride which is added to remove the excess of stannous chloride, will be reduced to the black metallic state.

An alternative procedure for the stannous chloride reduction has been described by Hume and Kolthoff,[29] in which cacotheline is added to indicate the presence of a slight excess of stannous chloride. This slight excess is then titrated with ceric sulphate solution before the titration of the ferrous iron with the same reagent.

METHOD BASED UPON STANNOUS CHLORIDE REDUCTION

In this method stannous chloride is used to reduce any ferric iron present in the solution to the divalent state. In the procedure described in detail below, the excess stannous chloride is removed by titration with standard ceric sulphate solution. The older mercuric chloride procedure is suggested as an alternative, for which a dichromate titration is described.

Method

Reagents: *Stannous chloride solution*, dissolve 7·5 g of stannous chloride dihydrate in 100 ml of 6 N hydrochloric acid. Prepare freshly at frequent intervals.

Ceric sulphate 0·03 N *standard solution*, dissolve 10 g of ceric sulphate in water containing 50 ml of 20 N sulphuric acid and dilute to volume in a 1 litre volumetric flask. Standardise by titration against a pure iron solution, or preferably against pure arsenious oxide in the presence of osmium tetroxide as catalyst.

Cacotheline reagent, grind together 0·5 g of the solid reagent and 0·5 ml of water. Add 50 ml of water. Shake before using.

Ferroin indicator solution, dissolve 0·742 g of 1,10-phenanthroline monohydrate in 50 ml of a solution containing 6·95 g of ferrous sulphate heptahydrate per litre.

Mercuric chloride solution, saturated.

Potassium dichromate 0·03 N *standard solution*, dry a small amount of the pure reagent at a temperature of 150° for 4 hours, and then dissolve approximately 1·5 g in water and dilute to volume in a 1 litre volumetric flask. The strength of this solution can be calculated from the weight of material taken, but for higher accuracy, the solution obtained should be standardised by titration against a standard iron solution.

Diphenylamine indicator solution, dissolve 10 mg of diphenylamine sulphonate (sodium salt) in a mixture of 50 ml of water and 50 ml of syrupy phosphoric acid.

Procedure. Accurately weigh approximately 1 g (note 1) of the finely powdered silicate rock material into a platinum dish, moisten with water and add 5 ml of concentrated perchloric acid and 10 ml of concentrated hydrofluoric acid. Transfer the dish to a hot plate and evaporate first to fumes of perchloric acid and then to complete dryness. Allow to cool, moisten the dry residue with a little perchloric acid and again evaporate to dryness on a hot plate. Add 5 ml of water and 5 ml of concentrated hydrochloric acid to the residue and warm to bring all soluble material into solution. With some silicate rocks, complete solution will be obtained at this stage. With others a small residue, insoluble in hydrochloric acid, will remain. Collect this residue on a small filter, wash with water, dry and ignite in a small silica crucible. Add a small amount of potassium pyrosulphate to the residue and fuse at a dull red heat until a quiescent melt is obtained. Allow to cool, extract the melt into a little dilute hydrochloric acid and add to the main rock solution (note 2).

Dilute the solution to about 75 ml, add 5 ml of concentrated hydrochloric acid, heat almost to boiling and add stannous chloride solution drop by drop until the yellow colour of the solution is completely discharged. Add 1 drop of stannous chloride solution in excess. Allow to cool, and add 5 drops of the cacotheline suspension which turns the solution deep violet in colour. Titrate the solution, drop by drop, with ceric sulphate solution until the violet colour fades, giving first a brown colour, then a pure yellow. Near the end-point it may be necessary to add 2 or 3 drops of the cacotheline indicator suspension to observe the end-point.

Add 125 ml of water, 6 ml of 20 N sulphuric acid and 3 drops of ferroin indicator solution, and titrate with the standard ceric sulphate solution to the disappearance of the red colour given by the indicator with ferrous iron. There is a small indicator and reagent blank value which should be determined and subtracted from the sample titration value before calculation of the total iron content.

The excess of stannous chloride can also be removed by adding 5 ml of mercuric chloride solution and allowing the solution to stand—preferably under a blanket of inert gas such as nitrogen or carbon dioxide, for 5 minutes. Then add 50 ml of water and 10 ml of the diphenylamine indicator solution and titrate with 0·03 N potassium dichromate solution until a faint purple colour is obtained.

Notes: 1. This weight of sample material should give a titration of about 30 ml of 0·03 N oxidant with samples containing about 6 per cent Fe. For basic and other rocks rich in iron, the sample weight should be reduced. Alternatively, 0·1 N oxidant solution can be used.

2. Most iron minerals will be decomposed by this treatment. If any residue remains it should be collected and fused with a little sodium hydroxide, the melt extracted with water containing a little hydrochloric acid, and the solution added to the main rock solution.

3. The total iron content of carbonate rocks can be determined in the same way as that of silicate rocks, except that for the initial stage of the decomposition it is sufficient to evaporate the sample material to dryness with hydrochloric acid. After dissolution of the residue in dilute hydrochloric acid, any insoluble material can be recovered and decomposed by fusion with a little anhydrous sodium carbonate.

METHODS BASED UPON METALLIC REDUCING AGENTS

Although a number of metals have been suggested as suitable for reducing ferric to ferrous iron, only two, zinc and silver, have found extensive use in the analysis of silicate rocks. A Jones reductor of amalgamated zinc is less suited to this application as there is considerable interference from titanium which is reduced to Ti^{3+}. Chromium and vanadium which may be present to a minor extent, are reduced to Cr^{2+} and V^{2+} respectively. Many other elements are reduced by passage through the reductor, but are unlikely to be present in normal silicate rocks in amounts sufficient to interfere.

Neither titanium nor chromium (Cr^{3+}) are reduced by passage through a silver reductor, although vanadium (V^{5+}) is reduced to V^{4+}. The formation of hydrogen peroxide in silver reductors has been reported by Miller and Chalmers,[30] and this prevents the complete reduction of ferric iron. This problem has been overcome by the use of solutions saturated with carbon dioxide. Platinum, derived from the apparatus used, is reduced (Pt^{4+} to Pt^{2+}) in a silver reductor, and may also interfere with the titration of iron by catalysing the reduction of titanium. The introduction of platinum can be avoided by conducting the fusion in crucibles of gold or silver.

Method

The method described in detail is based upon that given by Mercy and Saunders[20] using a silver reductor and titrating with standard potassium dichromate solution. The silicate rock material is decomposed by fusion with a mixture of sodium carbonate and borax glass in a silver crucible.

Apparatus. *Silver crucibles*, approximately 40 ml capacity, with silver lids. Silver melts at a temperature of 960°, so particular care must be taken during the fusion stages and on no account should they be heated over the full flame

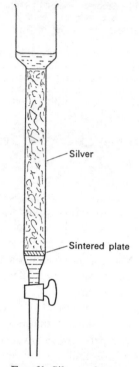

FIG. 61. Silver reductor.

of a gas burner. They may be cleaned by fusing in them a little sodium hydroxide and extracting the melt with water. After removal of the extract, the crucibles should be boiled with 6 N hydrochloric acid.

Silver reductor, Fig. 61. This consists of a small glass column of about 15 cm in length and 1·5 cm in diameter, with a stop-cock and reservoir. The silver metal is prepared by precipitation from silver nitrate solution with metallic copper and, after transfer to the column, is washed several times with N hydrochloric acid saturated with carbon dioxide. The silver metal is left covered with this acid at all times, and must not be allowed to run dry. Before use, pass a solution containing about 5 mg of ferrous sulphate in about 50 ml of N hydrochloric acid through the reductor. This serves to remove all traces of hydrogen peroxide from the column, which is then washed with about 120 ml of N hydrochloric acid saturated with carbon dioxide.

Reagents: *Fusion mixture*, mix 30 g of anhydrous sodium carbonate with 20 g of borax glass.

Procedure. Accurately weigh approximately 0·5 g of the finely powdered silicate rock material into a silver crucible. Add approximately 4 g of the fusion mixture and mix carefully with a small nickel or platinum spatula. Dust any particles adhering to the spatula back into the crucible and cover with a thin layer of the fusion mixture. Cover the crucible with a silver lid and transfer to an electric furnace. Slowly raise the temperature from ambient to about 880° over a period of at least 1 hour, and maintain at this temperature for a further hour. Remove the crucible and allow to cool.

Transfer the crucible and lid to a 150-ml beaker and extract the melt with 30 ml of water and 20 ml of 6 N hydrochloric acid, by warming gently on a hot plate or steam bath. When the disintegration is complete and all soluble material has dissolved, rinse the crucible and lid and remove them. If any residue remains it should be collected, washed with water, dried and fused with a smaller quantity of the fusion mixture.

Pass the whole of the solution through the silver reductor and rinse the column with about 120 ml of N hydrochloric acid previously saturated with carbon dioxide and added in small portions at a time. Collect the eluate and washings in a conical beaker containing a small piece of marble and a few ml of N hydrochloric acid to provide an atmosphere of carbon dioxide. Add 10 ml of 20 N sulphuric acid and 10 ml of the diphenylamine indicator solution and titrate the ferrous iron with standard 0·03 N potassium dichromate solution. Determine also the reagent blank value and subtract this from the titre before calculating the total iron content of the silicate rock material.

THE TITRATION OF FERRIC IRON

The difficulties that arise during the reduction of ferric iron in the rock solution can be avoided by titrating not the ferrous to ferric but the ferric to ferrous. Thornton and Chapman[31] have described a procedure in which potassium permanganate is used to ensure that all the iron present is in the higher valency state, and they then titrated the ferric iron with a standard solution of titanous sulphate. Then end-point was indicated by the disappearance of the red colour given by ferric iron with thiocyanate. Some experience is necessary to judge the exact point at which this occurs, as the reaction is rather slow near the end-point.

Nitric acid, free hydrofluoric acid, vanadium, molybdenum and a number of other metallic elements interfere with the determination. The reagent can be prepared by dissolving pure titanium sponge in sulphuric acid. Once prepared it should be stored under a layer of inert hydrocarbon and away from direct sunlight. Titanous sulphate solutions are very readily oxidised, and special precautions are necessary for the delivery of the solution in the course of the analysis. Experience with titanous sulphate has shown the necessity of standardisation at frequent intervals, if accurate results are to be obtained.

Mercurous nitrate can also be used for this titration,[3] and possibly also a number of other titrants such as vanadous sulphate or chromous sulphate solutions. These reagents would also need to be protected from atmospheric oxidation, and are unlikely to offer any significant advantage over titanous sulphate solutions.

Ferric iron can also be titrated with solutions of EDTA, neither ammonium thiocyanate nor salicylic acid are suitable for indicating the end points of the reaction, and a redox indicator such as variamine blue B is preferred.[32] An alternative procedure is the indirect titration suggested by Pribil and Vesely,[33] involving the addition of an excess of EDTA solution and a titration of this excess with bismuth, thorium or lead solution using xylenol orange as indicator.

References

1. EASTON A. J. and LOVERING J. F., *Geochim. Cosmochim. Acta* (1963) **27**, 753.
2. FRIEDHEIM C., *S.B. Akad. Wiss. Berlin* (1888) 345.
3. TRUSOV YU. P., *Zhur. Anal. Khim.* (1959) **14**, 139.
4. PRATT J. H., *Amer. J. Sci.* (1894) **48**, 149.
5. HARRIS F. R., *Analyst* (1950) **75**, 496.
6. TREADWELL F. P., *Kurzes Lehrbuch der Analytischen Chemie* (1913) **2**, p. 425, (6th ed.).
7. LO-SUN JEN, *Anal. Chim. Acta* (1973) **66**, 315.
8. WILSON A. D., *Bull. Geol. Surv. Gt. Brit.* (1955) (9), 56.
9. WILSON A. D., *Analyst* (1960) **85**, 823.
10. REICHEN L. E. and FAHEY J. J., *U.S. Geol. Surv. Bull.* 1144-B, 1962.
11. ROWLEDGE H. P., *J. Roy. Soc. W. Aust.* (1934) **20**, 165.
12. VINCENT E. A., *Geol. Mag.* (1937) **35**, 86.
13. GROVES A. W., *Silicate Analysis*, Allen & Unwin, London, 1951 (2nd ed.).
14. MEYROWITZ R., *Analyt. Chem.* (1970) **42**, 1110.
15. SHAPIRO L., *U.S. Geol. Surv. Research* 1960, B-496 (1961).
16. FRENCH W. J. and ADAMS S. J., *Analyst* (1972) **97**, 828.
17. NICHOLLS G. D., *J. Sed. Petrol.* (1960) **30**, 603.
18. HEISIG G. B., *J. Amer. Chem. Soc.* (1928) **50**, 1687.
19. HEY M. H., *Amer. Mineral.* (1949) **34**, 769.
20. MERCY E. L. P. and SAUNDERS M. J., *Earth, Planet. Sci. Lett.* (1966) **1**, 169.
21. KISS E., *Anal. Chim. Acta* (1967) **39**, 223.
22. YOE J. H. and ARMSTRONG A. R., *Analyt. Chem.* (1947) **19**, 100.
23. RIGG T. and WAGENBAUER H. A., *Analyt. Chem.* (1961) **33**, 1347.
24. ARCHER K., FLINT D. and JORDAN J., *Fuel, London* (1958) **37**, 421.
25. TRUSELL F. and DIEHL H., *Analyt. Chem.* (1959) **31**, 1979.
26. CLULEY H. J. and NEWMAN E. J., *Analyst* (1963) **88**, 3.
27. SCHILT A. A., SMITH G. F. and HEIMBUCH A., *Analyt. Chem.* (1956) **28**, 809.
28. RICHARDSON J. and JEFFERY P. G., Unpubl. Rept., 1962.
29. HUME D. S. and KOLTOFF I. M., *Analyt. Chem.* (1957) **16**, 415.
30. MILLER C. C. and CHALMERS R. A., *Analyst* (1952) **77**, 2.
31. THORNTON W. M. Jr. and CHAPMAN J. E., *J. Amer. Chem. Soc.* (1921) **43**, 91.
32. FLASCHKA H., *Mickrochim Acta* (1954) 361.
33. PRIBIL R. and VESELY V., *Talanta* (1963) **10**, 361.

CHAPTER 28

LEAD

Occurrence

The most extensive account of the occurrence and distribution of lead in igneous, metamorphic and sedimentary rocks is that of Wedepohl.[1] The values given, Table 26, are similar to those reported by Sandell and Goldich[2] in North American rocks. It is unfortunate that the value given by Wedepohl for the granite G-1 (26 ppm Pb) is only about half that of the recommended value.[3] The original lead values for G-1 suggested a bimodal distribution in the several sample portion, with peak values at about 27 and 50 ppm. This distribution now appears to have arisen as a result of analytical errors,[4] and a value near to 50 ppm is now accepted.

TABLE 26. THE AVERAGE LEAD CONTENT OF
IGNEOUS AND OTHER ROCK TYPES[1]

Rock type	Pb, ppm
Ultrabasic	3
Basic	6
Intermediate (calc-alkaline)	10
Intermediate (alkaline)	12
Granodiorite	15
Granite	20
Argillaceous sediments	20
Arenaceous sediments	7
Calcareous and dolomitic sediments	9

Lead is a chalcophilic element, appearing as the sulphide galena, PbS as a primary mineral in magmatic rocks, and as an ore mineral deposited at a later date in both silicate and carbonate rocks. From the work of Sandell and Goldich,[2] it is clear that lead can also appear in the silicate mineral fraction of silicate rocks, substituting for potassium which has a comparable ionic radius. Correlations have been noted between the lead content of silicate rocks and the content of both potassium and silicon.

Galena is the most important mineral and ore of lead. Within a mineral deposit a number of secondary lead minerals are usually also noted, particularly from the zone of oxidation. These include the sulphate anglesite $PbSO_4$, carbonate cerussite $PbCO_3$, and sometimes smaller amounts of phosphate, vanadate, molybdate, etc. Lead also occurs as an essential constituent of some uranium minerals, and is usually reported in other radioactive minerals such as euxenite, thorite and samarskite.

The Determination of Lead in Silicate Rocks

SEPARATION PROCEDURES

Most of the procedures for the determination of lead in silicates require the removal of silica as the first step in obtaining a solution of the rock material. Mixtures of hydrofluoric acid with either perchloric or nitric acid are commonly used, although hydrofluoric and hydrochloric acids have been reported. In general sulphuric acid has been avoided, probably because of the possibility that lead sulphate may be co-precipitated with barium or calcium from rocks rich in these elements.

Procedures based upon ion-exchange separation have not been extensively used for the separation of lead in silicate rocks, although for the determination of lead in marine sediments Korkisch and Feik[5] have described the use of an anion exchange separation, based upon elution with tetrahydrofuran–nitric acid mixtures from a strongly basic resin (Dowex 1 × 8).

Many analysts prefer to make this initial separation of lead from other elements present in silicate rocks by solvent extraction. Dithizone has been described for this,[6] but difficulties have been experienced with rocks rich in iron.[7] Diethyldithiocarbamate has been proposed[8] for this purpose and its use was considered to be satisfactory by Baskova,[9] who compared a number of methods for the separation and determination of lead in silicate rocks.

Traces of lead can be precipitated with suitable metal carriers. Methods suggested include precipitation as sulphide with mercury or zinc and as sulphate with barium. None of these methods gave a good recovery of lead[9] at the concentration levels encountered in some silicate rocks, but could form the basis of an isotope dilution procedure.

One procedure for separating lead that does not require either the removal of silica or an acid dissolution of the rock material, is based upon the ease with which the lead can be sublimed or distilled from the powdered material. The method as described by Iordanov and Kocheva[10] involved sublimation *in vacuo*, but a simpler procedure by Marshall and Hess[11] involved sublimation in a stream of nitrogen at a temperature of 1400°. As little as 0·01 ppm of lead can be determined in this way.

The application of atomic absorption spectroscopy to the determination of lead in silicate rocks is limited more by the low sensitivity than by other factors. Moldan *et al.*[12] calculated that the determination of about 2 ppm Pb in solid samples should be possible with a reasonably sensitive spectrometer, a hydrogen–air flame and a three-slot burner. To improve this, Moldan *et al.* used a long path absorption tube as described by Rubeska.[13] A somewhat simpler procedure was described by Jenkins and Moore[14] involving a prior extraction with diethyl-ammonium diethyldithiocarbamate. Poor recoveries of lead are obtained unless the iron present is reduced, for which ascorbic acid is suggested. Blank values can be high.

METHODS OF DETERMINATION

In acetic acid–ammonium acetate solution lead has a well defined polarographic reduction wave at $-0·61$ V. This enables a few micrograms of lead to be determined, preferably after an initial separation from the bulk of the rock material. This technique has been reported for the analysis of silicates, but has not been widely applied. Atomic absorption has also been suggested,[15] but does not appear to be sufficiently sensitive, at least at the present time, for application to normal silicate rocks.

In view of the importance of determining small quantities of lead in a variety of matrices, it is surprising that so little attention has been given to developing new reagents for lead. However, for most purposes dithizone is a reagent of adequate sensitivity and, provided that certain complexing agents are present, is of reasonable selectivity. The reagent itself is dark green, almost black, in colour and gives green solutions in chloroform and carbon tetrachloride that deteriorate slowly with time. It reacts readily with a large number of metallic ions in solution to give highly coloured—mostly brown, orange or red—complexes that are soluble in organic solvents. In the presence of cyanide ions, only lead, bismuth, thallium, tin(II) and possibly indium are extracted as dithizonates. Bismuth, thallium, tin and indium are present in only very small amounts in silicate rocks, and are not likely to interfere. All four elements are, however, separated from lead in a preliminary concentration procedure involving extraction of the lead complex with diethyldithiocarbamate into organic solution.

Solutions of lead dithizonate in carbon tetrachloride (the solvent most frequently used) have a maximum absorption at a wavelength of 520 nm and obey the Beer–Lambert Law up to about 3 ppm Pb, although at this concentration the solutions are probably supersaturated. 0 to 1 ppm Pb is a convenient concentration range to use for the determination in silicate rocks.

A Solvent Extraction—Spectrophotometric Method

This method is based upon methods given in outline by Baskova[9] and in greater detail by Gage.[16] A mixture of hydrofluoric and nitric acid is used to remove the silica and to obtain the lead and other metallic constituents in solution. The initial separation from iron and certain other metals is by extraction of the lead complex of diethyldithiocarbamate into an organic solvent consisting of a mixture of pentanol and toluene. The lead is transferred to aqueous solution by shaking with dilute hydrochloric acid, and is added to an ammoniacal solution of dithizone containing potassium cyanide and sodium metabisulphite. The red-coloured lead dithizonate is extracted into carbon tetrachloride and the optical density of the extract measured at 520 nm.

Method

Reagents: *Pentanol-toluene mixture,* mix equal volumes of pentanol and toluene. Some batches of this mixture have given low recoveries of lead, even when "sulphur-free" solvents have been used.

These mixtures can be purified by treating with sufficient bromine to give a deep yellow colour, allowing to stand for 30 minutes, decolorising with bisulphite solution and washing with water.

Citrate–bicarbonate solution, dissolve 25 g of sodium citrate dihydrate and 4 g of sodium bicarbonate in water and dilute to 100 ml.

Sodium diethyldithiocarbamate.

Ammonia–cyanide–sulphite solution, mix 95 ml of 10 M aqueous ammonia with 5 ml of a solution of potassium cyanide containing 10 g per 100 ml, and dissolve 5 g of sodium metabisulphite in the mixture.

Dithizone solution, dissolve 2 mg of the solid reagent in 100 ml of pure carbon tetrachloride. Store in a refrigerator.

Standard lead stock solution, dissolve 0·160 g of lead nitrate in water and dilute to 1 litre. This solution contains 100 μg Pb per ml.

Standard lead working solution, dilute 10 ml of the stock solution to 1 litre with water. This solution contains 1 μg Pb per ml.

Procedure. Accurately weigh from 0·5 to 1 g of the finely powdered silicate rock material into a small platinum basin and evaporate to dryness with 10 ml of concentrated hydrofluoric acid and 5 ml of concentrated nitric acid. Add 5 ml of hydrofluoric acid and 5 ml of nitric acid to the dry residue and again evaporate to dryness. Moisten the residue with 2 ml of concentrated nitric acid and evaporate to dryness, and repeat this evaporation with two further portions of nitric acid to expel hydrofluoric acid. Moisten the dry residue with nitric acid and rinse the solution into a small beaker (note 1). Evaporate the solution to dryness twice with concentrated hydrochloric acid to convert nitrates to chlorides, and finally evaporate the chloride solution to give a moist chloride residue. Dissolve this residue in 10 ml of 0·6 N hydrochloric acid, by warming if necessary, and set aside.

Transfer 25 ml of the citrate–bicarbonate solution to a 100-ml separating funnel and add approximately 10 mg of sodium diethyldithiocarbamate. Swirl to dissolve the solid, allow to stand for 15 minutes and then add 25 ml of the pentanol–toulene solvent. Stopper the separating funnel and shake for 1 minute, allow the layers to separate and then run the lower, aqueous layer into the beaker containing the rock solution, previously set aside. Shake the organic layer remaining in the funnel with two successive 10-ml portions of 0·6 N hydrochloric acid and discard these washings.

Now transfer the rock solution to the separating funnel, rinsing the beaker with a few millilitres of the pentanol–toulene solvent mixture. Allow to stand for 15 minutes, shake for 2 minutes, allow the layers to separate and then discard the lower, aqueous layer. Wash the organic layer with 25 ml of redistilled water and discard the washings. Extract the organic layer with two successive 10-ml portions of 0·6 N hydrochloric acid, shaking for 2 minutes and· 1 minute respectively, and transfer the acid extracts to a 50-ml volumetric flask. Rinse the funnel with 0·6 N hydrochloric acid, add the washings to the solution in the flask and dilute to volume with 0·6 N hydrochloric acid. Mix well.

Transfer 25 ml of the ammonia–cyanide–sulphite reagent to a clean 100-ml separating funnel, add 10 ml of the dithizone solution, stopper and shake for 1 minute. Allow the layers to separate and discard the lower, organic layer. Wash the ammoniacal layer by shaking with 10 ml of carbon tetrachloride and discard the washings. Transfer to the separating funnel an aliquot of the rock solution from the volumetric flask, containing not more than 10 μg of lead, and extract the lead into organic solution by shaking for 2 minutes with 10 ml of carbon tetrachloride. Run off the lower, organic layer, filter through a dry paper into a spectrophotometer cell and measure the optical density at a wavelength of 520 nm.

Determine also the optical density of a reagent blank solution similarly prepared but omitting the sample material.

Calibration. To effect the calibration it is necessary to take aliquots of the standard lead solution containing 2–10 μg lead through the procedure, including the extraction with diethyldithiocarbamate, and finally measuring the optical density of each of the dithizone extracts. Plot the relation of the optical densities of the extracts to lead concentration.

Note: 1. The small residue that is usually obtained can often be neglected. Where an appreciable residue is obtained it should be collected, fused with sodium carbonate, extracted with water and filtered. Discard the filtrate. Dissolve the residue in a little 0·6 N hydrochloric acid and add to the main rock solution.

A Sublimation Procedure

For those rocks containing 1 ppm of lead or less, the sublimation procedure of Marshall and Hess[11] can be used for the determination. Approximately 20 g of the silicate rock material are used for each measurement. This sample weight is transferred to a tall-form carbon crucible made from high purity graphite that had previously been cleaned by soaking in concentrated hydrochloric acid, rinsing, drying and baking at 1300–1400° for an hour. This temperature is achieved using an induction heater and is measured with an optical pyrometer.

The apparatus required is shown in Fig. 62. It consists of a water-cooled quartz glass tube as the furnace chamber, with the graphite crucible containing the sample material supported on quartz rings, themselves on a graphite pedestal.

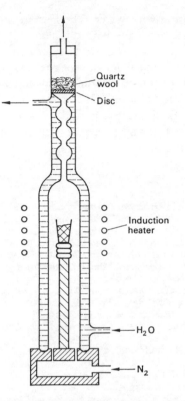

FIG. 62. Apparatus for the recovery of lead by distillation.

A slow stream of nitrogen is used to sweep the gaseous products including volatilised lead out of the hot zone, through a sintered quartz-glass disc and to a plug of quartz-glass wool. The lead is condensed on the walls of the chamber as well as in the disc and quartz-glass wool plug. Marshall and Hess heated their samples for 1 hour, but this is probably longer than is necessary. Some silicates tend to froth excessively, and the rate of increase of temperature must be carefully controlled to keep the melt within the graphite crucible.

After cooling and dismantling of the apparatus, the lead from the sample together with other metals that are also sublimed onto the walls of the apparatus are recovered by washing with concentrated nitric acid, follówed by rinsing with water. The lead in this solution can then be determined photometrically with dithizone after a preliminary separation with diethyldithiocarbamate as described above.

The recovery of lead from silicate rocks is not always quantitative, but this loss of lead can be determined by an isotope dilution technique involving the addition of lead-212 to the rock material in the crucible before ignition. Equilibration is said to take place during the fusion, and the losses at this and subsequent stages of the analysis can then readily be determined.

References

1. WEDEPOHL K. H., *Geochim. Cosmochim. Acta* (1956) **10**, 69.
2. SANDELL E. B. and GOLDICH S. B., *J. Geol.* (1943) **51**, 99 and 167.
3. FLEISCHER M., *Geochim. Cosmochim. Acta* (1965) **29**, 1263.
4. FLANAGAN F. J., *U.S. Geol. Surv. Bull.* 1113, p. 113, 1960.
5. KORKISCH J. and FEIK F., *Analyt. Chem.* (1964) **36**, 1793.
6. SANDELL E. B., *Colorimetric Determination of Traces of Metals*, Interscience, New York, 3rd ed., 1959, p. 572.
7. THOMPSON C. E. and NAKAGAWA H. M., *U.S. Geol. Surv. Bull.* 1084-F, 1960.
8. MAYNES A. D. and MCBRYDE W. A. E., *Analyt. Chem.* (1957) **29**, 1259.
9. BASKOVA Z. A., *Zhur. Anal. Khim.* (1959) **14**, 75.
10. IORDANOV N. and KOCHEVA L., *Bulgar. Akad. Nauk, Izv. Khim. Inst.* (1956) **4**, 327.
11. MARSHALL, R. R. and HESS D. C., *Analyt. Chem.* (1960) **32**, 960.
12. MOLDAN B., RUBESKA I. and MIKSOVSKY M., *Anal. Chim. Acta* (1970) **50**, 342.
13. RUBESKA I., *Anal. Chim. Acta* (1968) **40**, 187.
14. JENKINS L. B. and MOORE R., *U.S. Geol. Surv. Prof. Paper* 700-D, D222 (1970).
15. IIDA C., TANAKA T. and YAMASAKI K., *Bunseki Kagaku* (1966) **15**, 1100.
16. GAGE J. C., *Analyst* (1955) **80**, 789 and (1957) **82**, 453.

CHAPTER 29

MAGNESIUM

Occurrence

Magnesium is usually reported as the oxide MgO, and as such can constitute as much as 51 per cent of certain silicate rocks—varieties of dunite for example, 30 to 40 per cent MgO is fairly common, being encountered in certain ultrabasic rocks such as picrites and peridotites. The distribution of magnesium in silicate rocks is shown in Fig. 63, which illustrates both the association of magnesium with basic rocks, and the increasing depletion of magnesium in residual magmas as differentiation proceeds.

Magnesium and iron form isomorphous crystals with the substitution of Fe^{2+} ($R = 74$ pm) for Mg^{2+} ($R = 66$ pm). In the course of the crystallisation of ferromagnesian minerals, the first solid fractions to appear are always enriched in magnesium relative to the composition of the magma, and conversely the last fractions to crystallise are always enriched in ferrous iron. The end members of these ferromagnesian mineral series, although petrographically important, are of less frequent occurrence than the intermediate members. These include members of the forsterite (Mg_2SiO_4)–fayalite (Fe_2SiO_4) series in common olivine, enstatite ($MgSiO_3$)–ferrosilite ($FeSiO_3$) in the hypersthene orthopyroxenes, and similar series in the remaining rock-forming silicate mineral groups.

Other silicate minerals containing magnesium as an essential component include talc, chlorite, chrysotile and serpentine. Carbonate minerals include magnesite $MgCO_3$, dolomite $CaMg(CO_3)_2$ and ankerite in which iron and manganese have substituted to some extent for magnesium in dolomite.

Rarer minerals of magnesium include spinel $MgAl_2O_4$, periclase MgO, brucite $Mg(OH)_2$ and sellaite MgF_2. Evaporite deposits may contain carnallite $K_2MgCl_4.6H_2O$, bischofite $MgCl_2.6H_2O$, kieserite $MgSO_4$.

307

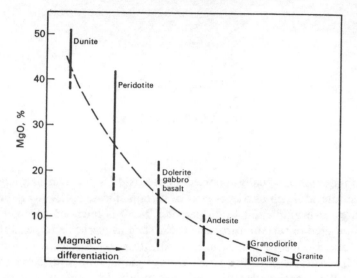

FIG. 63. Magnesium content of silicate rocks.

H_2O, hexahydrite $MgSO_4.6H_2O$, epsomite $MgSO_4.7H_2O$, and mixed chlorides and sulphates.

Gravimetric Determination in Silicate Rocks
PYROPHOSPHATE METHOD

This method has long formed part of the classical procedure for the main fraction in silicate and carbonate rock analysis. In this scheme it is applied after the separation of almost all other constituents of the sample material including silicon, iron, aluminium, titanium, vanadium, chromium, calcium and part of the manganese. The remaining part of the manganese is precipitated with magnesium as the ammonium phosphate $(Mg,Mn)(NH_4)PO_4$, which is then ignited to pyrophosphate $(Mg,Mn)_2P_2O_7$ as the weighing form. This method is described on p. 45, Chapter 4.

8-HYDROXYQUINOLINE METHOD

Experience with the pyrophosphate method has shown that with rocks containing very little magnesium, the recoveries by the classical method are both poor and erratic. With some rocks it was found impossible to recover any magnesium at all until the excess of ammonium salts had been destroyed, and then the recoveries were incomplete. For these rocks the gravimetric procedure based upon precipitation with 8-hydroxyquinoline is preferred. This is not suited to the determination of large amounts and for rocks rich in magnesium, the solution should be diluted to volume and a suitable aliquot taken for the precipitation. As with the pyrophosphate method, it is necessary to remove most other constituents of silicate rocks by ammonia precipitation, and preferable also to remove the bulk of ammonium salts before precipitating magnesium. This is usually accomplished by evaporating the rock solution and heating the residue with concentrated nitric acid. This step is not necessary where only a portion of the solution is taken for the recovery of magnesium.

Method

Reagents: *8-Hydroxyquinoline solution*, dissolve $2 \cdot 5$ g of the reagent in 100 ml of 2 N acetic acid.

Procedure. Combine the filtrates and washings from the precipitation of calcium oxalate, make just acid with hydrochloric acid and evaporate to near dryness on a steam bath. Allow to cool, cover with a watch glass and add 50 ml of concentrated nitric acid. Transfer the beaker to a steam bath and gradually raise the temperature until a vigorous reaction sets in, with evolution of brown nitrous fumes. When the reaction has subsided, rinse and remove the cover, wash down the sides of the beaker and again evaporate to dryness. If any appreciable ammonium salt residue remains, repeat the evaporation with a further quantity of concentrated nitric acid, and then evaporate to dryness. Moisten the residue with a little concentrated hydrochloric acid and evaporate to dryness once more, to expel the remaining traces of nitric acid.

Dissolve the residue in a little hot water containing 1 ml of concentrated hydrochloric acid and filter if necessary into a clean 400-ml beaker. Dilute to a volume of about 100 ml with water, heat to boiling and add concentrated ammonia solution until a small excess is present. If any precipitate forms at this stage it should be collected, washed with a little dilute ammonia and discarded (note 1). Allow the combined filtrate and washings to cool to a temperature of 65–70°, then add 10 ml of 8-hydroxyquinoline solution (note 2) and 8 ml of concentrated ammonia solution. Stir the solution, transfer to a steam bath for 5 minutes, stir again and then set aside for 30 minutes.

Collect the precipitate on a weighed, sintered glass or porcelain crucible and wash well with warm $0 \cdot 5$ N ammonia solution. Transfer to an electric oven set at a temperature of $140°$, and dry to a constant weight. The precipitate of $Mg(C_9H_6ON)_2$ contains $12 \cdot 91$ per cent magnesium oxide, MgO (note 3).

Notes: 1. Traces of alumina that have not been collected in the main ammonia precipitate are usually noted at this stage.

2. This quantity of reagent is sufficient to precipitate about 30 mg of magnesium oxide, corresponding to 3 per cent MgO from a 1 g sample portion. For rocks of higher magnesia content, the amount of reagent should be increased proportionately, or the rock solution diluted to volume and an aliquot taken for the precipitation.

3. If a large excess of reagent has been added, the magnesium oxinate precipitate will be contaminated with reagent. This can be removed by dissolving the precipitate in dilute acid and reprecipitating as described above.

4. Any manganese present in the solution will be recovered with the magnesium. It can be determined photometrically (Chapter 30, p. 326) in the precipitate after destruction of the organic matter with sulphuric and nitric acids.

Titrimetric Determination in Silicate Rocks

DETERMINATION WITH 8-HYDROXYQUINOLINE

The precipitation of magnesium as the complex with 8-hydroxyquinoline can also form the basis of a titrimetric determination of magnesium. As with the gravimetric method, aluminium, iron and other elements of the ammonia group must be absent, whilst any manganese that is not collected in the ammonia precipitate will be collected with the magnesium as the organic complex with 8-hydroxyquinoline. This precipitate is collected, dissolved in warm dilute hydrochloric acid, and a known excess of a standard potassium bromide–potassium bromate solution added to brominate the organic reagent. The excess of oxidising agent is then determined by adding potassium iodide and titrating the liberated iodine with standard sodium thiosulphate solution.

DETERMINATION WITH EDTA

Magnesium and calcium both form stable complexes with EDTA at pH 10, whereas at pH $7 \cdot 6$ only that of calcium is stable. It is therefore theoretically possible, by choosing a pH just above $7 \cdot 6$ and a suitable indicator, to titrate calcium in the presence of magnesium, whereas by using a pH greater than $10 \cdot 0$, calcium and magnesium will be titrated together. In practice, however, calcium is usually titrated at pH of 12 or more, where magnesium is precipitated as hydroxide and then does not interfere. The titrimetric determination of magnesium consists therefore of two separate titrations, that of calcium alone and that of calcium plus magnesium, giving the magnesium content by difference.

Many other metals present in the solution will be similarly titrated and some, such as iron, react irreversibly with the recommended indicators. Iron, aluminium and manganese can be complexed by adding triethanolamine but the presence of more than traces of these elements, even when masked in this way can give rise to colour changes in the indicator, making end-point detection difficult. Even when titrating pure magnesium and calcium solutions the exact end-points are subjective, and for best results it is necessary to complete each titration under the same conditions of dilution, indicator concentration, lighting and viewing.

Indicators that have been suggested for the titrimetric determination of magnesium plus calcium include eriochrome black T (solochrome black T, C.I. 14645),[1] eriochrome blue black B (solochrome black 6B, C.I. 14640),[2] calmagite,[3] metalphthalein with naphthol green B screening,[4] and 4-[bis(carboxymethyl)aminomethyl]-3-hydroxy-2-naphthoic acid (DHNA)[5]. Where magnesium is present in only very small amounts (as, for example, in many limestones), acid alizarin black S.N. (C.I. 21725)[6] and methyl thymol blue[7] can be used to titrate calcium present in solution. In the more general case where calcium and magnesium are both present in quantity, calcein,[8] calcon (eriochrome blue black R, C.I. 15705)[9] or HSN[10] ("Patton and Reeder's Indicator") can all be used for the titration of calcium.

In the procedure described below, magnesium plus calcium are determined with DHNA as described by Clements et al.[5] Calcium is determined by titration of a further aliquot of the same rock solution with EGTA at a pH of 13 in the presence of triethanolamine, with calcein as indicator.[11]

An alternative approach to the titrimetric determination of both calcium and magnesium is that described by Abdullah and Riley[12] in which these elements are separated from other elements present in silicate rocks and from each other by elution from a column of cation exchange resin. This procedure is also described in detail below.

Method

Reagents: *Triethanolamine solution*, equal volumes of triethanolamine reagent and water.

pH 10 buffer solution, dissolve 67·5 g of ammonium chloride in water, add 570 ml of ammonia solution (s.g. 0·880) and dilute to 1 litre with water.

1-Dicarboxymethylaminomethyl-2-hydroxy-3-naphthoic acid reagent (DHNA), mix together 0·05 g of the solid material with 10 g of sodium chloride.

Diaminocyclohexanetetraacetic acid solution (CyDTA), 0·01 M, dissolve 3·3 g of the solid reagent in about 200 ml of water with small additions of M sodium hydroxide solution as necessary. Add acetic acid to bring the pH to 10 and dilute to 1 litre with water. Standardise by titration with the standard magnesium solution using the procedure given below.

Ethylenebis(oxyethylenenitrilo) tetraacetate solution (EGTA), 0·02 M, dissolve 3·81 g of the free acid in 25 ml of 1 M sodium hydroxide solution and dilute to 1 litre with water. Standardise by titration with aliquots of standard calcium chloride solution using the procedure given below.

Potassium hydroxide solution, 1 M, dissolve 14·0 g of the solid material in 250 ml of water.

Calcein reagent, grind together 0·1 g of the solid reagent with 10 g of solid potassium nitrate.

Standard magnesium solution, clean a length of magnesium ribbon and dissolve 0·603 g in about 100 ml of water to which 10 ml of concentrated perchloric acid have been added. Dilute to 1 litre with water. This solution contains 1 mg MgO per ml.

Standard calcium solution, 0·02 M, dissolve 2·002 g of pure, dried calcium carbonate in the minimum quantity of very dilute hydrochloric acid and dilute to 1 litre with water. This solution contains 1·12 mg CaO per ml.

Procedure for the determination of magnesium plus calcium. Accurately weigh approximately 1 g of the finely powdered silicate rock material into a platinum of PTFE basin, moisten with water and add 1 ml of concentrated nitric acid, 5 ml of concentrated perchloric acid and 20 ml of concentrated hydrofluoric acid and allow the basin to stand overnight. Complete the decomposition of the rock material by evaporating to dryness. Add 2 ml of 20 N sulphuric acid and 2 ml of concentrated perchloric acid and evaporate almost to dryness. Repeat this evaporation almost to dryness twice more with intervening additions of small quantities of perchloric acid. Finally add 5 ml of concentrated perchloric acid, 50 ml of water and heat on a water bath until all solid material has dissolved, allow to cool and dilute to volume with water in a 200-ml volumetric flask.

Transfer by pipette 10 ml of the rock solution to a titration vessel and add 80 ml of water, 5 ml of the triethanolamine solution, 10 ml of the prepared buffer solution and about 30 mg of the DHNA indicator reagent. Using a fluorimetric titrator, titrate the total calcium plus magnesium with CyDTA solution under filtered ultraviolet illumination to the end point given by the disappearance of the blue-green fluorescence (note 1).

Procedure for determining calcium. Transfer by pipette 20 ml of the rock solution (note 3) to a titration vessel, add 4 ml of triethanolamine solution (note 4) and 2 ml of 1 M potassium hydroxide solution (note 5). Add about 30 mg of the prepared calcein reagent and titrate slowly with the 0·02 M EGTA solution to the disappearance of the green fluorescence (note 6).

Notes: 1. The precise end point is difficult to distinguish visually under normal lighting conditions due to a slight residual fluorescence, hence the use of a fluorimetric titrator.

2. Barium and strontium are partially titrated in the presence of calcium and magnesium, but do not themselves form fluorescent complexes with DHNA.

3. The solution should contain not more than 40 mg total magnesium plus calcium, with not more than 50 mg of iron plus aluminium.

4. This quantity can be increased up to 8 ml if necessary to complex the amounts of iron, aluminium, etc., present in the sample solution.

5. i.e. one-tenth of the original sample volume.

6. When much magnesium is present, there is a tendency to obtain slightly high results, possibly by as much as 5 per cent. Where magnesium is present in quantity and more accurate results are required then a back titration procedure, as described by Pribil and Vesely,[11] should be used.

The Separation and Determination of Calcium and Magnesium

This procedure, based upon that described by Abdullah and Riley,[12] uses a cation exchange resin to separate iron, aluminium and other elements by elution with EDTA solution at a pH of 4·5. Calcium and magnesium are not eluted with EDXA, but can be recovered successively by elution with ammonium chloride solution. The complete procedure takes several days, but this does not involve a great deal of manipulation and the separations continue whilst other determinations are completed.

Method

Apparatus. *Ion exchange column,* 34 cm in length, 0·8 mm in diameter, filled with 52–100-mesh cation exchange resin such as Zeo-Carb 225. Wash with 250 ml of 2 N hydrochloric acid and 50 ml of water. Convert to the ammonium form by washing with 200 ml of 2 N ammonium chloride solution. Just before use wash with 50 ml of water.

Reagents: *EDTA solution* (for elution), for each separation prepare 250 ml of EDTA(NH₄) solution by titrating 0·55 g of the free acid suspended in 200–225 ml of water with 1 N aqueous ammonia to a pH of 4·5 and dilute to 250 ml with water.

Ammonium chloride solutions, dissolve 26·8 g, 53·5 g and 107 g of ammonium chloride reagent separately in 1 litre volumes of water to give 0·5 M, 1 M and 2 M solutions respectively.

EDTA standard solution, dissolve approximately 1 g of the disodium salt of EDTA in water, add about 0·1 g of magnesium chloride hydrate, transfer to a 1-litre flask and dilute to volume with water. Standardise by titration against standard calcium and magnesium solutions.

Eriochrome black T solution, dissolve 0·25 g of indicator in 50 ml of methanol. This solution is not completely stable and should be discarded after 7 to 10 days.

Procedure. Prepare a solution of the rock material in dilute perchloric acid by evaporating 0·5 g of the sample material with hydrofluoric and perchloric acids as described above. Dilute the solution to 250 ml in a volumetric flask.

Pipette 50 ml of this solution into a 400-ml beaker, add 150 ml of water and transfer to the ion-exchange column. Allow to percolate at a rate of about 40 ml per hour. When all the rock solution has been added, rinse the column with 200 ml of water and then elute the iron, aluminium, etc., at the same rate with 250 ml of the EDTA solution. Rinse the column with 1 litre of water to ensure that all the EDTA is removed and then wash with 200 ml of 0·5 M ammonium chloride solution. Discard all washings.

Elute magnesium with 240 ml of the 0·5 M ammonium chloride solution, collecting the eluate in a 250-ml volumetric flask, test for complete elution of magnesium (note 1) and then elute the calcium with 240 ml of the 1 M ammonium chloride solution, again collecting the eluate in a 250-ml volumetric flask. When this elution is complete, pass 100 ml of 2 M ammonium chloride solution through the column to prepare the resin for the next analysis. Dilute the calcium and magnesium solutions to volume with water, mix well and transfer 100-ml aliquots to separate 250-ml conical flasks. Add to each 10 ml of concentrated aqueous ammonia followed by 2–3 drops of the eriochrome black T indicator solution and then titrate with the standard EDTA solution to a blue-green end-point.

Note: 1. Pass a further 50 ml of 0·5 M ammonium chloride solution through the column and test for magnesium in the eluate by adding 10 ml of concentrated aqueous ammonia and a few drops of the indicator solution. If any magnesium is present, the solution, otherwise blue, will be tinged with pink and should be titrated with the standard EDTA solution.

Photometric Determination in Silicate Rocks

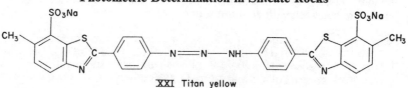

XXI Titan yellow

Very few photometric reagents have found general application for the determination of magnesium in silicate or carbonate rocks, partly because of the general lack of good photometric reagents for magnesium, and partly because other elements tend to interfere excessively. Two reagents that have been applied to this determination are titan yellow and magon—trivial names for the reagents XXI and XXII ($R = H$).

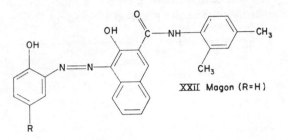

XXII Magon (R=H)

A great deal more is known concerning the application of titan yellow, C.I. 19540 known also as titan yellow 2GS, Clayton yellow, thiazole yellow, acridingelb 5G, azidingelb 5G, and brilliant yellow. This compound is a water soluble triazole dye which forms a reddish coloured lake with freshly formed colloidal magnesium hydroxide. The colloid is protected from precipitation by the addition of other colloids such as agar, starch, gum arabic, and the more recently suggested polyvinyl alcohol, polyacrylate and glycerol. In colloidal solution, the lake formed between magnesium hydroxide and titan yellow has maximum light absorption at a wavelength of 530 nm, although if polyvinyl alcohol has been added, a somewhat higher wavelength of 540 nm is used to prevent interference from a polyvinyl-alcohol complex which has maximum absorption at 490 nm.

Titan yellow is produced by coupling dehydrothio-*p*-toluidine sulphonic acid with its diazonium salt, and commercial samples have been shown [13, 14] to differ considerably in their reactivity towards magnesium. The complexity of the organic product is well known, and inorganic salts (especially sodium chloride) have been known to form the major part of some batches of the reagent. King and Pruden[14] have suggested fractionation of the acetone-soluble material on a Sephadex G-10 (a dextran gel) column, but in a more recent paper[15] a new synthesis is suggested for a titan yellow product that is much superior to that previously available in its reactivity towards magnesium. A calibration graph obtained with this material is shown in Fig. 64.

Elements that interfere with the determination of magnesium include aluminium and other elements of the ammonia group and also calcium and phosphorus. These interferences may be prevented by the addition of suitable masking agents, but this has been found to affect the stability of the magnesium lake and to impair the reproducibility of the results. The optical density of the colloid is also affected by the ammonium salt

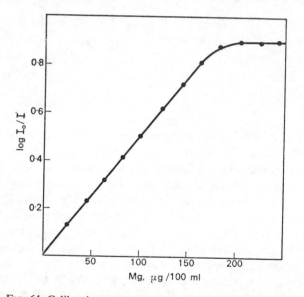

FIG. 64. Calibration graph for magnesium using titan yellow.

concentration, and any scheme for removing the elements of the ammonia group, together with calcium by precipitation with ammonia and ammonium oxalate, must include also a subsequent stage in which the added ammonium salts are removed or destroyed. An alternative procedure for the removal of the elements of the ammonia group described by Evans[16] is based upon their precipitation with sodium succinate at a pH of 6. The quantity of succinate has little effect upon the determination, except to increase slightly the reagent blank value. Calcium is not removed, but interference from it is avoided by adding sucrose.

Magon ($R = $ H) and magon sulphate ($R = $ SO$_3$H) (known also as xylidyl blue) form soluble pink coloured complexes with magnesium.[17, 18] These reagents were used in a procedure for magnesium described by Abbey and Maxwell,[19] in which iron and aluminium were removed by precipitation with ammonia, and an aliquot of the filtrate was evaporated with concentrated hydrochloric and nitric acids to decompose ammonium salts. Triethanolamine was added to complex any residual aluminium and the pink colour was developed in a borax-buffered solution.

The presence of calcium has a slight effect upon the magnesium determination, although interference from this source can be avoided by adding an excess, equivalent to 40 per cent CaO, to give a constant calcium effect.

Magon reagent is blue in colour, and has an appreciable absorption at the wavelength used to measure the absorption of the magnesium complex. A certain instability of the magnesium complex has also been noted.[17] In the recommended procedure two or three standards are analysed with each batch of samples, and the results are calculated by linear interpolation.

Method

This procedure is based upon that described by Evans,[16] it involves the decomposition of the sample material by evaporation with hydrofluoric and perchloric acids, precipitation of iron, aluminium and other elements by boiling with sodium succinate solution, and finally photometric determination with titan yellow.

Reagents: *Sodium succinate.*

Sodium succinate wash solution, dissolve 0·25 g of sodium succinate in 500 ml of water.

Sodium hydroxide 2 M *solution,* dissolve 8 g of the reagent in 100 ml of water.

Titan yellow stock solution, dissolve by boiling 50 mg of polyvinyl alcohol in 50 ml of water, add 100 mg of pure titan yellow and shake to dissolve.

Titan yellow working solution, dissolve by boiling 100 mg of polyvinyl alcohol in 50 ml of water, add 5 ml of the titan yellow stock solution, 4 g of sucrose and 100 ml of glycerol. Transfer to a 200-ml volumetric flask and dilute to volume with water.

Standard magnesium stock solution, dissolve 0·151 g of clean fresh magnesium ribbon in a mixture of 10 ml of perchloric acid and 90 ml of water. Transfer to a 1-litre volumetric flask and dilute to volume with water. This solution contains 250 μg MgO per ml.

Standard magnesium working solution, dilute 10 ml of the stock solution with water to give a volume of approximately 450 ml, bring the pH to 6 by adding sodium succinate wash solution, transfer to a 500-ml volumetric flask and dilute to volume with water. This solution contains 5 μg MgO per ml.

Procedure. Accurately weigh approximately 1 g of the finely powdered silicate rock material into a platinum dish, moisten with water and add 5 ml of concentrated perchloric acid and 20 ml of concentrated hydrofluoric acid. Transfer the dish to a hot plate and evaporate, first to fumes of perchloric acid and then to complete dryness. Allow to cool, rinse down the sides of the dish with a little water, add 5 ml of concentrated perchloric acid and again evaporate to dryness on a hot plate. Repeat the evaporation with perchloric acid twice more, finally dissolving the residue in 5 ml of concentrated perchloric acid and about 40 ml of water.

With many rocks this treatment will give complete dissolution. If any residue remains, it may contain or be contaminated with magnesium fluoride, and should be collected, washed with water, ignited in a silica crucible and fused with a little potassium pyrosulphate. Allow to cool, extract the melt into water containing a few drops of perchloric acid, and add to the main rock solution. Transfer to a 100-ml volumetric flask and dilute to volume with water.

Pipette 10 ml of the rock solution into a 150-ml beaker, add 30 ml of water and, using a pH meter, bring the pH of the solution to a value of 2 by adding 2 M sodium hydroxide solution drop by drop. Add 2 g of solid sodium succinate and a little macerated filter paper, boil the solution for a few minutes and then allow the precipitate to settle. Filter the solution through a close-textured paper into a 200-ml volumetric flask and wash the residue with the sodium succinate wash solution. Rinse the residue back into the beaker, add 1 ml of concentrated perchloric acid and dilute with water to a volume of about 40 ml. Now readjust the pH to a value of 2 and re-precipitate with sodium succinate as before. Allow the precipitate to settle and filter into the 200-ml volumetric flask containing the earlier filtrate. Discard the residue and dilute the combined filtrates and washings to volume with water.

Pipette an aliquot of the solution containing from 10 to 50 μg of magnesium oxide into a 50-ml volumetric flask (note 1) and dilute to 30 ml with water. Now add by pipette 10 ml of the titan yellow working solution, mix by shaking and then add 5 ml of the 2 M sodium hydroxide solution. Dilute to volume with water, mix well and allow to stand for 1 hour. Measure the optical density of the solution in 2-cm cells using the spectrophotometer set at a wavelength of 540 nm.

Prepare also a reagent blank solution in the same way as the sample solution, but omitting the sample material. The colour development and measurement of optical density of the sample solution, reagent blank solution and a set of standard solutions should preferably be made at the same time.

Calibration. Transfer aliquots of 0–10 ml of the standard magnesium solution containing 0 to 50 μg MgO to separate 50-ml volumetric flasks and dilute each solution to a volume of about 30 ml. Add titan yellow and sodium hydroxide solutions as described above, dilute to volume with water and measure the optical densities in 2-cm cells at a wavelength of 540 nm. Plot the relation of optical density to magnesium concentration (Fig. 64).

Notes: 1. All volumetric flasks used for this determination should be acid-washed with 6 N hydrochloric acid, rinsed with water and allowed to drain before use.

2. This procedure can be used to determine the magnesium content of limestones and carbonate rocks with the exception of those containing magnesium as a major constituent (dolomite, magnesite, magnesian limestone, ankerite, etc.). The initial decomposition with hydrofluoric and perchloric acids should be replaced by a digestion with dilute perchloric acid.

Determination of Calcium and Magnesium by
Atomic Absorption Spectroscopy

Flame photometric methods have not achieved any degree of popularity for the determination of magnesium, probably because other elements occurring in silicate rocks seriously interfere with the magnesium emission. Aluminium, silicon, phosphate and sulphate in particular have been implicated, although some improvement can be obtained by adding an excess of calcium or strontium as a releasing agent, and also by working in an aqueous acetone medium.[20]

A similar although very much less severe interference has been noted in the determination of calcium and magnesium by atomic absorption spectroscopy, again principally from the elements silicon, aluminium, phosphate and sulphate. The interference from phosphate and sulphate is considerably reduced by using a high temperature (e.g. air–acetylene) flame, and provided that the sample material is decomposed with a mixture of perchloric and hydrofluoric acid, rather than with sulphuric and hydrofluoric acid, the effect of sulphate and phosphate can be ignored in the analysis of most silicate rocks. Silicon is removed at this stage and the only serious interference is therefore that from aluminium.

The addition of calcium has been shown to lessen the depressing effect of aluminium on the magnesium absorption, but very large amounts of calcium are required for the complete release of magnesium and the determination of calcium is then no longer possible. Lanthanum has also been recommended for this purpose, as in the procedure given below. Rubeska and Moldan[21] who investigated the determination of magnesium, obtained their best results by adding both calcium and 8-hydroxyquinoline dissolved in methanol to the solution before spraying. The 8-hydroxyquinoline served to reduce (although not eliminating completely) the depressing effect of aluminium, whilst the methanol enhanced the magnesium absorption. The standards used for plotting the working curve should have the same concentration of acid, added calcium, 8-hydroxyquinoline and particularly of methanol. Aluminium oxide is reported as forming a mixed compound with calcium and magnesium which is not dissociated in an air–acetylene flame. The hotter flame obtained with a nitrous oxide burner, is however, sufficient for this[22] and the interference from aluminium can thus be avoided.

An interesting method for determining calcium and magnesium by atomic absorption spectroscopy was described by Govindaraju.[23] The rock sample was fused with a mixture of lithium and strontium borates and the fusion product dissolved in dilute citric acid. Under these conditions the interference from the inorganic acids usually used for this determination is avoided and interference from other major elements present in the silicate rock greatly reduced.

Method

The procedure given below is based upon that described by Esson,[24] in which a mixture of perchloric and hydrofluoric acids is used to decompose the silicate sample material, and lanthanum chloride solution is added to serve as releasing agent.

Reagents: *Lanthanum chloride solution,* dissolve 58·6 g of lanthanum oxide by heating with 1 litre of 1·2 N hydrochloric acid. Store in a polyethylene bottle.

        Standard magnesium stock solution, dissolve 0·151 g of pure fresh magnesium ribbon in dilute perchloric acid, transfer to a 1 litre volumetric flask and dilute to volume with water. This solution contains 250 μg MgO per ml.

        Standard magnesium working solution, pipette 10 ml of the stock solution into a 250-ml volumetric flask and dilute to volume with water. This solution contains 10 μg MgO per ml.

        Standard calcium stock solution, dissolve 0·446 g of pure calcium carbonate in dilute perchloric acid, transfer to a 1-litre volumetric flask and dilute to volume with water. This solution contains 250 μg CaO per ml.

Standard calcium working solution, pipette 10 ml of the stock solution into a 250-ml volumetric flask and dilute to volume with water. This solution contains 10 μg CaO per ml.

Procedure. Accurately weigh 0·1 g of the finely powdered silicate rock material into a small platinum crucible, moisten with water and add 5 ml of concentrated hydrofluoric acid and 4 ml of concentrated perchloric acid. Transfer the crucible to a hot plate and decompose the sample material by repeated evaporation with perchloric acid as described above for the photometric determination of magnesium. Dilute the rock solution containing 1 ml of concentrated perchloric acid to volume with water in a 100-ml volumetric flask.

Pipette a volume of from 1 to 10 ml (depending upon the calcium and magnesium content) of the rock solution into a clean 100-ml volumetric flask, add 20 ml of the lanthanum solution, dilute to volume with water and mix well. With the instrument operating in accordance with the manufacturers instructions, spray the solution into the air-acetylene flame of an atomic absorption spectrometer fitted in succession with a calcium and a magnesium lamp. Spray also a reagent blank solution, similarly prepared but omitting the sample material.

Calibration. Prepare a set of standard solutions containing 0–200 μg CaO and 0–200 μg MgO each with 20 ml of the lanthanum solution in separate 100-ml volumetric flasks. Spray in turn into an air–acetylene flame of the spectrometer as for the sample and reagent blank solutions, and plot the absorption against the calcium and magnesium concentration. The calibration graphs may be slightly convex to the concentration axis.

Determination of Magnesium in Carbonate Rocks

Magnesium can be determined in most samples of carbonate rock by methods that are very similar to those used for the corresponding determination in silicate rocks. The addition of hydrofluoric acid is, however, seldom required as complex silicate minerals present in these rocks can usually be converted to acid-soluble calcium silicate by igniting the sample material at about 1000°, prior to acid treatment. Carbonatite rocks containing a wide variety of accessory minerals will probably require further treatment, usually by fusion of an acid-insoluble residue. The composition of the flux required will depend upon the mineralogical composition of the residue.

The gravimetric pyrophosphate and the titrimetric EDTA methods are not ideal for the determination of the small quantities of magnesium that occur in many carbonate rocks, but both photometric and atomic absorption methods are readily applicable. The interferences noted in these two methods for magnesium are not likely to have any serious effects upon its determination, and special methods are not required for application to limestones, except for those rich in magnesium. For these and EDTA titrimetric procedure is the best alternative to the classical gravimetric determination as magnesium pyrophosphate.

Methyl thymol blue and some other indicators commonly used for the EDTA titration of calcium plus magnesium, do not have a sufficiently sharp end-point for the titration of magnesium alone, when present in quantity, and indicators such as eriochrome black T (solochrome black T, C.I. 14645) eriochrome blue-black B (solochrome black 6B, C.I. 14640) are preferred. These are, however, particularly sensitive to the presence of iron and manganese, even when triethanolamine has been added, making the end-point difficult if not impossible to detect. Both iron and manganese can be removed from the solution by a chloroform extraction of the metal complexes with diethyldithiocarbamate. Aluminium is not extracted, but can be complexed with triethanolamine.

Unless a considerable amount of ammonium chloride is added, some magnesium may precipitate as hydroxide when the pH of the solution is adjusted prior to titrating with the EDTA. This precipitation can be avoided by adding the larger part of the EDTA required for the titration before adjusting the pH. This can conveniently be done using a more concentrated EDTA solution than that suggested for the titration.

Under these conditions any calcium present in the sample solution will also be titrated. A separate EDTA titration is therefore also required using a lower pH as for silicate rocks and employing calcein as indicator. Whilst this double titration procedure is adequate for dolomites, ankerites and magnesium limestones, difficulties arise in determining the small amounts of calcium present in magnesites. For these rocks an atomic absorption procedure is recommended for calcium.

Method

The procedure given in detail below is for magnesites that are completely soluble in hydrochloric acid. Rocks containing acid-resistant silicates should be strongly ignited before acid treatment. If a small residue is obtained after digestion with acid, it can usually be decomposed by fusion with a little anhydrous sodium carbonate in the usual way.

Reagents: *Chloroform.*
Diethyldithiocarbamate solution, dissolve 2 g of the sodium salt in 20 ml of water as required.
Ammonium chloride.

Triethanolamine solution, dilute 10 ml of the reagent with 40 ml of water.

Eriochrome black T indicator solution, dissolve $0 \cdot 5$ g of the reagent in 100 ml of methanol. Prepare freshly every 7 to 10 days.

Standard EDTA, $0 \cdot 2$ M *solution,* and *Standard EDTA,* $0 \cdot 05$ M *solution,* prepare by dissolving $37 \cdot 2$ g and $9 \cdot 3$ g respectively of the disodium salt, dihydrate in 500 ml of water, and standardise by titration against a pure magnesium solution.

Procedure. Accurately weigh approximately $0 \cdot 25$ g of the finely powdered magnesite rock material into a 250-ml tall-form or conical beaker and add 50 ml of water. Cover the beaker with a clock glass and carefully add 5 ml of concentrated hydrochloric acid in small portions at a time, down the sides of the beaker. When all the acid has been added and the effervescence has practically ceased, transfer the beaker to a steam bath and heat for about an hour or until complete dissolution is obtained. If any residue remains, collect on a small filter, wash with water and discard (see above if more than a trace of material remains undissolved). Combine the filtrate and washings in a 250-ml volumetric flask and dilute to volume with water, mix well.

Pipette 100 ml of this solution into a 250-ml separating funnel and add aqueous ammonia solution drop by drop until the first appearance of an ammonia group precipitate that does not re-dissolve on shaking. Just clear this precipitate with a drop of concentrated hydrochloric acid and add 1 ml in excess. Add 20 ml of chloroform and 5 ml of diethyldithiocarbamate solution, stopper the funnel and shake for a few seconds before releasing the pressure. Continue shaking for about 1 minute then allow the layers to separate and run off the organic layer. Add 20 ml of chloroform and a further 5 ml of diethyldithiocarbamate solution and repeat the extraction, again removing the organic extract. Repeat the extraction with smaller volumes of chloroform until the extracts are completely colourless.

Discard the organic extracts and transfer the aqueous solution to a conical flask. Rinse the separating funnel with water and add the washings to the conical flask. A small amount of chloroform is usually retained in the aqueous solution, and should be removed by heating the solution almost to boiling. Allow to cool and add 10 ml of $0 \cdot 2$ M EDTA solution followed by 2 g of ammonium chloride, 5 ml of triethanolamine solution and 20 ml of concentrated aqueous ammonia. Add a few drops of eriochrome black T indicator solution and titrate with standard $0 \cdot 05$ M EDTA solution to a colour change from red to a pure blue.

References

1. BIEDERMANN, W. and SCHWARZENBACH G., *Chimia* (1948) **2**, 56.
2. SCHNEIDER F. and EMMERICH A., *Zucker Beih.* (1951) **1**, 53.
3. LINDSTROM F. and DIEHL H., *Analyt. Chem.* (1960) **32**, 1123.
4. TUCKER B. M., *J. Austr. Inst. Agri. Sci.* (1955) **21**, 100.

5. CLEMENTS R. L., READ J. I. and SERGEANT G. A., *Analyst* (1971) **96**, 656.
6. BELCHER R., CLOSE R. A. and WEST T. S., *Talanta* (1958) **1**, 238.
7. KORBL J. and PRIBIL R., *Chem. and Ind.* (1957) p. 233.
8. DIEHL H. and ELLINGBOE J., *Analyt. Chem.* (1956) **28**, 882.
9. HILDEBRAND G. P. and REILLEY C. N., *Analyt. Chem.* (1957) **29**, 258.
10. PATTON J. and REEDER W., *Analyt. Chem.* (1956) **28**, 1026.
11. PRIBIL R. and VESELY V., *Chemist-Analyst* (1966), **55**, 82.
12. ABDULLAH M. I. and RILEY J. P., *Anal. Chim. Acta* (1965) **33**, 391.
13. MIKKELSEN D. S. and TOTH S. J., *J. Amer. Soc. Agron.* (1947) **39**, 165.
14. KING H. G. C. and PRUDEN G., *Analyst* (1967) **92**, 83.
15. KING H. G. C., PRUDEN G. and JANES N. F., *Analyst* (1967) **92**, 695.
16. EVANS W. H., *Analyst* (1968) **93**, 306.
17. MANN C. K. and YOE J. H., *Analyt. Chem.* (1956) **28**, 202.
18. MANN C. K. and YOE J. H., *Anal. Chim. Acta* (1957) **16**, 155.
19. ABBEY S. and MAXWELL J. A., *Anal. Chim. Acta* (1962) **27**, 233.
20. DINNIN J. I., *U.S. Geol. Surv. Prof. Paper* 424-D, p. 391, 1961.
21. RUBESKA I. and MOLDAN B., *Acta Chim., Hung.* (1964) **44**, 367.
22. WALSH J. N. and HOWIE R. A., *Inst. Min. Metall. Trans. B* (1967) **76**, 119.
23. GOVINDARAJU K., *Appl. Spectrosc.* (1970) **24**, 81.
24. ESSON J., Unicam Instruments, Atomic Absorption Methods No. Ca-3 and Mg-4.

CHAPTER 30

MANGANESE

Occurrence

In the igneous silicate rocks, manganese is present in the divalent state, associated largely with the ferromagnesian and accessory iron minerals. In the process of weathering, progressive oxidation leads to the formation of minerals containing tri- and quadrivalent manganese. The highest manganese contents are recorded in the earliest rocks to crystallise, particularly peridotites, basalts and gabbros (Table 27), whilst granitic rocks contain only very small amounts.

TABLE 27. THE MANGANESE CONTENT OF SOME
IGNEOUS ROCKS*

Rock type	MnO, %
Granite and granophyre	0·05
Felsite, rhyolite and obsidian	0·05
Tonalite and granodiorite	0·07
Diorite	0·12
Andesite and trachyandesite	0·15
Gabbro, basalt and dolerite	0·17
Picrite and peridotite	0·17

* From published analyses of igneous rocks of the U.K.[1]

There is a sufficient difference between the ionic radii of magnesium (Mg^{2+}, $R = 66$ pm), iron (Fe^{2+}, $R = 74$ pm) and manganese (Mn^{2+}, $R = 80$ pm) to cause a very definite priority among these three elements with regard to their entry into ionic silicate lattices, manganese being concentrated relative to the other two elements in the later crystalates. Thus, as individual iron and manganese concentrations decrease in the course of magmatic differentiation, the ratio Mn:Fe actually increases.

325

Clays and shales contain similar amounts of manganese to those found in igneous rocks, whilst sandstones and residual sediments in general contain rather less. Carbonate rocks are very variable, ranging from chalk and other limestones which are poor in manganese, to the manganese-rich ankeritic carbonatites. One curious feature of the geochemistry of manganese is the occurrence of accretions containing manganese(IV) in certain oxidate sediments; these serve to concentrate a number of rarer elements.

The principal ore of manganese is the dioxide pyrolusite, MnO_2. Other oxide ores include hausmannite Mn_3O_4, braunite Mn_2O_3, psilomelane and wad as well as the carbonate rhodochrosite $MnCO_3$ and silicate rhodonite $MnSiO_3$.

The Determination of Manganese in Silicate Rocks

Although there are a number of volumetric methods for determining manganese in manganese ores, these are unlikely to be required for the analysis of silicate rocks. Atomic absorption spectroscopy has been shown to be an acceptable alternative to those methods based upon spectrophotometry.

PHOTOMETRIC DETERMINATION

The method most commonly employed is based upon the oxidation of manganese(II) to permanganate with either potassium periodate or ammonium persulphate in the presence of silver ions as catalyst. Attempts to use silver(II) oxide for this purpose were unsuccessful, giving low results for a number of standard samples. It seems likely that some reduction of the permanganate had occurred after the destruction of excess silver(II) oxide. Sodium perxenate has been suggested[2] for the oxidation of manganese to permanganate but in view of the cost of the reagent, is unlikely to be widely used in the analysis of silicate rocks.

Oxidation with potassium periodate proceeds fairly rapidly in nitric or sulphuric acid solution at or near the boiling point and, provided that more than a trace of manganese is present, complete oxidation can be obtained in about 1 hour. According to Nydahl,[3] under these conditions the oxidation is incomplete if only trace amounts of manganese are present, and for this reason he prefers to use ammonium persulphate which gives a much more rapid oxidation. This occurs smoothly in nitric-phosphoric acid solution provided that a catalytic amount of silver ion is present. Just as Nydahl obtained complete oxidation of manganese with persulphate and not periodate, so Langmyhr[4] obtained complete oxidation only with periodate. In the author's laboratory both reagents were found to give complete oxidation, but fading occurred when persulphate solutions were boiled.

Solutions of permanganate obey the Beer-Lambert Law (Fig. 65) and slight variations in the reagent concentrations do not affect the optical density values of the solution. The absorption spectrum consists of a series of maxima, in the range 500 to 575 nm (Fig. 66) and a wavelength of 525 nm is recommended for the determination.

In the determination of manganese, the yellow colour of the ferric ion is removed by adding phosphoric acid. This can introduce difficulties if the sample material contains a great deal of titanium, when titanium phosphate may be precipitated. This can be avoided by increasing the sulphuric acid concentration. Of the elements remaining in the rock solution, only chromium has an appreciable absorption at 525 nm. Interference from chromium can be avoided by making the photometric measurement at a wavelength of 575 nm,[5] but at this wavelength, the absorption curve falls steeply, and the wavelength settings must be made and reproduced very accurately. Since chromate solutions are not reduced by sodium nitrite, chromium interference can be avoided by measuring relative to a reagent blank produced by adding a little sodium nitrite to a separate aliquot of the permanganate solution.

All permanganate solutions should be diluted to volume with very dilute nitric acid that has previously been boiled with either potassium periodate or ammonium persulphate and allowed to cool. If this is not done, and distilled water is used, some fading of the permanganate is likely to be noted. Water from polyethylene wash bottles should be avoided, as this has been implicated as one cause of fading.[6]

Sulphuric or perchloric acid solutions are commonly used for the photometric determination of manganese. These are readily obtained by evaporation of the sample material with hydrofluoric and either sulphuric or perchloric acid, as described on p. 22. Any accessory minerals that remain after this initial attack may contain a large proportion of the total manganese present in the sample material, and provision must be made to recover the manganese in this residual

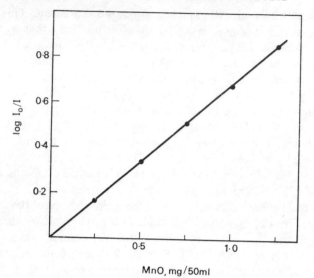

FIG. 65. Calibration graph for manganese as permanganate (1-cm cells, 525 nm).

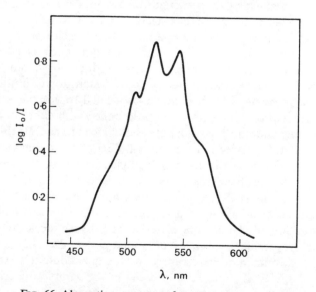

FIG. 66. Absorption spectrum of a permanganate solution.

fraction. With many rocks complete decomposition will be obtained by evaporating the sample portion to dryness with hydrofluoric and sulphuric acids and fusing the dry residue with a little potassium pyrosulphate.

Determination with Potassium Periodate

In the procedure described in detail below, sulphuric and hydrofluoric acids are used to decompose the sample material and, after decomposition of any remaining accessory minerals, the rock solution is diluted to volume for the determination of manganese. This sample solution can be used also for the determination of total iron, titanium and phosphorus if required.

Method

Reagents: *Potassium periodate.*

Sodium nitrite.

Nitric acid solution, boil 1 litre of $0 \cdot 2$ N nitric acid with approximately $0 \cdot 1$ g of potassium periodate, allow to cool and store in an all-glass wash-bottle.

Standard manganese stock solution, accurately weigh $0 \cdot 155$ g of pure manganese into a small beaker, dissolve in 50 ml of $0 \cdot 5$ N sulphuric acid, transfer to a 1-litre volumetric flask and dilute to volume with water. This solution contains 200 μg MnO per ml.

Standard manganese working solution, transfer 25 ml of the stock solution to a 100-ml volumetric flask and dilute to volume with water. This solution contains 50 μg MnO per ml and should be used for the calibration of the 1-cm spectrophotometer cells. If 4-cm cells are to be used, the working solution can be prepared by diluting 5 ml of the stock solution to 100 ml, giving 10 μg per ml.

Procedure. Accurately weigh approximately $0 \cdot 5$ g (note 1) of the finely powdered silicate rock material into a platinum crucible or dish, moisten with a little water and add 1 ml of concentrated nitric acid, 5 ml of 20 N sulphuric acid and 10 ml of hydrofluoric acid. Transfer the dish to a hot plate and evaporate to fumes of sulphuric acid. Allow the dish to cool, rinse down the sides of the dish with a few ml of water, add 5 ml of concentrated hydrofluoric acid and again evaporate, this time to complete dryness. Allow the dish to cool.

Add a small quantity of potassium pyrosulphate to the residue and fuse gently to give a completely fluid melt. Allow to cool, add a little water to the dish and warm to detach the melt from the dish. Rinse the melt into a 100-ml beaker, add 5 ml of concentrated nitric acid, $2 \cdot 5$ ml of syrupy phosphoric acid and approximately 0·2 g of potassium periodate and dilute to a volume of approximately 45 ml. Add a few granules of fused alumina (or other "anti-bump" device, such as a Gernez boiling tube), cover the beaker with a clock glass and gently boil until the purple permanganate colour forms and then no longer deepens in intensity. If the colour does not develop after boiling for about 30 minutes, add a further 0·2 g of potassium periodate and continue boiling for a further period of 30 minutes.

Allow the solution to cool, transfer to a 50-ml volumetric flask and dilute to volume with the dilute nitric acid, previously boiled with a little potassium periodate.

Fill two matched spectrophotometer cells with the coloured sample solution and add to one cell only a small crystal of sodium nitrite. Stir gently with a thin glass rod to complete the decomposition of the permanganate ion. Measure the optical density of the coloured solution, relative to the solution to which the sodium nitrite has been added, using the spectrophotometer set at a wave-length of 525 nm, and hence determine the manganese content of the sample material by reference to the calibration graph or by using a calibration factor (note 2).

Calibration. Transfer aliquots of 5 to 25 ml of the standard manganese solution containing 0·25–1·25 mg MnO (note 3) to separate 100-ml beakers, dilute each aliquot to about 40 ml, add concentrated nitric acid, syrupy phosphoric acid and potassium periodate and continue as described above for the sample solution. Measure the optical density of each solution and plot these values against concentration of manganese.

An optical density of 1·000 corresponds to a manganese concentration of about 1·48 mg MnO per 50 ml, when measured in 1-cm cells, and to about 0·37 mg MnO per 50 ml in 4-cm cells.

Notes: 1. This quantity of material is recommended for acid rocks such as granites and rhyolites and other similar materials low in manganese. A 0·1-g sample weight is adequate for intermediate and basic rocks.

2. If the measured optical density of the sample solution is beyond the range covered by the 1-cm cell calibration graph, transfer 5 ml of the permanganate solution to a 100-ml beaker, add further amounts of concentrated nitric acid, syrupy phosphoric acid and potassium periodate, dilute to about 45 ml and repeat the oxidation as described.

3. These aliquots are suggested for the calibration of the 1-cm matched spectro-photometer cells. When using 4-cm cells, use the more dilute working standard when these aliquots will contain 50–250 μg MnO.

Determination with Ammonium Persulphate

As with the procedure described above, the following procedure with ammonium persulphate can be used for the determination of manganese in the same sample solution as that prepared for the determination of total iron, titanium and phosphorus. Alternatively, a separate, somewhat smaller sample weight can be taken for the determination of manganese only. Mercuric sulphate is added to complex any chloride ion present in the solution. This can be omitted with silicate rocks that contain only traces of chlorine.

Method

Reagents: *Acid reagent solution*, dissolve $36 \cdot 5$ g of mercuric sulphate in a mixture of 200 ml of concentrated nitric acid with 100 ml of water. Add 100 ml of syrupy phosphoric acid and $0 \cdot 017$ g of silver nitrate. When cold, dilute to 500 ml with water.

Dilute acid reagent solution, add about $0 \cdot 5$ g of ammonium persulphate to 25 ml of the acid reagent solution and dilute to 1 litre with water. Boil this solution for 5 minutes, allow to cool and store in a glass wash-bottle.

Ammonium persulphate, use only fresh, good-quality reagent.

Procedure. Decompose a $0 \cdot 5$-g sample portion of the finely powdered rock material by evaporation with nitric, hydrofluoric and sulphuric acids, and fuse the residue with potassium pyrosulphate as described above. Extract the melt with water, dilute to volume in a volumetric flask and transfer an aliquot containing not more than $1 \cdot 25$ mg MnO to a 100-ml beaker. Dilute to about 40 ml with water. Add 3 ml of the acid reagent solution and about $0 \cdot 5$ g of ammonium persulphate. Using a few granules of fused alumina or some other "anti-bump" device, bring the solution to the boil and boil rapidly for 2 minutes but not longer. Cool the solution rapidly and dilute to volume in a 50-ml volumetric flask with the dilute acid reagent solution.

Prepare also a reagent blank solution by diluting a second aliquot of the rock solution, together with 3 ml of the acid reagent solution to volume in a second 50-ml volumetric flask.

Fill a spectrophotometer cell with the coloured permanganate solution, and measure the optical density relative to the reagent blank in a separate cell. Prepare also the calibration solutions as described above, but using the acid reagent and ammonium persulphate as oxidant.

Determination with Formaldoxime

Manganese(II) reacts with formaldoxime in alkaline solution to give an orange-red complex that can be used for photometric measurement. The sensitivity of the determination is approximately five times that using the red-purple colour of permanganate. The most serious interference appears to be that from iron(III) which also forms a highly coloured complex with the reagent. This interference may be overcome by adding ascorbic acid, hydroxylamine hydrochloride and EDTA, as described by Abdullah.[7]

Procedure. Decompose a 0·5 g sample of the finely powdered silicate rock material by evaporation with hydrofluoric and perchloric acids in the usual way, and dilute the solution of metallic perchlorates to 500 ml with water. Transfer by pipette, a 20-ml aliquot of the solution to a 50-ml volumetric flask and add in succession 3 ml of a freshly prepared 0·4 M ascorbic acid solution, 3 ml of a 0·4 M formaldoxime solution and 4 ml of a buffer solution containing 70 g of ammonium chloride and 600 ml of ammonia (s.g. 0·880) per litre. Allow to stand for 2 minutes.

Add 3 ml of a 0·1 M EDTA solution and 4 ml of 2·2 M hydroxylamine hydrochloride solution. Dilute to volume with water, stopper the flask and mix well. Allow to stand for 5–10 minutes, then measure the optical density in 1-cm cells at a wavelength of 490 nm, relative to water. Prepare also a reagent blank and a series of calibration solutions containing up to 0·5 mg Mn in 50 ml.

Photometric Determination of Manganese in Carbonatite Rocks

Most carbonate rocks can be decomposed by warming the finely ground sample material with dilute nitric acid. A suitable aliquot of the solution can then be taken for a photometric determination by either of the methods given above for silicate rocks. When this method is applied to carbonatite rocks, the carbonate mineral fraction is decomposed together with certain of the accessory minerals (sulphides for example), leaving most of the oxide and silicate minerals which may contain an appreciable proportion of the total manganese content of the sample. For these materials it has been found possible to obtain a complete decomposition by fusing a small sample weight with sodium peroxide in a zirconium crucible. After extraction with water, the material is usually completely soluble in dilute nitric acid. After dilution to volume, a suitable aliquot can then be taken for photometric determination of manganese following oxidation with potassium periodate.

The Determination of Manganese by Atomic Absorption Spectroscopy

Flame emission methods have been suggested for determining manganese in silicate material,[8, 9] but have not been extensively employed, possibly because of direct interference from the potassium line at 404 nm with the manganese emission at 403 nm. Phosphate and sulphate ions tend to decrease the manganese emission, whereas chloride and perchlorate ions enhance the flame emission.

In contrast, atomic absorption spectrophotometric methods appear to be increasingly used for the determination of manganese in silicate materials. Allan[10] has reported poor sensitivity by absorption at 403 nm, and recommended a wavelength of $279 \cdot 5$ nm as the most sensitive. No depression or enhancement effects were noted by Trent and Slavin,[11] who used both evaporation with a mixture of sulphuric and hydrofluoric acids and fusion with sodium carbonate to decompose silicate samples, and then sprayed the solution in hydrochloric acid directly into the flame. The rock solution in perchloric acid, prepared as described on p. 82 for the determination of the alkali metals, can also be used for the determination of manganese by atomic absorption spectroscopy.

References

1. GUPPY E. M., *Chemical Analyses of Igneous Rocks, Metamorphic Rocks and Minerals*, Geol. Surv. Gt. Brit., H.M.S.O., 1931 and 1956.
2. BANE R. W., *Analyst* (1965) **90**, 756.
3. NYDAHL F., *Anal. Chim. Acta* (1949) **3**, 144.
4. LANGMYHR F. J. and GRAFF P. R., *Norges Geol. Undersokelse* No. 230, 1965, p. 17.
5. SANDELL E. B., *Colorimetric Determination of Traces of Metals*, Interscience, 1950, 2nd ed., p. 433.
6. RILEY J. P., *Anal. Chim. Acta* (1958) **19**, 421.
7. ABDULLAH M. I., *Anal. Chim. Acta* (1968) **40**, 526.
8. DIPPEL W. A. and BRICKER C. E., *Analyt. Chem.* (1955) **27**, 1484.
9. ROY N., *Analyt. Chem.* (1956) **28**, 34.
10. ALLAN J. E., *Spectrochim. Acta* (1959) (10), 800.
11. TRENT D. and SLAVIN W., *Atomic Absorption Newsletter* (Perkin-Elmer Corp.), No. 19. March 1964.

CHAPTER 31

MERCURY

Occurrence

Very little is known concerning the distribution of mercury in silicate and other rocks. Stock and Cucuel[1] have reported from $0 \cdot 6$ to $0 \cdot 1$ ppm in some magmatic rocks, and the average abundance in igneous rocks is given by Winchester[2] as $0 \cdot 06$ ppm. Somewhat higher values were reported for G-1 and W-1,[3] although it is possible that some contamination had occurred. More recent work by Ehmann and Lovering[4] has indicated higher mercury abundances in the more acid rock types than in the basic or ultrabasic types. Values of $0 \cdot 004$–$0 \cdot 04$ ppm Hg in a series of six standard rocks from the U.S.A. suggest that many of the earlier results were too high. Rocks from mineralised areas may contain high concentrations of mercury. Kaspar and Kral,[5] for example, have reported 1–3 ppm in the average rocks of the mercury province in the Presor Mountains of Eastern Slovakia, and greater concentrations in certain of the rocks.

The Determination of Mercury in Silicate Rocks

The normal techniques of optical spectroscopy and spectrophotometry are not sufficiently sensitive to determine mercury in silicate rocks. Neutron activation analysis can be used,[4, 5] but wherever chemical processing is undertaken care must be exercised to prevent or restrict loss of mercury by volatilisation.

For rocks associated with mineral deposits containing high concentrations of mercury, Popea and Jemăneanu[6] have proposed a method based upon dithizone extraction into carbon tetrachloride solution, 1–5 g of the sample are heated under reflux with sulphuric and nitric acids, and the mercury dithizonate extracted at pH 4–5 from acetic acid solution containing both EDTA and potassium thiocyanate. This method is useful for recovering 2–20 μg mercury. A combination of

334

dithizone extraction with direct atomic absorption spectroscopy using the organic extract was described by Pyrih and Bisque;[7] the detection limit was given as 0·05 ppm Hg in the rock sample.

Vapour Absorption for the Determination of Mercury

This method of determining mercury, known also as flameless atomic absorption and as cold vapour absorption, was used initially[8, 9] to determine mercury in rocks down to about 0·05 ppm. However, its usage has been extended [10, 11] to rocks containing no more than a few ppb (1 in 10⁹). Interference from sulphur dioxide and organic compounds which absorb at the mercury resonance line of 253·7 nm, can be prevented by using a double-beam spectrophotometer and measuring the difference in absorption between the vapour from the rock material and the same vapour from which the mercury has been stripped.

An alternative procedure is to use a purification stage involving the trapping of the evolved mercury in a low temperature (e.g. liquid nitrogen) trap as described by Aston and Riley,[12] or a metallic gold trap which removes the mercury as an amalgam, as described by Warren et al.[13] The procedure given below is based upon that described by Head and Nicholson[14] and Omang and Paus.[11]

Method

In order to ensure that all the mercury present in the rock sample is released, a decomposition procedure based upon the use of hydrofluoric acid is included. A sealed PTFE vessel is used for this, and nitric acid is added to ensure that oxidising conditions are maintained, as mercury is readily lost from reduced solutions. Boric acid is used to complex fluoride ion and stannous chloride to provide the reduced condition at the appropriate moment.

The solution is aerated to remove the mercury which is collected on gold wire. Once collection is complete, the gold wire is heated in a furnace to remove the mercury and the absorption of the vapour measured at 253·7 nm using a standard atomic absorption spectrophotometer with a quartz-window gas cell and a pen recorder.

Apparatus. For the sample decomposition a PTFE-lined bomb with an internal volume of about 110 ml is used.

The additional apparatus required is shown in Fig. 67. It consists of a sample bubbler A, fitted with a magnesium perchlorate drying tube B, a gold wire collector C, furnace D, and quartz-window gas cell E. The gas flow system incorporates a mercury trap F, in the form of a coil of silver wire, a flow meter and a furnace by-pass system. The silica tube furnace, described in detail by Nicholson and Smith,[15] is wire-wound and gives a temperature of 850°. The infrared gas cell is fitted to a standard atomic absorption spectrophotometer in the usual way.

The gold wire collector consists of about 10 g of gold wire of 0·5 mm diam. cut into lengths of about 2 mm packed between gold spirals into the quartz tube furnace.

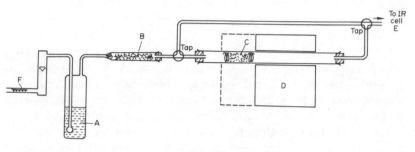

FIG. 67

Reagents: *Boric acid*, saturated aqueous solution.

Stannous chloride solution, dissolve 50 g of stannous chloride in 250 ml of 4 N sulphuric acid.

Standard mercury stock solution, dissolve 0·0540 g of mercuric oxide (HgO) in 250 ml of N sulphuric acid. This solution contains 200 μg Hg per ml.

Standard mercury working solution, dilute 10 ml of the stock solution to 500 ml with N sulphuric acid. This solution contains 4 μg Hg per ml. As required, prepare from this a further dilution with N sulphuric acid to give 0·02 μg Hg per ml.

Procedure. Accurately weigh approximately 0·2 g of the finely powdered rock material into the PTFE-lined decomposition vessel, moisten with water and add 5 ml of concentrated hydrofluoric acid and 0·5 ml of concentrated nitric acid. Seal the bomb, transfer to an electric oven and heat at a temperature of 120° for 10 minutes. Allow to cool to room temperature, open the bomb and add 50 ml of saturated boric acid and heat to dissolve any precipitated fluorides.

Set up the apparatus as shown in Fig. 67, with the 250-ml bubbler containing 50 ml of water. Adjust the air flow through the apparatus to a rate of about 2·5 litres per minute and heat the furnace to a temperature of 850°. Push the silica tube with the gold wire collector into the furnace for 2 or 3 minutes and then divert the air flow over the collector. Any absorbed mercury will be removed and appear as a response on the spectrophotometer recorder. Repeat this operation two or three times, when no further mercury response should be obtained. When this point has been reached, move the silica tube so that the gold wire collector is outside the hot zone of the furnace and allow it to cool. Divert the air stream to bypass the furnace.

Replace the bubbler containing water with a fresh bubbler containing the sample solution to which has been added 2 ml of 20 N sulphuric acid and 2 ml of the stannous chloride solution immediately before the replacement. At the same time, divert the air flow back over the gold wire collector and through the furnace, allowing the evolved mercury to be collected on the gold wire.

After about 2 minutes, divert the air flow to bypass the furnace once again and push the silica tube so that the collector is once again in the hot zone of the furnace, heating for 1 minute. Again divert the air flow and sweep the evolved mercury into the gas cell and record the mercury absorption. For calibration use aliquots of the lowest concentration of working standard solution containing up to 100 ng mercury. The calibration is linear up to at least this quantity, which gives an optical density of about $0 \cdot 215$ in a 150-mm gas cell.

Notes: 1. The silver coil is necessary to remove the small quantity of mercury that is usually found in the air supply.

2. The flow meter, which can be of the Rotameter type, should be calibrated for air in the range $0 \cdot 5$ to $2 \cdot 5$ litres per minute.

References

1. STOCK A. and CUCUEL F., *Naturwiss.* (1934) **22**, 319.
2. WINCHESTER J. W., *Progr. Inorg. Chem.* (1960) **2**, 1.
3. MORRIS D. F. C. and KILLICK R. A., *Talanta* (1964) **11**, 781.
4. EHMANN W. D. and LOVERING J. F., *Geochim. Cosmochim. Acta* (1967) **31**, 357.
5. KASPAR J. and KRAL R., *Sbornik Vysoke Skoly Chem.-Technol. v Praze, Oddil Fak. Anorg. a Org. Techol.* (1958) 281–288.
6. POPEA F. and JEMĂNEANU M., *Acad. R.P.R., Stud. Cercet. Chim.* (1960) **8, 607**.
7. PYRIH R. Z. and BISQUE R. E., *Econ. Geol.* (1969) **64**, 825.
8. VAUGHN W. W. and McCARTHY J. M., *U.S. Geol. Surv. Prof. Paper* 501-D (1964) p. 123.
9. JAMES C. H. and WEBB J. S., *Bull. Inst. Min. Metall.* (1964) **691**, 633.
10. HATCH W. R. and OTT W. L., *Analyt. Chem.* (1968) **40**, 2085.
11. OMANG S. H. and PAUS P. E., *Anal. Chim. Acta* (1971) **56**, 393.
12. ASTON S. R. and RILEY J. P., *Anal. Chim. Acta* (1972) **59**, 349.
13. WARREN H. V., DELAVAULT R. E. and BARAKSO J., *Econ. Geol.* (1966) **61**, 1010.
14. HEAD P. C. and NICHOLSON R. A., *Analyst* (1973) **98**, 53.
15. NICHOLSON R. A. and SMITH J. D., *Lab. Practice* (1972) **21**, 638.

CHAPTER 32

MOLYBDENUM AND TUNGSTEN

Occurrence

Studies of the occurrence and distribution of molybdenum has been published by Ishimori,[1] Kuroda and Sandell,[2] Vinogradov *et al.*[3] and Studennikova *et al.*[4] and of tungsten by Jeffery[5] and Vinogradov *et al.*[3] These studies indicate that both molybdenum and tungsten occur in very small amounts in almost all sedimentary, metamorphic and igneous rocks. Typical values for a number of silicate rock types are given in Table 28, compiled from data from a number of published and unpublished sources.

TABLE 28. THE MOLYBDENUM AND TUNGSTEN
CONTENT OF SOME SILICATE ROCKS

Rock type	Mo, ppm	W, ppm
Granitic rocks	1·5	2·1
Intermediate rocks	1·0	2·0
Basalts and diabases	2·0	1·0
Gabbros	1·9	0·6
Ultrabasic rocks	0·4	0·7
Shales	3	4
Limestones	0·4	0·6

It is possible that regional variation exists in both the molybdenum and tungsten distributions. Both elements are present to an increased extent in shales rich in organic matter and sulphides. Although molybdenum and tungsten occur in many rocks to about the same extent, some occasional large differences are to be expected from the greater chalcophilic nature of molybdenum. Thus, for example, molybdenum tends to accumulate in the residual magma and then crystallise as the disulphide molybdenite, MoS_2 (as, for example, in the adamellite from Shap), whereas tungsten usually crystallises as the iron–manganese tungstate, wolframite.

Molybdenum is frequently associated with chalcopyrite in "porphyry ores", noted as being poor in metal values, but very large in extent. A number of rare molybdates are also known including wulfenite, $PbMoO_4$ and powellite $Ca(Mo,W)O_4$. The chief ores of tungsten are scheelite $CaWO_3$ and wolframite ranging in composition from ferberite $FeWO_4$ to hubnerite $MnWO_4$. Tungstenite WS_2 resembles molybdenite, but is a very rare mineral. Rare tungstates include raspite and stolzite, both $PbWO_4$, and cuproscheelite $(Ca,Cu)WO_4$.

The Determination of Molybdenum and Tungsten

These two elements occur in very small amounts in most silicate rocks. They are difficult to determine by emission spectrography, although a combined solvent extraction-spectrographic technique has been proposed for molybdenum by Edge et al.[6] The abundance data so far recorded have been obtained largely by spectrophotometric methods, with some contribution from neutron activation analysis. Atkins and Smales[7] have described the application of this latter technique for tungsten in rocks and meteorites, whilst Hamaguchi et al.[8] have used it to determine both tungsten and molybdenum in silicates. In both procedures the pure separated tungsten and molybdenum products were determined by their β-activity. The limits of detection were given as $0 \cdot 0021$ ppm W for the procedure of Atkins and Smales[7] and as $0 \cdot 01$ ppm W and about 1 ppm Mo for that described by Hamaguchi et al.[8]

A polarographic method for determining molybdenum in rocks and ores has been described by Holten.[9] The limit for this determination appears to be about $0 \cdot 01$ per cent Mo, which is considerably higher than the molybdenum content of most silicate and other rock types.

Atomic absorption spectroscopy has been used for the determination of trace amounts of molybdenum in silicate materials by Butler and Mathews[10] and by Hutchison.[11] The sensitivity of the technique is insufficient for a direct determination using an acid solution of the rock material after removal of silica, but can be increased by a concentration step involving the solvent extraction of the molybdenum complex with

8-hydroxyquinoline[10] or α-benzoinoxime.[11] Even with these concentration steps, the sensitivity of the method is still insufficient for most normal silicate rocks.

PHOTOMETRIC METHODS

There are only a small number of reagents that have been suggested for the photometric determination of molybdenum and tungsten, and of these only thiocyanate and dithiol have been widely used for the analysis of silicate rocks.

Thiocyanate method. Both molybdenum and tungsten react with alkali thiocyanate to give intense yellow-to-orange colours that can be used for the photometric determination of either element. The reactions occur only in acid solution in the presence of strong reducing agents such as stannous chloride. The coloured species can be measured directly in the aqueous solution, or after extraction into a water-immiscible solvent such as isopentanol or isopropyl ether. If a heavier-than-water solvent is required, a mixture of isopentanol and carbon tetrachloride can be used.

Titanium, vanadium and chromium also form coloured products and can interfere with the determination of both molybdenum and tungsten. In the presence of stannous chloride, iron is reduced to the ferrous state, which does not then react to give a coloured product. In the presence of much iron, some air-oxidation may occur, resulting in the formation of red ferric thiocyanate which colours the organic extracts. However, the presence of a little iron is known to be beneficial, assisting with the colour development. Molybdenum interferes with the determination of tungsten to a greater extent than tungsten interferes with that of molybdenum.

A procedure using this reaction for the determination of molybdenum in silicate rocks was described by Sandell.[12] The rock material was decomposed by fusion with alkali carbonate, and the melt extracted with water. Chromium, vanadium and molybdenum were all determined in aliquots of the filtrate. Difficulties encountered in using this procedure include the precipitation of silica, leading to low recoveries of molybdenum, fading of the organic extracts, also leading to low recoveries and the temperature dependence of the colour-forming reaction, leading to erratic values.

For tungsten, Sandell[13] has described a more detailed procedure based upon a decomposition of the silicate rock material with hydro-fluoric and sulphuric acids. Tungsten was separated from iron and titanium by precipitation with aqueous alkali and from molybdenum by precipitation of the latter as sulphide with antimony as carrier. This separation has been criticised by Chan and Riley[14] who found that at low tungsten levels, some of the tungsten was co-precipitated as sulphide with the molybdenum and antimony. Fading of the organic extracts has also been noted with this procedure. A lower limit of 0·5 ppm (using a 1-g sample weight) has been suggested for this procedure, which is not as sensitive as that for molybdenum, and barely adequate for many basic rocks.

XXIII   Toluene−3,4−dithiol

Method using toluene-3,4-dithiol. Toluene-3,4-dithiol (XXIII), gener-ally referred to by most analysts as "dithiol", forms green-coloured complexes with both molybdenum and tungsten. These are soluble in many organic solvents giving green solutions that can also be used for photometric measurement. Tin, bismuth, copper and other metals form coloured complexes with the reagent, but these are insoluble in most organic solvents. The reagent itself is unstable, and solutions deteriorate slowly on standing, even when thioglycollic acid has been added. For this reason, the solutions should be kept in a refrigerator and discarded after about 14 days. The keeping qualities of the reagent itself can be improved by conversion to the zinc salt or the diacetyl derivative.

The conditions necessary for the quantitative formation of the complexes of molybdenum and tungsten have been the subject of con-siderable investigation, and the reports available are to some extent contradictory. It is, however, clear that the molybdenum complex is formed under conditions of high acidity and that reducing agents are not required. For tungsten, two quite different sets of conditions have been proposed, based respectively upon reaction in hot, strongly acid solution in the presence of a reducing agent, and upon reaction in hot, weakly acid solution without additional reducing agent. Straight-line calibration curves are obtained for both complexes (Fig. 68). The

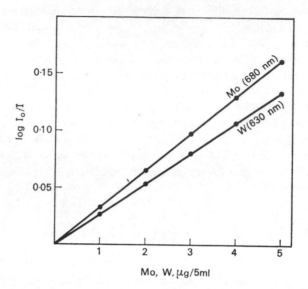

FIG. 68. Calibration graphs for molybdenum and tungsten using dithiol
(1-cm cells, 680 nm (Mo) and 630 nm (W)).

maximum absorption occurs at wavelengths of 630 nm (tungsten) and
680 nm (molybdenum) (Fig. 69).

Chloroform, carbon tetrachloride, light petroleum, isopentyl acetate
and n-butyl acetate have all been used as solvents for the molybdenum
and tungsten complexes, but pronounced fading has been noted with
certain hydrocarbon solvents. This can be avoided by allowing the
solvent to stand over concentrated sulphuric acid until no further
darkening of the acid occurs, and then washing the solvent with aqueous
alkali followed by water.

The formation of the tungsten complex is completely suppressed by
the addition of citric acid,[15] in contrast to the molybdenum complex
which is not suppressed and can be extracted into a suitable organic
solvent. This enables molybdenum to be separated quantitatively from
tungsten, which is not extracted, but can be determined with dithiol
after the destruction of the citric acid remaining in solution. This procedure for separating these two elements gives results that are more

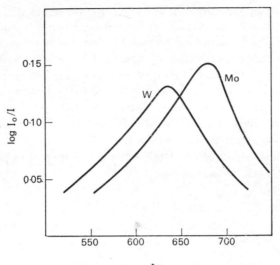

FIG. 69. Absorption spectra of the dithiol complexes of molybdenum
and tungsten (1-cm cells, 5 μg MO (W) per 5 ml extract).

precise than those obtained by an earlier procedure,[16] based upon
simultaneous photometric determination.

The Determination of Tungsten and Molybdenum
in Silicate Rocks

Earlier procedures for molybdenum and tungsten in silicate rocks
were based upon decomposition of the sample by fusion with alkali
carbonate, followed by an aqueous extraction of the melt. It has since
been observed that erratic recoveries of molybdenum and tungsten are
due in part to the failure to recover these two elements quantitatively
in the alkaline filtrate. For this reason an acid decomposition procedure
based upon evaporation with hydrofluoric and sulphuric acids is now
preferred.

Stepanova and Yakumina,[17] who have also advocated this acid
decomposition, suggested that for some silicate rocks it is possible to
extract the molybdenum and tungsten complexes with dithiol directly

from the acid sulphate solution remaining after the removal of silica. For more complex rocks, however, a separation stage is recommended. This involves the extraction of the molybdenum and tungsten complexes with α-benzoinoxime as previously described by Jeffery.[16] Alternative schemes for the separation of molybdenum and tungsten were described by Chan and Riley,[14, 18] involving coprecipation with manganese dioxide followed by ion-exchange separation and by Kawabuchi and Kuroda[19] involving anion exchange separation from acid sulphate solution containing hydrogen peroxide.

The procedure described in detail below is based upon published and unpublished work by the author.

Method

Reagents: *α-Benzoinoxime solution*, dissolve 2 g of the reagent in 100 ml of ethanol.

Chloroform.

Citric acid solution, dissolve 25 g of the reagent in water and dilute to 100 ml.

Toluene-3,4-dithiol solution, dissolve 1 g of the reagent and 5 g of sodium hydroxide in 500 ml of water. When solution is complete, add 5 ml of thioglycollic acid. Store in a refrigerator and discard after 14 days. This reagent can also be prepared from $1 \cdot 4$ g of the zinc complex or from $1 \cdot 5$ g of the diacetyl derivative by dissolution in dilute sodium hydroxide solution.

Carbon tetrachloride.

Standard molybdenum and tungsten stock solutions, dissolve $0 \cdot 150$g of pure molybdic oxide and $0 \cdot 126$ g of pure tungstic oxide separately in small volumes of aqueous sodium hydroxide solution and dilute each to a volume of 1 litre with water. These solutions contain 100 μg Mo and W per ml, respectively.

Standard molybdenum and tungsten working solutions, using a pipette, transfer 10 ml of each of the stock solutions to separate 1 litre volumetric flasks and dilute to volume with water. These solutions contain 1 μg Mo and W per ml respectively.

Procedure. Accurately weigh approximately 1 g of the finely ground silicate rock material into a platinum dish, moisten with water and add 1 ml of concentrated nitric acid, 3 ml of 20 N sulphuric acid and 10 ml of concentrated hydrofluoric acid. Transfer the dish to a hot plate and evaporate to fumes of sulphuric acid. Allow to cool, rinse down the sides of the dish with a little water, add 5 ml of concentrated hydrofluoric acid and again evaporate to fumes of sulphuric acid. Allow to cool, again rinse down the sides of the dish with a little water and evaporate, this time to dryness. Allow to cool and fuse the dry residue with 2 g of potassium pyrosulphate to give a completely fluid melt.

Extract the melt with approximately 100 ml of N hydrochloric acid (note 1), and transfer the solution to a 250-ml separating funnel. Add 2 ml of the α-benzoinoxime solution and mix by shaking. Add 10 ml of chloroform, stopper the funnel and shake for about $1\frac{1}{2}$ minutes to extract the molybdenum and tungsten into organic solution. Allow the phases to separate, remove the organic layer and repeat the extraction with a 5 ml portion of chloroform. Remove the chloroform layer and repeat the extraction with a further 5-ml portion of chloroform. Discard the aqueous layer.

Combine the chloroform extracts in a platinum crucible, add about 50 mg of sodium carbonate and allow the chloroform to evaporate. Dry the residue, ignite and fuse gently over a burner and allow to cool. Extract the melt with 10 ml of water (note 2), transfer the solution to a 100-ml conical flask, add 1 ml of citric acid solution and sufficient 20 N sulphuric acid to bring the final acid concentration to 10 N. Add 5 ml of the dithiol solution, mix and allow to stand for 30 minutes. Add 5 ml of carbon tetrachloride, stopper the flask and shake for 1 minute to extract the molybdenum–dithiol complex into the organic solution.

Transfer the contents of the conical flask to a separating funnel and run off the organic layer. Filter if necessary through a small dry paper into a 1-cm cell and measure the optical density using the spectrophotometer set at a wavelength of 680 nm.

After the extraction of the molybdenum, rinse the aqueous solution with 2 ml of carbon tetrachloride, remove and discard the organic layer. Run the aqueous layer into a 100-ml beaker, add 3 ml of concentrated nitric acid and 2 ml of concentrated perchloric acid and evaporate to dryness of a hot plate. A white residue should remain after the sulphuric acid has evaporated. If the residue is not white, add 1 ml of 20 N sulphuric acid, 3 ml of concentrated nitric acid and 2 ml of concentrated perchloric acid and repeat the evaporation. This oxidation should be repeated again if necessary to give a white residue.

Dissolve the residue in 50 ml of water and, using a pH meter, adjust the pH of the solution to a value of 2·0 by the addition of dilute ammonia or dilute sulphuric acid as necessary. Transfer the solution to a 100-ml conical flask and add 3 ml of the dithiol reagent solution. Place the flask on a hot plate and heat to just below boiling for a period of 30 minutes. Allow to cool, add 5 ml of carbon tetrachloride, stopper the flask and shake for 1 minute to extract the tungsten into the organic layer. Allow the layers to separate and filter the green organic solution through a small dry paper into a 2-cm cell. Measure the optical density using the spectrophotometer set at a wavelength of 630 nm.

Calibration. Transfer aliquots of 0–5 ml of the standard molybdenum solution containing 0–5 μg Mo to a series of 100-ml conical flasks and dilute each solution to 10 ml with water. Add 1 ml of citric acid solution to each followed by sufficient 20 N sulphuric acid to bring the final acid concentration to 10 N. Add 5 ml of the dithiol reagent solution, mix by swirling and allow to stand for 30 minutes. Extract the molybdenum complex into 5 ml of carbon tetrachloride and measure the optical density at a wavelength of 680 nm as described above. Plot the relation of optical density to molybdenum concentration (Fig. 68).

Similarly transfer aliquots of 0–5 ml of the standard tungsten solution containing 0–5 μg W to a series of 100-ml beakers, dilute each solution to about 50 ml with water and adjust the pH to a value of 2 by adding dilute sulphuric acid. Rinse each solution into a 100-ml conical flask with a little water and add 3 ml of the dithiol reagent to each. Transfer the flasks to a hot plate and heat to near boiling for a period of 30 minutes. Allow to cool, extract the tungsten complex into 5 ml of carbon tetrachloride and measure the optical density at a wavelength of 630 nm as described above. Plot the relation of optical density to tungsten concentration (Fig. 68).

Notes: 1. With most silicate rocks little or no residue will be obtained at this stage. If a residue is observed it should be collected, washed with water and ignited and fused in a platinum crucible with a little anhydrous sodium carbonate. Extract the melt with water, acidify with a little hydrochloric acid and add to the main rock solution.

2. Many silicate rocks contain less than 5 ppm of molybdenum and tungsten. With these samples the whole of the rock solution should be taken for the formation and extraction of the complexes with dithiol. Rocks containing larger amounts of molybdenum or tungsten should be diluted to volume and an aliquot taken for the subsequent stages of the analysis.

References

1. ISHIMORI T., *Bull. Chem. Soc. Japan* (1951) **24**, 251.
2. KURODA P. K. and SANDELL E. B., *Geochim. Cosmochim. Acta* (1954) **6**, 35.
3. VINOGRADOV A. P., VAINSHTEIN E. E. and PAVLENKO L. I., *Geokhimiya* (1958) (5), 497.
4. STUDENNIKOVA E. V., GLINKINA M. I. and PAVLENKO L. I., *Geokhimiya* (1957) (2), 136.
5. JEFFERY P. G., *Geochim. Cosmochim. Acta* (1959) **16**, 278.
6. EDGE R. A., DUNN J. D. and AHRENS L. H., *Anal. Chim. Acta* (1962) **27**, 551.
7. ATKINS D. H. F. and SMALES A. A., *Anal. Chim. Acta* (1960) **22**, 462.
8. HAMAGUCHI H., KURODA R., SHIMIZU T., SUGISITA R., TSUKAHARA I. and YAMAMOTO R., *J. Atomic Energy Soc. Japan* (1961) **3**, 800.
9. HOLTEN C. H., *Acta Chem. Scand.* (1961) **15**, 943.
10. BUTLER L. R. P. and MATHEWS P. M., *Anal. Chim. Acta* (1966) **36**, 319.
11. HUTCHISON D., *Analyst* (1972) **97**, 118.
12. SANDELL E. B., *Ind. Eng. Chem., Anal. Ed.* (1936) **8**, 336.
13. SANDELL E. B., *Ind. Eng. Chem., Anal. Ed.* (1946) **18**, 163.
14. CHAN K. M. and RILEY J. P., *Anal. Chim. Acta* (1967) **39**, 103.
15. JEFFERY P. G., *Analyst* (1957) **82**, 558.
16. JEFFERY P. G., *Analyst* (1956) **81**, 104.
17. STEPANOVA N. A. and YAKUMINA G. A., *Zhur. Anal. Khim.* (1962) **17**, 858.
18. CHAN K. M. and RILEY J. P., *Anal. Chim. Acta* (1966) **36**, 220.
19. KAWABUCHI K. and KURODA R., *Talanta* (1970) **17**, 67.

CHAPTER 33

NICKEL

Occurrence

Nickel has long been known as a frequent minor constituent of certain ultrabasic rocks. The ability of nickel ions to substitute for those of magnesium, results in a close association of these two metals, and as magnesium tends to appear in the earliest ferromagnesian minerals to crystallise, these minerals and the rocks in which they appear, are also those with the highest nickel content. Thus for example, Wager and Mitchell[1] reported 0·2 per cent nickel in the earliest olivine of the Skaergaard series, but only 10 ppm in the later, iron-rich olivine.

The depletion of the basic magma in nickel by the crystallisation of the magnesian minerals results in the impoverishment of successive rock types, as shown in Table 29, based on data by Vogt,[2] Goldschmidt[3] and Sandell and Goldich,[4] and in Fig. 41, where the nickel content of a number of silicate rocks has been plotted against the silica content from data given by Unksov and Lodochnikova.[5]

The relation of nickel to magnesium in some silicate rocks from the U.S.A. is shown in Fig. 70.

TABLE 29. THE NICKEL CONTENT OF SOME
SILICATE ROCKS

Rock type	Ni, ppm
Peridotite, dunite	1000–3000
Pyroxenite	800
Norite	350
Basalt, diabase, etc.	50– 300
Syenite, diorite, etc.	10– 60
Granite	2– 10

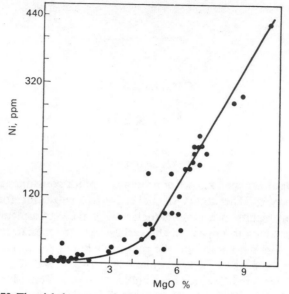

FIG. 70. The nickel content of silicate rocks in relation to the magnesium
content.

Olivine has already been noted as one mineral in which considerable
concentrations of nickel have been reported; others include hypersthene
and certain pyroxenes. The first iron minerals to crystallise from a basic
magma may contain nickel (up to 300 ppm), in contrast to those crystal-
lising later (less than 2 ppm). Nickel minerals include the rare sulphides
millerite NiS, and maucherite Ni_4S_8, and rare arsenides, sulpho-
arsenides and antimonides. Nickel occurs in sulphide minerals of iron
(pentlandite and bravoite) and cobalt (linnaeite and siegenite).

The Determination of Nickel in Silicate Rocks

XXIV Dimethylglyoxime

$$CH_3 - C = NOH$$
$$CH_3 - C = NOH$$

Early procedures for the gravimetric determination following precipitation as sulphide from an ammoniacal solution are now entirely

discarded in favour of either gravimetric or photometric determination with dimethylglyoxime (XXIV). A number of other α-dioximes have been suggested for the determination of nickel, but do not appear to have been widely applied to silicate rocks.

The gravimetric procedure given in detail below is based upon that described by Harwood and Theobald.[6] It can be recommended for those rocks that are rich in nickel, but cannot be applied to those rock materials containing only a few parts per million. For these, photometric methods have been devised based upon the colour given by nickel with dimethylglyoxime in the presence of oxidising agents.[7] This colour can be developed directly in the rock solution, or can preferably be formed after a preliminary separation from other elements present in silicate rocks—as, for example, by extraction of the nickel complex into chloroform.[8]

GRAVIMETRIC DETERMINATION WITH DIMETHYLGLYOXIME

This procedure is rather time consuming, but is preferred where only a small number of samples are to be examined, or where results obtained by other methods are to be checked.

The rock material is decomposed and silica removed by evaporation with hydrofluoric and sulphuric acids in the usual way. The nickel present in the sulphate solution is then precipitated with dimethylglyoxime in the presence of citric acid which minimises the co-precipitation of iron and aluminium hydroxides. The organic precipitate is destroyed by evaporation with nitric acid, and the nickel again precipitated with dimethylglyoxime, this time from a small volume, but again in the presence of citric acid. The red nickel complex is collected on a sintered glass crucible dried and weighed.

The reagent, dimethylglyoxime, is not very soluble in water, and is often added in alcoholic solution. When added in this way, some of the excess reagent may contaminate the precipitated nickel complex; for this reason an aqueous solution of the sodium salt is preferred.

Method

Reagents: *Citric acid.*

Sodium hydroxide solution, dissolve 250 g of the reagent in 250 ml of water.

Methyl red indicator solution.
Dimethylglyoxime solution, dissolve 2 g of the sodium salt in 100 ml of water.

Procedure. Accurately weigh 1–2 g of the finely powdered silicate rock material into a platinum dish and moisten with a little water. Add 10 ml of 20 N sulphuric acid, followed by 1 ml of concentrated nitric acid and 25 ml of concentrated hydrofluoric acid. Transfer the dish to a hot plate or sand bath and evaporate just to fumes of sulphuric acid, stirring as necessary with a platinum rod and taking care that loss by spitting does not occur, particularly in the latter stages of the evaporation. Allow to cool. Add 10 ml of water to the dish and wash down the sides with a further quantity of water. Stir the solution and again evaporate on a hot plate, this time to copious fumes of sulphuric acid. Allow to fume for at least 10 minutes and then allow to cool.

Dilute the cold solution with water and rinse the solution and any residue into a 600-ml beaker using a total of 200–300 ml of water. Heat on a steam bath to bring all soluble material into solution, and filter if necessary (note 1). Wash the residue with water and combine the filtrate and washings.

Add approximately 3 g of solid citric acid to the solution and stir to dissolve. Add concentrated aqueous sodium hydroxide until the solution is alkaline to methyl red. If iron and aluminium hydroxides are precipitated, make acid with sulphuric acid, add an additional 3 g of citric acid, stir to dissolve and again make the solution alkaline to methyl red. Now add dilute sulphuric acid until the solution is just acid. Heat the solution to a temperature of about 50°, then add 25 ml of dimethylglyoxime solution followed by dilute ammonia drop by drop until a test with methyl red indicator paper indicates that the solution is just alkaline, and add 2 or 3 drops of ammonia in excess. Transfer the beaker to a steam bath for 30 minutes, then allow to cool. Collect the precipitate on a close-textured filter paper and wash thoroughly with cold water (note 2).

Using a fine jet of water, rinse the residue into a small beaker and dissolve in a little 8 N nitric acid. Moisten the filter paper with 8 N nitric acid and filter the solution back through the paper, finally washing the paper with small quantities of water. Collect the filtrate and washings in a small beaker.

Add 5 drops of 20 N sulphuric acid to the solution and evaporate on a water bath. Transfer the beaker to a hot plate and heat to fumes of sulphuric acid to complete the expulsion of all nitric acid. If any darkening of the solution occurs, clear this with concentrated nitric acid and again evaporate to fumes of sulphuric acid. Dissolve the residue in a little water and filter if necessary into a 250-ml beaker. The volume at this stage should be about 100 ml. To the solution add approximately 0·1 g of solid citric acid, 10 ml of dimethylglyoxime solution and a few drops of methyl red indicator solution. Now add dilute ammonia until the solution is just alkaline (note 3), followed by 2 to 3 drops in excess. Allow the solution to stand overnight and then collect the red precipitate of the nickel salt on a sintered glass crucible of medium porosity previously dried and weighed. Dry the crucible and precipitate in an electric oven set at a temperature of 120–130°, and weigh. The precipitate contains 20·31 per cent Ni or 25·8 per cent NiO.

Notes: 1. Some silicate rocks are completely attacked by this evaporation and no residue will be observed, with others only a small, pure white residue consisting largely of barium sulphate will be obtained. This is unlikely to contain nickel and can be discarded. Dark-coloured residues may contain an appreciable proportion of the total nickel content of the rock, and should be ignited, fused with a little sodium carbonate (or peroxide if the rock contains chromite,) the melt extracted with water, acidified with sulphuric acid and added to the main solution.

2. Small amounts of iron are usually precipitated with nickel at this stage. With less than about 0·04 per cent Ni, it is not always possible to distinguish the slight precipitate of the nickel complex and for this reason the second precipitation should be made, before reporting the absence of nickel at this concentration level.

3. The indicator colour change may be obscured in the presence of much nickel. Methyl red test paper should be used if any difficulty is experienced in neutralising the solution.

PHOTOMETRIC DETERMINATION FOLLOWING A PRELIMINARY SEPARATION BY CHLOROFORM EXTRACTION CTION

This determination is based upon the formation of a deep red-coloured complex of nickel with dimethylglyoxime in strongly alkaline solution in the presence of an oxidising agent. This reaction was first noted by Feigl,[9] who used bromine as the oxidising agent. Other agents that have been used include persulphate,[10, 11] hypochlorite, periodate and iodine.

Two coloured compounds can be formed in solution. Hooreman[12] has suggested that in these two compounds nickel and dimethylglyoxime are present in the ratios 1:2 and 1:4 respectively. The 1:2 complex is unstable and for this reason analytical procedures for nickel are designed to favour the formation of the 1:4 complex. This is done by using an excess of the reagent, by restricting the quantity of ammonia added and by using sodium hydroxide to increase the alkalinity of the solution. In the procedure described below, broadly based upon that of Rader and Grimaldi,[13] potassium persulphate is used as oxidising agent, and a prior extraction of the nickel complex into chloroform is used to separate the nickel from other elements such as iron, that interfere. Citric acid is added to the rock solution to prevent the precipitation of iron and aluminium hydroxides, and hydroxyammonium chloride added to prevent the oxidation of manganese. Any traces of copper that are extracted with the nickel are removed by shaking the chloroform extract with dilute aqueous ammonia.

Method

Reagents: *Hydroxyammonium chloride.*

Sodium hydroxide solution, dissolve 250 g of the reagent in 250 ml of water, stirring as necessary.

Dimethylglyoxime solution, dissolve 1 g of the reagent in 100 ml of ethyl alcohol. Store in a stoppered bottle.

Sodium citrate solution, dissolve 10 g of the dihydrate in water and dilute to about 100 ml.

Phenolphthalein indicator solution.

Potassium persulphate solution, dissolve 5 g of the pure fresh reagent in water and dilute to 100 ml in a volumetric flask. This solution deteriorates on keeping, and should be prepared as required.

Chloroform.

Standard nickel stock solution, dissolve 0·405 g of nickel chloride hexahydrate in water containing a few ml of concentrated hydrochloric acid, and dilute to 1 litre. This solution contains 100 μg Ni per ml.

Standard nickel working solution, dilute 25 ml of the stock solution to 250 ml with water in a volumetric flask. This solution contains 10 μg Ni per ml.

Procedure. Accurately weigh 1–2 g of the finely powdered rock material into a platinum dish and moisten with a little water. Add 5 ml of concentrated nitric acid, 5 ml of concentrated perchloric acid and 15 ml of concentrated hydrofluoric acid. Transfer the dish to a hot plate or sand bath and evaporate to copious fumes of perchloric acid. Allow to cool, wash down the sides of the dish with a little water, add a further 5 ml of water to the dish and again evaporate on a hot plate or sand bath—this time until the residue remains just moist. Do not allow to dry completely.

Add 1–2 ml of concentrated hydrochloric acid and 20 ml of water and warm, adding a further quantity of water if necessary to complete the dissolution of all soluble material. If any unattacked residue remains, collect on a small filter paper and wash well with cold water (note 1). Dilute the combined filtrate and washings to 100 ml in a volumetric flask.

Transfer 25 ml of this solution (for rocks containing more than about 200 ppm Ni use a smaller aliquot containing not more than 100 μg Ni and dilute to 25 ml with water) to a 60- or 100-ml separating funnel and add 10 ml of sodium citrate solution (note 2). Now add concentrated ammonia to bring the pH of the solution to 9 (just pink to phenolphthalein), making the final adjustments with dilute ammonia or hydrochloric acid. Add approximately 0·1 g of hydroxyammonium chloride and 3 ml of dimethylglyoxime solution to the separating funnel, shake well and allow the solution to stand for 5 minutes.

Add 10 ml of chloroform to the solution and shake for 2 minutes. Allow the phases to separate and then transfer the lower, organic layer to a second separating funnel. Repeat the extraction of the aqueous solution twice more with 5-ml aliquots of chloroform and combine the three extracts. Discard the aqueous layer. Now add 10 ml of the dilute ammonia to the chloroform solution, shake for 2 minutes and again separate the organic layer and transfer it to a third separating funnel. Add 3 ml of chloroform to the aqueous phase remaining in the second funnel, shake for 1 minute and add the chloroform layer to the main extract in the third funnel. Discard the aqueous phase.

Add 10 ml of $0 \cdot 5$ N hydrochloric acid to the chloroform solution and shake vigorously for 2 minutes. Transfer the lower, organic layer to yet another separating funnel and shake it with a further 10-ml portion of $0 \cdot 5$ N hydrochloric acid. Discard the chloroform layer. Combine the acid extracts and filter through a small close-textured filter paper into a 50-ml volumetric flask. Using no more than 10 ml of water rinse the separating funnels, filter paper and filter funnel, adding the washings to the main solutions. Add 2 ml of the sodium citrate solution to the flask, followed by $0 \cdot 6$ ml of concentrated sodium hydroxide solution to give a pH of 12 or just over. Add 10 ml of potassium persulphate solution followed by 3 ml of dimethylglyoxime solution. Dilute to volume, mix well and allow to stand for between 30 and 60 minutes for the colour to develop.

Measure the optical density of the solution in 4 or 5 cm cells, using the spectrophotometer set at a wavelength of 530 nm. The optical density of a reagent blank extract should also be measured. This extract is prepared in the same way as the sample extract, but from 25 ml of water in place of 25 ml of the rock sample solution. Calculate the nickel content of the sample material by reference to the calibration graph.

Calibration. Transfer aliquots of 0–12 ml of the nickel solution containing 0–120 μg Ni, to separate 50-ml volumetric flasks and add to each 2 ml of sodium citrate, $0 \cdot 6$ ml of sodium hydroxide solution, 10 ml of potassium persulphate solution and 3 ml of dimethylglyoxime solution. Dilute each to volume with water, mix well, allow to stand and then measure the optical densities as described above. Plot the relation of optical density to nickel concentration (Fig. 71).

Notes: 1. If the residue contains dark gritty particles of unattacked sample material, further treatment to recover any contained nickel is imperative. The dried and ignited residue should be fused with sodium carbonate, leached with water, acidified with hydrochloric acid and added to the main solution. Where the rock sample contains chromite, it may be necessary to sinter or fuse the unattacked residue with sodium peroxide in a zirconium crucible.

2. An increased quantity of sodium citrate solution may be needed to keep all the iron and aluminium of some basic rocks in solution.

PHOTOMETRIC DETERMINATION AFTER
ION-EXCHANGE SEPARATION

In this procedure bromine water is used to oxidise nickel in aqueous solution prior to forming the soluble red-coloured complex with

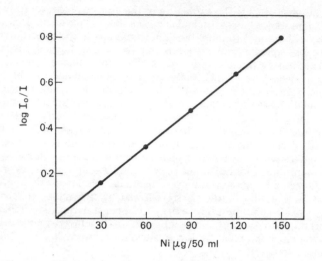

Fig. 71. Calibration graph for nickel as complex with dimethylglyoxime
after oxidation with persulphate (2-cm cells, 530 nm).

dimethylglyoxime. Nickel solutions prepared in this way are unstable,
and the optical densities should be measured as soon as possible after
preparation.

The separation given below is based upon the removal of iron and
other interfering elements by anion exchange. The chloride complexes
of iron, copper, cobalt, cadmium and zinc can be retained on a column
of strongly basic anion exchange resin, enabling a separation to be made
from nickel, aluminium, titanium, calcium, magnesium and the alkali
metals which are not retained. The procedure given is similar to that
described by Lieberman[14] for nickel in copper ores, and by Easton
and Lovering[15] for nickel in chondritic meteorites.

Method

Apparatus. *Strongly basic anion exchange resin*, such as Dowex 1 × 4, in
the form of a small column 12 cm in length, 1 cm in diameter. Wash the bed
in succession with four column volumes each of 10 M hydrochloric acid,
0·5 M hydrochloric acid, water and again with 10 M hydrochloric acid.

Reagents: *Bromine water*, saturated.

Sodium citrate and *dimethylglyoxime solutions*, as described above.

Procedure. Decompose a 1–2 g portion of the finely ground sample material by evaporation with concentrated hydrofluoric, nitric and perchloric acids as described above. Any insoluble residue must be collected, decomposed and added to the main solution, prior to the evaporation to give a moist residue. Dissolve this residue in the smallest quantity of 10 M hydrochloric acid and transfer the solution to the top of the ion-exchange resin bed.

Elute the nickel from the column with four column volumes of 10 M hydrochloric acid, collecting the eluate in a small beaker. (The resin column must now be washed in succession with 0·5 M hydrochloric acid, water and 10 M hydrochloric acid before it can be used again for the separation of nickel.) Transfer the beaker to a hot plate and evaporate the solution almost but not quite to dryness to remove the excess of hydrochloric acid. Transfer the remaining solution to a 100-ml volumetric flask and dilute to volume with water.

Transfer an aliquot of this solution containing not more than 60 µg Ni to a 50-ml volumetric flask and add 2 ml of sodium citrate solution and 1 ml of saturated bromine water. This quantity of bromine should be sufficient to impart a distinct yellow colour to the solution. Allow to stand for about 5 minutes. Add concentrated aqueous ammonia drop by drop until the colour is discharged, then add 1 ml of ammonia in excess and 1 ml of dimethylglyoxime solution. Dilute to volume with water, mix well and measure the optical density against water. For this measurement use 4-cm cells and the spectrophotometer set to a wavelength of 450 nm. Determine also a reagent blank value and calculate the nickel content of the sample material by reference to the calibration graph.

Calibration. Transfer aliquots of 0–12 ml of the nickel solution containing 0–120 µg of nickel, to separate 50-ml volumetric flasks. Add 2 ml of sodium citrate solution to each, followed by bromine water to give a yellow-coloured solution. Now add ammonia solution drop by drop, to discharge the colour, excess ammonia and dimethylglyoxime solutions as described above. Measure the optical density of each solution in 4-cm cells using the spectrophotometer set at a wavelength of 450 nm, and plot the relation of optical density to nickel concentration.

Determination of Nickel by Atomic Absorption Spectroscopy

Beccaluva and Venturelli[16] have reported the interference of iron with the determination of nickel by atomic absorption spectroscopy when using an air–acetylene flame. A slight enhancement effect was noted, and to prevent this 300 µg per ml of iron was added to each of the nickel standards. Iida and Nagura[17] reported that the major components of silicate rocks and also mineral acids do not interfere with the determination of nickel by this technique, but first removed iron by extraction with isopropyl ether. Absorption measurements were made at a wavelength of 232·0 nm. The technique can be used to determine a few parts per million, but for less than 1 ppm nickel, the nickel should be extracted with dithizone and transferred to hydrochloric acid solution for photometric measurement using an air–hydrogen flame.

References

1. WAGER L. R. and MITCHELL R. L., *Geochim. Cosmochim. Acta* (1951) **1**, 129.
2. VOGT J. H. L., *Econ. Geol.* (1923) **18**, 307.
3. GOLDSCHMIDT V. M., *J. Chem. Soc.* (1937) p. 655.
4. SANDELL E. B. and GOLDICH S. S., *J. Geol.* (1943) **51**, 99 and 167.
5. UNKSOV V. A. and LODOCHNIKOVA N. V., *Geokhimiya* (1961) (9), 732.
6. HARWOOD H. F. and THEOBALD L. S., *Analyst* (1933) **58**, 673.
7. ROLLET A., *Compt. Rend.* (1926) **183**, 212.
8. SANDELL E. B. and PERLICH R. W., *Ind. Eng. Chem., Anal. Ed.* (1939) **11**, 309.
9. FEIGL F., *Ber.* (1924) **57**, 758.
10. HAAR K. and WESTERVELD W., *Rec. Trav. Chim.* (1948) **67**, 71.
11. CLAASEN A. and BASTINGS L., *Rec. Trav. Chim.* (1954) **73**, 783.
12. HOOREMAN M., *Anal. Chim. Acta* (1949) **3**, 635.
13. RADER L. F. and GRIMALDI F. S., *U.S. Geol. Surv., Prof. Paper* 391-A, 1961.
14. LIEBERMAN A., *Analyst* (1955) **80**, 595.
15. EASTON A. J. and LOVERING J. F., *Geochim. Cosmochim. Acta* (1963) **27**, 753.
16. BECCALUVA L. and VENTURELLI G., *Atom. Absorp. Newsl.* (1971) **10**, 50.
17. IIDA C. and NAGURA M., *Japan Analyst* (1968) **17**, 17.

CHAPTER 34

NIOBIUM AND TANTALUM

Occurrence

The well-known association of niobium and tantalum minerals with those of lithium and beryllium in pegmatite deposits reflects the tendency of these elements to accumulate together in residual magmas and to crystallise as separate distinct minerals in the later stages of granite emplacement. Alkali undersaturated rocks contain appreciable quantities of niobium but only relatively small amounts of tantalum. The association of niobium minerals, particularly pyrochlore (a calcium niobate), with carbonatite rocks has been recorded from many localities, whilst pyrochlore has also been noted as an accessory mineral in certain rare granite masses. Basic rocks contain significantly less niobium than acidic rocks (Table 30). In almost all silicate rocks the niobium content exceeds that of tantalum, with ratios ranging from about 4:1 in basic rocks to 400:1 in certain nepheline syenites.

TABLE 30. THE NIOBIUM AND TANTALUM
CONTENT OF SOME SILICATE ROCKS *

Rock type	Nb, ppm	Ta, ppm
Nepheline syenite	200	0·8
Syenite	30	2·0
Granite†	20	4·2
Diorite	3·6	0·7
Gabbro	19	1·1
Ultrabasic rocks	16	1·0

* Table based upon results by Rankama[1, 2] and unpublished work by Dubois and Jeffery.

† Znamenskii et al.[3] give the abundances in granite rocks as 14 ppm Nb and 1 ppm Ta, Edge and Ahrens[4] give 11·5, 11·6, 13·4 and 15·9 ppm Nb (average 13·1 ppm) for four granitic rocks from South Africa.

The Determination of Niobium

A number of reagents have been suggested for the determination of niobium, although few have found general application to the analysis of rocks and minerals. Some of the reagents suggested are given in Table 31, which also gives the molar absorptivities of the complexes formed. The high sensitivities of procedures based upon thiocyanate and 4-(2-pyridylazo)-resorcinol (PAR) have made these two reagents of particular importance in the examination of silicate rocks. Procedures based upon the formation of a peroxy complex, although insufficiently sensitive for application to most silicates, can be used for certain rocks and minerals where some degree of niobium enrichment can be expected. As an example of this, details are given below of a procedure for determining niobium in niobium-bearing carbonatites; this is based upon a published procedure for pyrochlore soils.[5]

TABLE 31. MOLAR ABSORPTIVITIES OF
SOME NIOBIUM COMPLEXES

Reagent	
Hydrogen peroxide	950*
Thiocyanate	32,400†
4-(2-Pyridylazo)- resorcinol	35,600[7]
	35,500[8]
Xylenol orange	16.000[9]
Tribromopyrogallol	6,170[10]
Lumogallion	16,400[11]

* Charlot[6] gives 892.
† With ether extraction.

Determination of Niobium in Carbonatites with Hydrogen Peroxide

Niobium is one of a number of elements forming yellow-coloured complexes with hydrogen peroxide. The only serious interference, however, is from titanium which forms its peroxy complex under somewhat similar conditions to that of niobium. This interference from titanium can be avoided by adding phosphoric acid to the solution of niobium in concentrated sulphuric acid, and by making the measurement of optical density at a wavelength of 356 nm (Fig. 72).

The conditions required for maximum sensitivity to niobium coupled with minimum interference from titanium were investigated by Pickup,[12] who reported that whereas the niobium peroxy colour decreased in intensity with increasing water content of the solution, the corresponding titanium colour increased. Similarly, provided that sufficient hydrogen peroxide was added to form the niobium complex, further increase

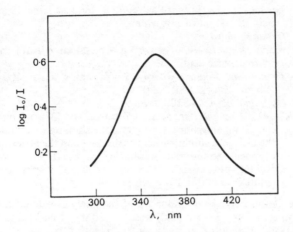

FIG. 72. Absorption spectrum of the niobium complex with hydrogen peroxide (1-cm cells, 10 mg Nb_2O_5/100 ml).

did not affect the intensity of the niobium colour, but increased the titanium colour. As phosphoric acid contains more water than sulphuric acid, an increase in the phosphoric to sulphuric acid ratio also adversely affected the niobium intensity and enhanced the titanium colour.

The following procedure has been devised for carbonatite rocks in which niobium occurs largely in the form of the mineral pyrochlore. Titanium minerals such as perovskite may also be niobium-bearing. These minerals are decomposed by the combination of evaporation with hydrofluoric acid and fusion with potassium pyrosulphate. The niobium present is separated by tannin precipitation using titanium as a carrier with cinchonine added to ensure complete precipitation of both

titanium and niobium, and the determination completed spectrophotometrically as the peroxy complex of niobium in a mixed sulphuric–
phosphoric acid solution.

Method

Reagents: *Hydroxyammonium chloride.*

     Titanium solution, dissolve 100 mg of titanium dioxide in 10 ml
of concentrated sulphuric acid, dilute to 100 ml with 2 N
sulphuric acid.

     Tannin solution, prepare as required by dissolving 10 g of tannic
acid in 100 ml of hot water.

     Cellulose powder.

     Cinchonine solution, dissolve 5 g of cinchonine in 100 ml of 0·2 N
hydrochloric acid.

     Tannin wash solution, dissolve 2 g of tannic acid in 100 ml of 0·2
N hydrochloric acid.

     Hydrogen peroxide, 100 volume.

     Standard niobium stock solution, fuse 0·143 g of pure ignited
niobium pentoxide with 1 g of potassium pyrosulphate. Extract
the melt into sulphuric acid as described in the procedure below,
evaporate to fumes of sulphuric acid and dilute to volume in a
100-ml volumetric flask with concentrated sulphuric acid. This
solution contains 1 mg Nb per ml.

Procedure. Accurately weigh about 2 g of the finely powdered sample into
a 400-ml beaker, cover with a clock glass, add 50 ml of water and with care
10 ml of concentrated nitric acid. When the reaction has largely subsided,
heat on a hot plate until no further action is apparent. Collect the residue on a
small close-textured filter paper, and wash well with water. Discard the filtrate
(note 1). Transfer the paper and residue to a small platinum dish, dry and burn
off the paper. Add 5–10 ml of concentrated hydrofluoric acid to the residue
and evaporate just to dryness on a hot plate (note 2). Add 3 g of potassium
pyrosulphate and fuse gently until a clear melt is obtained. Allow to cool and
dissolve the melt in 10 ml of 4 N sulphuric acid and add 150 ml of water.

Add 50 ml of concentrated hydrochloric acid and then make the solution
just alkaline by adding ammonia solution. Add 10 ml of the titanium solution
followed by 7·5 ml of concentrated hydrochloric acid and 2 g of hydroxyammonium chloride. Heat the solution nearly to boiling on a hot plate and
keep hot until the yellow colour of ferric iron in the chloride solution disappears. Dilute the solution to a volume of 450 ml with water, bring to the boil,
stir in 25 ml of the tannin solution and allow to stand on a water bath for a
period of 15 minutes. Stir in 1 g of cellulose powder, add 5 ml of cinchonine
solution and return the beaker to the water bath for a further period of 45
minutes.

While still hot, filter the solution through a hardened open-textured filter paper and wash well with the tannin wash solution. Transfer the paper and precipitate to a silica crucible (note 3), dry in an oven, burn off the paper and other organic material and ash over a burner. Fuse the residue with $3 \cdot 5$ g of potassium pyrosulphate. Cool, add 2 ml of 4 N sulphuric acid and warm to dislodge the melt. Transfer to a 100-ml beaker containing about 50 ml of 4 N sulphuric acid and 4 drops of hydrogen peroxide, and warm to complete the dissolution. Evaporate the solution to copious fumes of sulphuric acid and continue fuming for about 20 minutes.

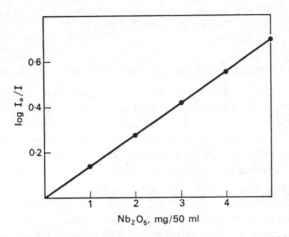

FIG. 73. Calibration graph for niobium using hydrogen peroxide (1-cm silica cells, 356 nm).

Cool the solution and transfer to a dry 50-ml volumetric flask containing 10 ml of phosphoric acid, rinsing the beaker with concentrated sulphuric acid until the solution is diluted almost to volume. Mix well, allow to cool, dilute to volume with concentrated sulphuric acid and again mix well. Fill a 25-ml volumetric flask with this solution, add exactly $0 \cdot 1$ ml of hydrogen peroxide and mix well. Measure the optical density of this solution relative to the solution remaining in the 50-ml flask, in 1-cm cells, with the spectrophotometer set at a wavelength of 356 nm.

Calibration. Prepare a calibration graph from aliquots of the standard niobium solution containing from 2–10 mg Nb. Transfer each aliquot to a 50-ml volumetric flask containing 10 ml of phosphoric acid and proceed as described above for the sample solution. Plot the relation of optical density to niobium concentration (Fig. 73).

Notes: 1. Niobium and titanium minerals are insoluble in dilute nitric acid, and only very small amounts of niobium are lost by discarding this filtrate.

2. In the presence of cerium, hydrogen peroxide is rapidly decomposed and fading of the yellow colour of the niobium complex then occurs. If rare earths are present, they should be removed by filtration as insoluble rare earth fluorides after treatment of this residue with hydrofluoric acid. Evaporate the filtrate just to fumes with sulphuric acid and then to dryness.

3. Platinum introduced into the solution from platinum apparatus also serves to catalyse the decomposition of hydrogen peroxide.

4. An adaptation of this procedure can be used for determining niobium in pyrochlore granites. A $0 \cdot 5$ g sample is decomposed by evaporation to dryness with hydrofluoric and sulphuric acids, the residue fused with potassium pyrosulphate and the melt dissolved in 4 N sulphuric acid. The niobium is precipitated with tannin and the determination completed as described above.

The Determination of Niobium in Silicate Rocks with Thiocyanate

Niobium forms a yellow-coloured thiocyanate that can be used as the basis of a spectrophotometric determination. Two general procedures have been developed involving extraction into organic solution,[13] and formation in a homogeneous acetone–water solution.[14, 15] The former has been used by a number of workers[16–18] for the examination of silicate rocks, and is given in detail below in the form proposed by Esson.[19]

The yellow complex has an absorption maximum at 385 nm (Fig. 74), and the calibration graph is a straight line over the range 0–10 μg Nb/25 ml ethyl acetate (Fig. 75). The intensity of the colour is time-dependent, although the change in optical density is small when the solutions are stood for 1 hour. The amount of niobium thiocyanate extracted from the aqueous phase into the organic is temperature-dependent, and Grimaldi has recommended that a standard niobium solution should accompany the sample solution through the colour formation and extraction stages when the room temperature differs by more than 2° from that at which the calibration was undertaken. The "ageing" of reagents may also affect the intensity of the colour, and it is recommended that the stannous chloride and thiocyanate solutions should be freshly prepared.

Even the small amount of platinum that can be removed from platinum apparatus in the course of a hydrofluoric–sulphuric acid attack of a rock sample is sufficient to interfere with the determination. For this reason Grimaldi recommends a decomposition procedure based upon fusion

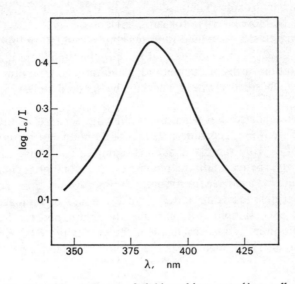

FIG. 74. Absorption spectrum of niobium thiocyanate (4-cm cells, 6 μg Nb_2O_5/25 ml).

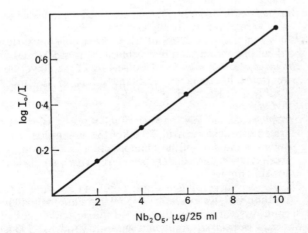

FIG. 75. Calibration graph for niobium using thiocyanate (4-cm cells, 385 nm).

with sodium hydroxide. The acid attack in PTFE dishes described by Esson may be more effective for some minerals.

Interfering elements include tungsten, molybdenum, uranium and a number of other rarer elements that are all separated from niobium in the precipitation stages of the method. Titanium and tantalum interfere but to a much smaller extent and can be ignored unless present in quantity.

Esson has reported that when the multi-stage separation and extraction procedure is used, considerable loss of niobium can occur and that this loss varies from sample to sample. Samples containing more iron tended to give better niobium recoveries, presumably because ferric hydroxide acts as a carrier for niobium during precipitation. To correct for these variable losses he added a small amount of radioactive niobium-95 to each sample and subsequently determined the "chemical yield" of niobium by measurement of the γ-activity of the final thiocyanate extract. This is included in the method described in detail below.

Method

Reagents: *Ammonium chloride solution*, dissolve 10 g of reagent in 500 ml of water.

Sodium hydroxide solution, dissolve 50 g of reagent in 500 ml of water.

Ammonium thiocyanate solution, dissolve 25 g in water and dilute to 100 ml. Prepare freshly each day.

Leach solution, mix 20 ml of 25 per cent aqueous tartaric acid, 65 ml of concentrated hydrochloric acid and 115 ml of water.

Stannous chloride solution, dissolve 40 g of the dihydrate in concentrated hydrochloric acid and dilute to 100 ml. Prepare freshly for each batch of samples.

Ethyl acetate.

Stripping solution, mix together 80 ml of water, 40 ml of concentrated hydrochloric acid, 30 ml of the ammonium thiocyanate solution and 3 ml of the stannous chloride solution. Prepare a fresh solution immediately before use. This quantity is sufficient for six samples.

Niobium-95 tracer solution, dilute 1 mCi of the carrier-free oxalate solution with water so that 2–5 ml gives approximately 10,000 cpm.

Standard niobium stock solution, accurately weigh 0·143 g of pure dried niobium pentoxide into a silica crucible and fuse with 2 g of potassium pyrosulphate. Allow to cool, dissolve in

10 ml of concentrated sulphuric acid and when cold, transfer to a 250-ml volumetric flask. Rinse the crucible with two 5-ml portions of sulphuric acid, and add these to the solution in the flask. Add 120 ml of 20 N sulphuric acid, dilute almost to volume with water and mix well. Allow the solution to cool to room temperature, make up to the mark with water and again mix thoroughly. This solution contains 400 μg Nb per ml.

Standard niobium working solution, prepare this as required by diluting 5 ml of the stock solution to 100 ml in a volumetric flask with sufficient sulphuric acid to give a final concentration of 12 N, and using the same diluting procedure as for the stock solution. This solution contains 20 μg Nb per ml.

Procedure. Accurately weigh from 0·3 to 0·5 g of the finely powdered rock material into a 40-ml PTFE dish. Moisten with water, add 2–5 ml of niobium tracer solution, 5 ml of concentrated hydrofluoric acid, 5 ml of 20 N sulphuric acid and 1 ml of concentrated nitric acid. Evaporate to strong fumes of sulphuric acid, cool, add 10 ml of water and 10 ml of concentrated hydrochloric acid and warm to dissolve most of the salts. Transfer the solution to a 150-ml beaker, rinsing the dish with dilute hydrochloric acid. Heat the solution to near boiling then make alkaline with ammonia. Digest the mixture for 15 minutes then collect the residue on a hardened open-textured filter paper and wash well with hot ammonium chloride solution. Rinse the precipitate into the original beaker with a jet of water. Wash the paper with 10 ml of 6 N hydrochloric acid and water, collecting the washings in the beaker. Heat to dissolve the precipitate.

Add 50 ml of aqueous sodium hydroxide to the hot solution, digest for 10 minutes and then collect the precipitate on the original hardened filter paper, washing with dilute aqueous ammonia. Dissolve this precipitate as before, and reprecipitate with ammonia, washing as before with ammonium chloride solution. Transfer the paper and precipitate to a silica crucible, dry and ignite.

Add 1 g of potassium pyrosulphate, weighed correct to the nearest 0·02 g, and fuse the mixture gently until a clear melt is obtained. Cool, add 0·4 ml of concentrated sulphuric acid and heat the mixture strongly on a sand bath to disintegrate the cake. Allow to cool. Dissolve the paste in the crucible by warming on a sand bath with 10 ml of the leach solution, and rinse into a 50-ml beaker with a further 20 ml of leach solution. Heat to incipient boiling and filter through a small open-textured filter paper into a 50- or 100-ml separating funnel. Prepare also a reagent blank solution from 1 g of potassium pyrosulphate, fused, dissolved and filtered in the same way as the rock sample.

To the cold filtered solutions add in the following order: 5 ml of ammonium thiocyanate solution, 0·5 ml of stannous chloride solution and 20 ml of ethyl acetate. Shake for 1 minute and allow to stand for the organic layer to separate. Prepare sufficient stripping solution as described above. Now run off and discard the aqueous layer. Shake the organic layer for 1 minute with 15 ml of the stripping solution, allow the phases to separate, discard the aqueous layer and repeat the shaking with a further 10 ml of the stripping solution. Transfer the ethyl acetate layer (which should be noticeably yellow if the sample contains more than about 10 μg of niobium) to a 25-ml volumetric flask and dilute to volume with ethyl acetate. Without delay (note 1) measure the optical density of the extract against the extracted reagent blank solution in 4-cm cells using the spectrophotometer at a wavelength of 385 nm.

Determine also the γ-activity of the organic extract and compare with the activity of a similar aliquot (2–5 ml) of the tracer solution diluted to 25 ml (note 2). Hence determine the chemical yield factor for niobium in the determination.

Calibration. Prepare a calibration graph by evaporating 0·1–0·5 ml aliquots of the dilute standard niobium solution containing 0–10 μg Nb together with added niobium tracer solution, in silica crucibles and fusing the residue with potassium pyrosulphate and continuing as described above. Both optical densities and chemical yield factors should be determined.

Notes: 1. The optical density decreases by about 2 per cent in 1 hour.
2. This tracer solution should be taken from the stock tracer solution at the same time as that added to the rock sample, a few mg of inactive niobium solution added and, after dilution to 25 ml with water, set aside until required for comparison.

The Determination of Niobium with 4-(2-Pyridylazo)-resorcinol

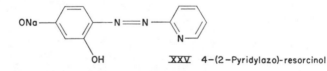

XXV 4-(2-Pyridylazo)-resorcinol

4-(2-Pyridylazo)-resorcinol (PAR, XXV), forms coloured complexes with many metals, but in the presence of EDTA or CyDTA the reagent is highly selective for niobium.[20, 21] The niobium complex, red in colour, has a maximum absorption at 550 nm (Fig. 76) whilst the reagent itself has a maximum at 410 nm and only negligible absorption at the higher wavelength. Close control of pH is required (Fig. 77), for which the addition of ammonium acetate has been recommended.

The time required for colour development, normally 25 minutes, is increased to 40 minutes in the presence of EDTA. After this period the optical density of both the reagent and the niobium solution increase slowly with time.

Interference from uranium, vanadium and phosphate can be avoided by making a prior separation, and from tantalum by masking with

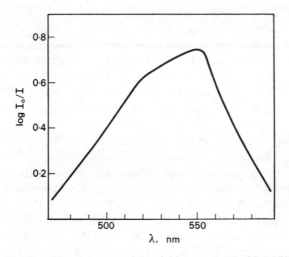

FIG. 76. Absorption spectrum of the niobium complex with PAR (4-cm cells, 55 μg Nb/100 ml).

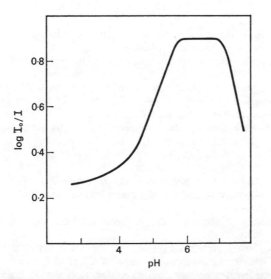

FIG. 77. The relation of optical density to pH in the formation of the niobium complex with PAR.

tartaric acid. In the examination of silicate rocks, the silica can conveniently be removed by evaporation with hydrofluoric and sulphuric acids in the usual way. The niobium present is readily separated from excess iron, aluminium and other elements by precipitation from the sulphate solution with a carefully controlled amount of cupferron. Vanadium is not precipitated if sulphurous acid is added to the solution, whilst any copper, uranium, tungsten and molybdenum are removed from the filter pad by washing with dilute aqueous ammonia.

The cupferrate residue is ignited in a silica crucible, fused with a little potassium pyrosulphate, extracted, and the coloured complex formed in tartaric-CyDTA solution. This procedure, described in detail below, has been adapted from that given by Jenkins[8] for niobium in mild steel.

A more elaborate procedure based upon the use of PAR, by Greenland and Campbell[22] uses an anion exchange column to separate niobium from interfering elements. This separation of niobium may not be quantitative and an isotope dilution technique is used to compensate for this.

Method

Reagents: *Cupferron*, dissolve 6 g in 100 ml of water as required. Use only fresh good-quality reagent.

Cupferron wash solution, add 1 ml of the above cupferron solution to 1 litre of 2 N sulphuric acid as required.

Sulphurous acid, saturate distilled water with sulphur dioxide.

Ammonia, dilute solution, dilute 25 ml of concentrated aqueous ammonia to 500 ml with water.

Tartaric acid solution, dissolve 50 g in 500 ml of water. Prepare also a 1 per cent solution by dilution as required.

Zinc sulphate solution, dissolve 0·75 g of zinc sulphate heptahydrate in 250 ml of water.

Cyclohexanediaminotetraacetic acid solution, dissolve 6·92 g of the reagent and 5 g of sodium hydroxide in water and dilute to 1 litre.

Sodium hydroxide solution, dissolve 12 g of reagent in 100 ml of water.

4-(2-Pyridylazo)-resorcinol (PAR) solution, dissolve 0·295 g of the monosodium salt in water and dilute to 1 litre.

Ammonium acetate buffer solution, dissolve 40 g of ammonium acetate in water, add 2·25 ml of glacial acetic acid and dilute to 500 ml with water.

Standard niobium stock solution, fuse 71·6 mg of pure dry niobium pentoxide with 2 g potassium pyrosulphate in a platinum crucible. Dissolve the melt in 10 per cent tartaric acid solution and dilute with tartaric acid solution to 500 ml. This solution contains 100 μg Nb/ml.

Standard niobium working solution, dilute 10 ml of the stock solution to 100 ml with water. Prepare this as required. This solution contains 10 μg Nb/ml in 1 per cent tartaric acid.

Procedure. Accurately weigh 0·5 g of the finely powdered sample material into a platinum crucible. Moisten with a little water and add 1 ml of 20 N sulphuric acid, 1 ml of concentrated nitric acid and 10 ml of concentrated hydrofluoric acid. Transfer the crucible to a hot plate and evaporate, first to fumes of sulphuric acid then to dryness in the usual way. Allow the crucible to cool, add 2 g of potassium pyrosulphate and fuse over a Bunsen burner until a quiescent melt is obtained. Allow to cool.

Disintegrate the melt by warming with a little water, then transfer the solution and any residue to a 250 ml beaker. Add 12 ml of 20 N sulphuric acid and warm on a hot plate until solution is complete. Dilute to about 90 ml with water, add 10 ml of sulphurous acid and cool the solution in a refrigerator if necessary, to bring the temperature into the range 10–15°. Precipitate any niobium present in the solution, together with some of the iron by adding 1 ml of cupferron solution. Stir in a little filter paper pulp and collect the precipitate on an open-textured paper and wash 5–6 times with the cupferron wash solution and 4–5 times with the dilute ammonia solution. Discard the filtrate.

Transfer the residue and paper to a platinum crucible, dry, burn off and ignite the residue at as low a temperature as possible, not exceeding 650°, then fuse the residue with 1 g of potassium pyrosulphate. Extract the melt into 10 ml of hot 1 per cent tartaric acid solution and transfer to a 100-ml beaker with a little water. Cool the solution to room temperature, add 1 ml of zinc sulphate solution and 10 ml of CyDTA solution. Using a pH meter adjust the pH of the solution to approximately 6 by adding sodium hyroxide solution (between 2 and 3 ml should be required) and add 10 ml of PAR solution and 0·5 ml of the ammonium acetate buffer solution. Transfer the solution to a 100-ml volumetric flask and dilute to volume with water. Mix well, stand for 40 minutes then measure the optical density in 4-cm cells using the spectrophotometer set at a wavelength of 550 nm.

Calibration. Transfer aliquots of 0–10 ml of the standard niobium working solution containing 0–100 μg Nb, to separate 100-ml beakers. Add sufficient tartaric acid solution to give a total volume of 10 ml and proceed as described above for the formation and measurement of the optical density of the niobium-PAR complex. Plot the relation of optical density to niobium concentration (Fig. 78).

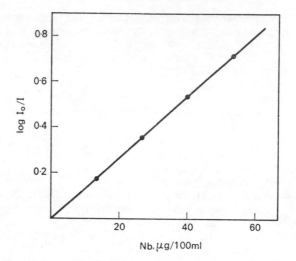

FIG. 78. Calibration graph for niobium using PAR (4-cm cells, 550 nm).

The Determination of Niobium by Atomic Absorption Spectroscopy

Atomic absorption spectroscopy has not been used extensively for the determination of niobium because of its inherent low sensitivity. Where an enrichment occurs, as for example in pyrochlore-bearing granites or alkaline rocks, the direct procedure of Husler[23] can be used. In this the sample is decomposed by evaporation with a mixture of concentrated hydrochloric and hydrofluoric acids. The fluoride solution, buffered by the addition of iron, aluminium and potassium, is used for direct absorption measurement without prior separation of niobium. The lower limit of determination by this method appears to be of the order of $0 \cdot 02$ per cent Nb_2O_5.

The Determination of Tantalum

The reagents that have been suggested for the photometric determination of tantalum are not as sensitive as thiocyanate or PAR for niobium. Patrovsky[22] has described a photometric procedure for 20 to 400 μg Ta, based upon the colour given with p-dimethylaminophenylfluorone. This method can be applied to certain pegmatites and other rare rocks appreciably enriched in tantalum, but is not applicable to normal silicate rocks. A somewhat more sensitive procedure has been described by Pavlova and Blyum,[25] involving first an extraction of tantalum as a complex with butylrhodanine S, and then photometric determination with rhodanine 6J.[26]

Unlike spectrophotometric and emission spectrographic methods procedures based upon thermal neutron activation analysis are particularly sensitive for this element. This is partly because the cross-section for thermal neutron capture is quite large, and partly because the convenient half-life of the product nuclide allows adequate time for the chemical processing of the irradiated material. Procedures for determining tantalum in silicate rocks by this method have been described by Atkins and Smales[27] and by Morris and Olya.[28]

References

1. RANKAMA K., *Bull. Comm. Géol. Finlande* (1944) **133**.
2. RANKAMA K., *Ann. Acad. Sci. Fenn.* (*A*) (1948) III, 13.
3. ZNAMENSKII E. B., RODIONOVA L. M. and KAKHANA M. M., *Geokhim.* (1957) (3), 267.
4. EDGE R. A. and AHRENS L. H., *Trans. Proc. Geol. Soc. S. Africa* (1963) **66**, 109.
5. BAKES J. M., GREGORY G. R. E. C. and JEFFERY P. G., *Anal. Chim. Acta* (1962) **27**, 540.
6. CHARLOT G., *Colorimetric Determination of Elements*, Elsevier, 1964, p. 315.
7. BELCHER R., RAMAKRISHNA T. V. and WEST T. S., *Talanta* (1963) **10**, 1013.
8. JENKINS N., *Metallurgia* (1964) **70**, 95.
9. CHENG K. L. and GOYDISH B. L., *Talanta* (1962) **9**, 987.
10. ACKERMANN G. and KOCH S., *Talanta* (1962) **9**, 1015.
11. ALIMARIN I. P. and HSI-I HAN, *Zhur. Anal. Khim.* (1963) **18**, 82.
12. PICKUP R., *Colon. Geol. Min. Res.* (1953) **3**, 358.
13. ALIMARIN I. P. and PODVAL'NAYA R. L. *Zhur. Anal. Khim.* (1946) **1**, 30.
14. FREUND H. and LEVITT A. E., *Analyt. Chem.* (1951) **23**, 1813.
15. MARZYS A. E. O., *Analyst* (1954) **79**, 327.
16. WARD, F. N. and MARRANZINO A. P., *Analyt. Chem.* (1955) **27**, 1325.
17. FAYE G. H., *Chem. in Canada* (1958) **10**, 90.
18. GRIMALDI F. S., *Analyt. Chem.* (1960) **32**, 119.
19. ESSON J., *Analyst* (1965) **20**, 488.
20. BELCHER R., RAMAKRISHNA T. V. and WEST T. S., *Talanta* (1962) **9**, 943.
21. ELINSON S. V. and POBEDINA L. I., *Zhur. Anal. Khim.* (1963) **18**, 199.
22. GREENLAND L. P. and CAMPBELL E. Y., *Anal. Chem. Acta* (1970) **49**, 109.
23. HUSLER J., *Talanta* (1972) **19**, 863.
24. PATROVSKY V., *Chim. Listy.* (1965) **59**, 1464.
25. PAVLOVA N. N. and BLYUM I. A., *Zavod. Lab.* (1966) **32**, 1196.
26. PAVLOVA N. N. and BLYUM I. A., *Zavod. Lab.* (1962) **28**, 1305.
27. ATKINS D. H. F. and SMALES A. A., *Anal. Chim. Acta* (1960) **22**, 462.
28. MORRIS D. F. C. and OLYA A., *Talanta* (1960) **4**, 194.

CHAPTER 35

NITROGEN

Occurrence

Appreciable quantities of nitrogen gas are released from many rocks by heating under vacuum. Silicate rocks are not likely to contain metallic nitrides (but see, for example, Baur[1]), but as the nitrogen is not so easily released as other gases,[2] it is probably not present as free nitrogen either. Whatever method of extraction is employed, the larger part of the nitrogen detected is in the form of ammonia. It is possible that this is the way in which most of it occurs, although the possibility that mixtures of nitrogen and hydrogen could be converted to ammonia at the time of their release from silicates has been noted.[3]

The ammoniacal–nitrogen content of silicates is very variable, although average values of $0 \cdot 00138$ per cent for dunites (thirteen determinations), $0 \cdot 00485$ per cent for mafic rocks (eleven determinations) and $0 \cdot 00267$ for granitic rocks (ten samples) were reported by Vinogradov et al.[2] Similar values were indicated by Stevenson[4] and Wlotzka,[5] with somewhat higher values for certain shales.

The ammoniacal–nitrogen appears to be concentrated in the mica fraction, where it may replace potassium (K^+, $R = 133$ pm; NH^+, $R = 143$ pm)—an indication of this was obtained by Stevenson,[4] who noted a correlation between nitrogen and potassium in mica minerals.

Wlotzka[5] examined a number of rocks for nitrate–nitrogen. He reported small amounts (5–20 ppm) in surface sediments, some in saline clays and limestones, but none in average clays, sandstones and magmatic rocks.

The Determination of Nitrogen

Some nitrogen may be extracted from silicate rocks by leaching with water, but this is unlikely to give any real indication of the total amount of ammonia present in the specimen. Methods reported for this determination include comminution in vacuum, ignition in vacuum and chemical decomposition with both acids and alkalis. Vinogradov *et al.*[1] reported that fusion of silicate samples in vacuum did not give concordant results, and preferred decomposition with a mixture of ortho- and pyrophosphoric acids in a partially sealed quartz ampoule. A temperature of 240° was used, and after the decomposition was complete, the ammonia liberated with alkali was absorbed in dilute hydrochloric acid and determined with Nessler's reagent.

A similar procedure was employed by Stevenson[6] using sulphuric acid in a completely sealed tube at a temperature of 420° for a minimum period of 90 minutes. A micro-Kjeldahl distillation apparatus was used to recover the ammonia, which was determined with Nessler's reagent. For those samples containing more than 100 μg of nitrogen, the ammonia distillate was collected in boric acid and titrated with 0·01 N sulphuric acid solution.

Wlotzka[5] noted that with an alkaline fusion in a closed system, a part of the ammonia was converted to nitrate. This was reduced with Devarda's alloy, and the total ammonia determined with Nessler's reagent. Wlotzka also decomposed silicate rocks and minerals with hydrofluoric acid in a closed polythene bottle, added sodium hydroxide and distilled the ammonia from the alkaline solution. By adding Devarda's alloy to the solution after removal of ammonia, it was possible to determine also the nitrogen originally present as nitrate.

A high-temperature fusion technique using helium as an inert carrier gas was described by Gibson and Moore.[7] Samples of silicate rocks and meteorites were fused in a graphite crucible using a Leco induction furnace and a temperature of 2400°. Any carbon monoxide produced was converted to carbon dioxide by passage over a copper oxide–rare earth oxide catalyst, and removed from the system, together with water vapour, by passage through an Ascarite-magnesium perchlorate trap. The remaining gases were collected in a molecular sieve trap at liquid nitrogen temperature and subsequently separated chromatographically using molecular sieve 5A. The components were detected by thermal conductivity.

The reagent blank was of the order of 4 to 6 μg N, and nitrogen contents of 27 to 59 ppm (μg N/g) were reported for seven USGS rock standards.

References

1. BAUR W. H., *Nature* (1972) **240**, 461.
2. VINOGRADOV A. P., FLORENSKII K. P. and VOLYNETS V. F., *Geokhimiya* (1963) (9), 875.
3. RAYLEIGH L., *Proc. Royal Soc.* (1939) **A170**, 451.
4. STEVENSON F. J., *Geochim. Cosmochim. Acta* (1960) **19**, 261.
5. WLOTZKA F., *Geochim. Cosmochim. Acta* (1961) **24**, 106.
6. STEVENSON F. J., *Analyt. Chem.* (1960) **32**, 1704.
7. GIBSON E. K. Jr. and MOORE B. M., *Analyt. Chem.* (1970) **42**, 461.

CHAPTER 36

PHOSPHORUS

Occurrence

Phosphorus, in the form of the mineral apatite, has long been recognised as a ubiquitous constituent of silicate rocks. Landergren[1] has indicated a gradual decrease of phosphorus content in rocks of increasing silica content in both the plutonic and volcanic series, Table 32.

TABLE 32. THE PHOSPHORUS
CONTENT OF SOME SILICATE
ROCKS

Rock type	P, ppm
Basalt	2440
Gabbro and norite	1700
Diorite	1400
Syenite	1330
Andesite	1230
Granite	870
Rhyolite and liparite	550

An inspection of published analyses, however, shows that these figures can be little more than a rough guide to the phosphorus content of any particular specimen. The overall abundance of phosphorus in igneous rocks has been given by Conway[2] as 1200 ppm. Limestones and sandstones contain rather less phosphorus than igneous rocks—a few hundred parts per million being a common value for many samples. Deep sea sediments contain rather more, a few thousand parts per million being typical.

Phosphorus occurs very largely as the phosphate ion, PO_4^{3-} in phosphate minerals, of which the apatite group is the most abundant. This group includes fluorapatite, hydroxyapatite and intermediate

375

members with the general composition $Ca_5X(PO_4)_3$, where X represents F and OH. Calcium can be replaced by alkali and other alkaline earth elements, and phosphate by sulphate and silicate. Other somewhat rarer phosphates include monazite, xenotime, amblygonite and triplite. There are a number of even rarer phosphates of lead (cerussite, pyromorphite, phosgenite and leadhillite) copper (turquoise, torbernite, libethenite) and other heavy metals, occurring in or associated with ore deposits.

The Determination of Phosphorus

When present in the amounts encountered in many silicate rocks, phosphorus can be precipitated from nitric acid solution as the complex phosphomolybdate and weighed as such or as magnesium pyrophosphate. These two procedures have been used for many years but they are tedious and not ideally suited to this phosphorus concentration range. They have now been largely displaced by photometric procedures based upon the yellow colour given by phosphorus with a vanadomolybdate reagent, or the blue colour of reduced phosphomolybdate solutions. The latter procedure is the more sensitive, whilst the former can be adapted for determining phosphorus in phosphate rock containing up to 30 per cent P_2O_5.

The Determination of Phosphorus using a Vanadomolybdate Reagent

The formation of a yellow-coloured complex of phosphorus with solutions of a vanadomolybdate reagent was used as early as 1908[3] for the determination of small amounts of phosphoric acid. More recently it has been established[4] that the maximum absorption of vanadomolybdophosphate solutions (some authors prefer "molybdovanadophosphate") occurs at a wavelength of 315 nm in the ultraviolet part of the spectrum. Most workers have however preferred to use a wavelength in the range 420–470 nm, where the interference from iron is much reduced or eliminated altogether.

The acid concentration required for the formation of this yellow colour has been given as 0·021–0·071 N hydrochloric,[4] and 0·2–1·6 N nitric acid.[5] Sulphuric and perchloric acids have also been used, particularly with sample materials containing large amounts of iron.

Interference has been reported from silicon which forms an analo-gous yellow-coloured vanadomolybdate[6] from titanium and zirconium when present in large amounts[7] and also from chromium[8] and arsenic. These latter metals are not likely to be present in phosphate rock in amounts sufficient to interfere with the determination, whilst silicon is removed in the early stages of the procedure described in detail below.

ACID-SOLUBLE PHOSPHATE IN PHOSPHATE ROCK
Method

Reagents: *Vanadomolybdate reagent solution*, prepare separate solutions by dissolving 35 g of ammonium molybdate in 500 ml of water, and 1·12 g of ammonium vanadate in a mixture of 240 ml of concentrated perchloric acid and 260 ml of water. These two solutions are quite stable and will keep for several months. Before use mix equal volumes of the two solutions.

Standard phosphate solution, dissolve 0·3835 g of dried potassium dihydrogen phosphate in water and dilute to 1 litre with water. This solution contains 0·2 mg P_2O_5 per ml.

Procedure. Accurately weigh from 0·2 to 1 g of the finely ground sample material (depending upon the phosphorus content) into a 150-ml beaker, add 25 ml of concentrated hydrochloric acid and 10 ml of concentrated nitric acid. Cover the beaker and heat gently on a hot plate or steam bath until all the immediate reaction has ceased, rinse and remove the cover glass, wash down the sides of the beaker with a little water and evaporate the solution to a volume of about 5 ml. Remove the beaker from the hot plate, allow to cool, and add 5 ml of concentrated perchloric acid and heat until the perchloric acid refluxes down the sides of the beaker. Allow this refluxing to continue for 10 minutes. Remove the beaker from the hot plate and allow to cool somewhat. Add 50 ml of water, heat almost to boiling and maintain at near boiling point for 20 to 30 minutes.

Collect any insoluble residue on a small close-textured filter paper and wash the residue and paper free of acid with water. Discard the residue. Combine the filtrate and washings and dilute to volume with water in a 250-ml volumetric flask.

Pipette an aliquot of from 5 to 25 ml of this solution into a 100-ml volu-metric flask. Using a pipette add 20 ml of the vanadomolybdate reagent solution and dilute to volume with water. Measure the optical density of the solution·in 1-cm cells at between 15 and 60 minutes after preparation. The spectrophotometer should be set at a wavelength of 460 nm, and the reference solution should be a reagent blank, prepared in the same manner as the sample solution, but omitting the sample material.

Calibration. Transfer aliquots of 0–30 ml of the standard phosphate solution containing 0–6 mg P_2O_5 to separate 100-ml volumetric flasks, add exactly 20 ml of the vanadomolybdate reagent solution to each and dilute to volume with water. Measure the optical densities relative to the solution containing no added phosphorus, as described above and construct a calibration curve by plotting the values obtained against phosphorus concentration.

Notes: 1. Only slight interference is encountered from arsenic, 1 mg As_2O_3 giving an optical density equivalent to about $0 \cdot 01$ mg P_2O_5. If large amounts of arsenic are present they can be removed by volatilisation as follows: After the sample has been attacked with hydrochloric and nitric acids and evaporated to a volume of 5 ml, add 10 ml of concentrated hydrochloric acid and again evaporate to a volume of about 5 ml. Repeat the evaporation with 10 ml of concentrated hydrochloric acid twice more. Add 10 ml of concentrated hydrochloric acid, 5 ml of water and 1 g of granulated zinc. When the reaction has subsided, evaporate to about 5 ml, cool slightly, add a few ml of water and 5 ml of concentrated perchloric acid. Evaporate to fumes of perchloric acid and continue as described above.

2. The interference from chromium can be removed by adding 1 g of sodium chloride immediately before adding the 5 ml of perchloric acid. Any chromium present is volatilised as chromyl chloride in the subsequent evaporation to fumes of perchloric acid.

TOTAL PHOSPHORUS IN SILICATE ROCKS

A number or workers have adapted the reaction of phosphorus with vanadomolybdate solutions to the determinations in silicate rocks. Bennett and Pickup[9] noted interference from titanium, and devised a procedure for separating both silica and titanium based upon a fusion of the rock sample with sodium carbonate. An aliquot of the rock solution prepared from 5 g of material, and used by them for the determination of barium, zirconium, chromium, vanadium, chlorine and total sulphur can be taken for the determination of phosphorus. It is essentially the procedure of Bennett and Pickup that is described below.

This procedure is adequate for many silicate rocks, but as the total phosphorus does not always appear in the aqueous leach from a carbonate fusion, low results may sometimes be obtained. This is particularly so for rocks containing appreciable amounts of calcium.

A somewhat simpler method that can be used for rocks containing not more than $1 \cdot 6$ per cent TiO_2 or $0 \cdot 1$ per cent ZrO_2 was described by Baadsgaard and Sandell.[7] Silica was eliminated by evaporation of the rock material with hydrofluoric and nitric acids, and a correction determined for the interference from iron. Shapiro and Brannock,[10] in their scheme for rapid rock analysis, determined phosphorus in an aliquot of their "solution B", prepared by evaporation of the rock material with hydrofluoric and perchloric acids.

Method

Reagents: *Sodium carbonate wash solution*, dissolve 10 g of solid reagent in 500 ml of water.

Nitric acid, dilute, dilute 22 ml of concentrated acid to 100 ml with water.

Vanadomolybdate reagent solution, dissolve 20 g of ammonium molybdate in water and pour slowly into 140 ml of concentrated nitric acid. Add 1 g of ammonium vanadate, stir until solution is complete and then dilute to one litre with water. This reagent is stable for several months.

Standard phosphate solution, dissolve $0 \cdot 192$ g of dried potassium dihydrogen phosphate in water and dilute to one litre in a volumetric flask. This solution contains 100 μg P_2O_5 per ml.

Procedure. Accurately weigh approximately $0 \cdot 2$ g of the rock powder into a small platinum crucible, add $1 \cdot 2$ g of anhydrous sodium carbonate and $0 \cdot 1$ g of potassium nitrate and fuse over a gas burner for an hour to complete the decomposition of all minerals present. Allow the crucible to cool, and then extract the melt with water. Collect the insoluble residue on a small, hardened, open-textured filter paper and wash well with small quantities of hot sodium carbonate wash solution. Discard the residue. Combine the filtrate and washings and transfer to a platinum basin. Cover the basin, cautiously acidify with concentrated nitric acid, and add approximately 5 ml in excess. Add 15 ml of concentrated hydrofluoric acid and evaporate the solution to dryness. Dissolve the residue in a little water, add 5 ml of concentrated nitric acid and 5 ml of concentrated hydrofluoric acid and repeat the evaporation to dryness. Add 2 ml of concentrated nitric acid and again evaporate to dryness.

Add 2 drops of concentrated nitric acid to the dry residue and dissolve in a little water. At this stage a clear solution should be obtained, but if the rock contains much magnesium, a certain amount of fluorine may be retained in the insoluble residue. This may be removed by filtration or by further evaporations with small quantities of concentrated nitric acid.

Transfer the clear aqueous solution to a 50-ml volumetric flask, add 5 ml of dilute nitric acid and 10 ml of the vanadomolybdate reagent solution and dilute to volume with water. Measure the optical density of the solution in 2-cm cells, using the spectrophotometer set at a wavelength of 470 nm, against a reagent blank solution prepared in the same way as the sample solution but omitting the sample material. Obtain the phosphate content of the sample by reference to the calibration graph.

Calibration. Transfer aliquots of 0–30 ml of the standard phosphate solution, containing 0–3 mg P_2O_5, to separate 50-ml volumetric flasks, add 5 ml of dilute nitric acid and 10 ml of vanadomolybdate reagent solution to each, and dilute to volume with water. Measure the optical densities against the solution containing no added phosphorus, in 2-cm cells as described above, and plot these values against the phosphate concentration to give the calibration graph (Fig. 79).

The solution containing $2 \cdot 5$ mg P_2O_5 per 50 ml should have an optical density of about $0 \cdot 8$.

DETERMINATION OF PHOSPHORUS IN CARBONATE ROCKS

There is no difficulty in applying the vanadomolybdate method to the determination of phosphorus in limestone rocks. A suitable weight of rock material is evaporated to dryness with nitric acid to remove carbon dioxide and oxidise organic matter, and twice to dryness with perchloric and hydrofluoric acids to remove any silica present in the sample. After further evaporation with perchloric acid to decompose any remaining

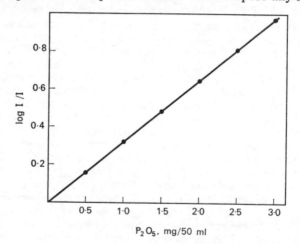

FIG. 79. Calibration graph for phosphorus by a molybdovanadate method (2-cm cells, 470 nm).

fluorides, the dry residue is moistened with perchloric acid and dissolved in water. This solution can be used directly for the determination of phosphorus by the addition of the vanadomolybdate reagent, as described above.

An alternative procedure of particular value in the analysis of those carbonates, where the iron content is appreciable is to separate phosphorus from iron and other elements present by using a cation exchange resin.[11] The phosphate ion is not retained by the resin and can be determined photometrically in the eluate with vanadomolybdate reagent. The sample solution should be just acid and, to prevent precipitation of iron and aluminium phosphates on the column, the phosphate should be eluted with 0·015 N hydrochloric acid.

Calcium, magnesium, iron and aluminium are retained by the cation exchange resin and can be subsequently recovered by elution with 4 N hydrochloric acid.

The Determination of Phosphorus as a Molybdenum Blue

Silica and phosphorus react with acid solutions of ammonium molybdate to give yellow-coloured complexes that can be reduced to blue-coloured solutions. These contain a colloidal complex known as molybdenum blue or heteropoly blue.[12] The intensity of the coloured product depends upon the pH of the solution, the reducing agent used, the molybdate–acid ratio, the amount of molybdate and the presence of other ions.[13, 14, 15] The procedure can be as much as twenty times as sensitive as that based upon the yellow vanadomolybdophosphate colour. A pH in the range 1·9–6·0 has been recommended.[16]

Many reducing agents have been suggested for the production of the blue species from the yellow phosphomolybdate, including ferrous sulphate, zinc dust, stannous chloride, hydrazine, 1-amino-2-naphthol-4-sulphonic acid, hydroquinone and many more.[17] Ascorbic acid is used in the procedure described below[18] this produces a stable molybdenum blue solution and provides a more sensitive procedure than those based upon many of the other recommended reductants. The disadvantage is that the colour is developed slowly (Fig. 80) and the solutions should preferably be allowed to stand overnight.

Using this reagent, the maximum absorption of molybdenum blue solutions occurs at a wavelength of about 830 nm (Fig. 81), measurements of optical density can also be made at 650 nm, although with much reduced sensitivity. In the procedure described, the calibration graph is a straight line (Fig. 82), indicating that the Beer–Lambert Law is valid over the concentration range used.

Interference from silica can be avoided by evaporation with hydrofluoric and either sulphuric or perchloric acid in the usual way. Arsenic

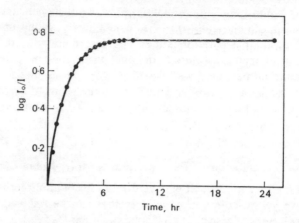

FIG. 80. Colour development of molybdenum blue in the determination of phosphorus (2-cm cells, 830 nm, 50 μg P_2O_5/50 ml).

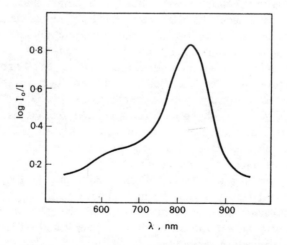

FIG. 81. Absorption spectrum of a molybdenum blue solution.

forms a yellow-coloured complex with ammonium molybdate that can similarly be reduced to a molybdenum blue. This element is unlikely to be present in silicate or carbonate rocks in amounts sufficient to interfere, and additional steps to volatilise arsenic are not necessary.

Because of the high sensitivity of the method to traces of phosphorus and silicon, all the glassware used for the determination must be scrupulously clean. Riley[18] has recommended that a separate set of volumetric flasks be set aside for this determination only. These flasks should be allowed to stand for several hours filled with concentrated sulphuric acid and then well washed out with distilled water. After use the flasks should be rinsed out, and then kept filled with distilled water.

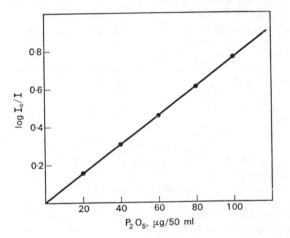

Fig. 82. Calibration graph for phosphorus by a molybdenum blue method (1-cm cells, 830 nm).

DETERMINATION IN SILICATE ROCKS
Method

Reagents: *Sulphuric acid,* 3 N.

Ammonium molybdate solution, dissolve 5 g of analytical grade reagent in water and dilute to 250 ml with water.

Ascorbic acid solution, dissolve 4·4 g of reagent in water and dilute to 250 ml. This solution deteriorates slowly on standing, and is best prepared freshly as required.

Reducing solution, dilute a mixture of 125 ml of 3 N sulphuric acid, 38 ml of ammonium molybdate solution and 60 ml of ascorbic acid solution to 250 ml with water. This reagent solution is faint-green in colour and should be prepared immediately before use.

Standard phosphate stock solution, dissolve 0·0959 g of dried potassium dihydrogen phosphate in water and dilute to 1 litre in a volumetric flask. This solution contains 50 μg P_2O_5 per ml.

Standard phosphate working solution, dilute 20 ml of the stock standard phosphate solution to 250 ml with water. This solution contains 4 μg P_2O_5 per ml. For the calibration of 4-cm spectrophotometer cells, prepare a phosphate solution containing 1 μg P_2O_5 per ml by diluting 5 ml of the stock solution to 250 ml.

Procedure. Accurately weigh 0·1 g of the finely powdered sample into a small platinum crucible and allow to stand overnight with 2 ml of concentrated perchloric acid and 5 ml of concentrated hydrofluoric acid. Evaporate on a water bath until no further fumes of hydrofluoric acid are visible, and then under an infrared heating lamp until most of the perchloric acid is removed. Do not allow the residue to become completely dry. Add 1 ml of concentrated perchloric acid to the moist residue and again evaporate almost but not quite to dryness.

Using a calibrated pipette, add 0·8 ml of concentrated perchloric acid to the residue, followed by about 20 ml of water. Warm gently on a hot plate or water bath until all soluble material has passed into solution, and then rinse into a small beaker. If there is no visible solid residue, transfer the solution to a 100-ml volumetric flask and dilute to volume with water.

If the rock material is not completely decomposed by this procedure and a small residue remains, collect this residue on a small, close-textured filter paper and wash it with a little water. Combine the filtrate and washings in a 100-ml volumetric flask. Dry and ignite the filter and residue in a small platinum crucible and fuse with a small quantity of anhydrous sodium carbonate. Allow the melt to cool, extract with water, acidify with the minimum quantity of diluted perchloric acid and add to the solution contained in the 100-ml volumetric flask. Dilute to volume with water.

Using a pipette, transfer 25 ml of the rock solution to a 50-ml volumetric flask, add 20 ml of the reducing solution, dilute to volume with water and mix well. Allow the solution to stand overnight and then measure the optical density in 1- or 4-cm cells with water as the reference solution, using the spectrophotometer set at a wavelength of 830 nm.

Measure also the optical density of a reagent blank solution prepared in the same way as the sample solution, but omitting the sample material. Calculate the phosphorus content of the sample by reference to the calibration graph, or by using a calibration factor.

Calibration. Pipette aliquots of 0–25 ml of the working standard phosphate solution containing 0–100 μg P_2O_5 (for calibration of 1-cm cells) or 0–25 μg P_2O_5 (for 4-cm cells) into separate 50-ml volumetric flasks, add 20 ml of reducing solution to each and dilute to volume. Allow the solutions to stand overnight and measure the optical densities in the appropriate cell using the spectrophotometer as described above. Plot the values obtained against the concentration of phosphate expressed as μg P_2O_5 per 50 ml of solution. The solution containing 100 μg P_2O_5 should have an optical density of about 0·780.

Notes: 1. This procedure can be used for the determination of phosphorus in carbonate rocks with only minor modifications. Dilute perchloric acid should be used for the sample decomposition, and if the residue contains silicate minerals it should be followed by the digestion with hydrofluoric and perchloric acids as described above for silicate rocks.

2. If arsenic is present in sufficient amount to interfere, the evaporation of the sample material with hydrofluoric and perchloric acids should be replaced by an evaporation with hydrofluoric, hydrochloric and nitric acids as described on p. 378. A few mg of metallic zinc is used to reduce the arsenic before volatilisation. Finally the solution is evaporated almost to dryness with perchloric acid before proceeding as described above.

References

1. LANDERGREN S. and GOLDSCHMIDT V. M., *Geochemistry*, Oxford, 1954, p. 457.
2. CONWAY E. J., *Amer. J. Sci.* (1945) **243**, 583.
3. MISSON G., *Chem.-Ztg.* (1908) **32**, 633.
4. MICHELSEN O. B., *Analyt. Chem.* (1957) **29**, 60.
5. KITSON R. E. and MELLON M. G., *Ind. Eng. Chem., Anal. Ed.* (1944) **16**, 379.
6. LEW R. B. and JAKOB F., *Talanta* (1963) **10**, 322.
7. BAADSGAARD H. and SANDELL E. B., *Anal. Chim. Acta* (1954) **11**, 183.
8. QUINLAN K. P. and DESESA M. A., *Analyt. Chem.* (1955) **27**, 1626.
9. BENNETT W. H. and PICKUP R., *Colon. Geol. Min. Res.* (1952) **3**, 171.
10. SHAPIRO L. and BRANNOCK W. W., *U.S. Geol. Surv. Circ.* 165, 1952, *Bull.* 1144-A, 1962.
11. SAMUELSON O., *Ion-Exchangers in Analytical Chemistry*, p. 146, Wiley, New York, 1953.
12. BELL R. D. and DOISY E. A., *J. Biol. Chem.* (1920) **44**, 55.
13. WILLARD H. H. and CENTER E. J., *Ind. Eng. Chem., Anal. Ed.* (1941) **13**, 81.
14. WOODS J. T. and MELLON M. G., *Ind. Eng. Chem., Anal. Ed.* (1941) **13**, 760.
15. MURPHY J. and RILEY J. P., *J. Marine Biol. Assoc. U.K.* (1958) **37**, 9.
16. KITSON R. E. and MELLON M. G., *Ind. Eng. Chem., Anal. Ed.* (1944) **16**, 466.
17. MACDONALD A. M. G., *Ind. Chemist* (1960) 88 and 134.
18. RILEY J. P., *Anal. Chim. Acta* (1958) **19**, 413.

CHAPTER 37

SCANDIUM, YTTRIUM AND THE LANTHANIDE RARE EARTHS

Occurrence

The lanthanide rare earth elements lanthanum to lutecium are considerably more plentiful than other well-known elements, and the term "rare" may well be considered as applicable to the individual members in a pure state. This group of elements is an interesting example of the Oddo-Harkins rule, that the elements with even atomic numbers are more plentiful than those with odd atomic numbers. They are of particular interest to geochemists and petrologists because under conditions of extreme fractionation, the steady decrease in ionic radius with increasing atomic number gives rise to preferential entry of the small cations, and leads to enrichment of the residual melt in the heavier elements. The ionic radius of yttrium is close to that of dysprosium, allowing yttrium to accumulate with and be camouflaged by the lanthanide earths. The state of oxidation of the magma may result in analogous values for certain lanthanides, particularly cerium and europium, that have more than one stable valency state.

There is some discrepancy between the abundance values reported by various authors. This may be due in part to the difficulty of obtaining reliable analytical results for individual rare earths and for scandium and yttrium, and in part to the variation existing between rocks of one area and those of another. Some typical values are given in Table 33. This has been compiled from data by Borisenkov,[1] Taylor,[2] Flanagan,[3] Fryklund and Fleischer,[4] Fleischer[5] and others.

Apart from the very rare occurrence of thortveitite $Sc_2Si_2O_7$ with about 40 per cent scandium oxide, scandium does not form independent minerals. Notable amounts of this element have been reported in wolframite, cassiterite and triplite. Yttrium occurs as yttrofluorite, xenotime, thalenite ($Y_2Si_2O_7$), gadolinite, as a minor constituent of zircon and thorite as well as in monazite, allanite and other rare earth minerals.

TABLE 33. SOME TYPICAL VALUES FOR A NUMBER
OF ROCK TYPES

Rock type	Sc, ppm	Y, ppm	TRE, *ppm
Ultrabasic	10		20
Basic	40	40	50
Intermediate	20	30	
Grandiorite	10	35	
Granite	5	50	200
Shale	15	25	100
Limestone	5	15	22

* Total rare earths.

In the course of magmatic evolution the residual liquors become enriched in all the rare earths, which appear to an increasing extent in the later crystallates and finally contribute characteristic phosphate, fluoride and oxide minerals to the pegmatite mineral assembly.

The geochemistry of the rare earth elements has been considered by many workers, and there is a very large number of recent papers describing work in this field. A review of certain aspects in the geochemistry of these elements has been published by Ahrens.[6]

The Determination of Scandium in Silicate Rocks

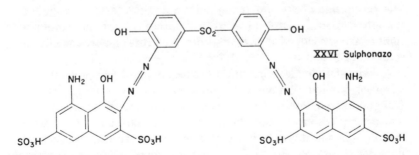

XXVI Sulphonazo

Most of the determinations of scandium in silicate rocks appear to have been made by emission spectrography. In recent years neutron activation analysis has also been successfully used.[7-9] A large number of photometric reagents have been suggested for the determination of scandium but none of these is specific and even the most selective requires extensive separation procedures to remove interfering elements. Brudz *et al.*[10] examined fourteen reagents that form coloured complexes with scandium, including arsenazo, thoron, alizarin, quinalizarin, carmine and murexide, but recommended sulphonazo(XXVI) as the reagent that combined high sensitivity with maximum selectivity to scandium.

Sulphonazo is dark-red in colour and readily soluble in water to give a violet-red coloured solution, which changes to pink on the addition of dilute hydrochloric acid. At pH 4·0–5·5 sulphonazo forms a stable violet-blue coloured water-soluble complex with scandium. Both this and the similar sulphonazo complex with yttrium have an absorption close to that of the reagent, and for this reason the measurements of optical density are made at a wavelength away from the absorption maximum. In an acetate buffered solution the absorption of the yttrium complex is similar to that of the reagent and does not interfere with the determination of scandium.

Elements that react with sulphonazo and interfere with the determination of scandium include vanadium, cobalt, gallium, copper, indium, nickel, uranium(VI), aluminium and zinc. Iron, titanium and zirconium interfere by hydrolysis. The separation procedure described by Brudz *et al.*[10] includes precipitation with sodium hydroxide to remove aluminium, extraction with ether to remove iron, and precipitation of the scandium with ammonium tartrate in the presence of added yttrium. The precipitate of yttrium–scandium ammonium tartrate is ignited to the mixed oxides, dissolved in hydrochloric acid and the colour with sulphonazo developed in acetate-urotropine (hexamethylenetetramine) solution at pH 5.

A somewhat similar separation procedure involving precipitation as a mixed yttrium-scandium ammonium tartrate was described by Belopol'ski and Popov,[11] who used xylenol orange to complete the photometric determination. Shimizu[12] used arsenazo to determine scandium in silicate rocks but used a more extensive separation procedure based on both cation and anion exchange. The same separation was advocated for the determination of scandium in silicate rocks[13] using 4-(2-thiazolyl)resorcinol as photometric reagent. A simpler separation procedure by Galkina and Strel'tsova[14] is based upon extraction of scandium into an isobutanol solution of butyric acid in the presence of sulphosalicyclic acid. The determination is completed photometrically using arsenazo III.

The Determination of Yttrium

As with scandium, yttrium is most frequently determined in silicate rocks by techniques of emission spectrography. The sensitivity is however poor, although some improvement can be obtained by using a cation exchange enrichment procedure as described by Edge and Ahrens.[15] Photometric reagents suggested for yttrium include methylthymol blue, alizarin red S, thoron, catechol violet and xylenol orange. These reagents do not have the selectivity really required for this purpose, and further work is necessary before they can be applied to the determination of yttrium in silicate rocks.

The Determination of the Lanthanide Rare Earths

Individual rare earth elements are usually determined by such physical methods as emission spectrography, or by a combination of chemical and physical methods such as neutron activation analysis, or chemical processing followed by X-ray fluorescence spectrography. Where the total rare earth content of rocks and minerals is required, gravimetric methods can be used.

GRAVIMETRIC DETERMINATION IN SILICATE ROCKS

The final determination step is a precipitation of the rare earths as mixed oxalates followed by ignition to mixed oxides as the weighing and reporting form. The major part of the cerium present will be in the higher valent state as CeO_2, whilst the weighed residue will include also any thorium present in the rock material, as ThO_2 and yttrium as Y_2O_3. If required, cerium and thorium can be separated chemically from the remaining earths, and determined individually.

Before the oxalates can be precipitated, it is necessary to separate the rare earths from most of the other components present in silicate rocks. Methods used for this include precipitation with sodium hydroxide to remove aluminium and the alkaline earth elements, precipitation with hydrofluoric acid to remove iron, titanium, zirconium and other elements forming soluble fluorides, and chlorination to remove elements that form volatile chlorides including iron, titanium, aluminium and zirconium. A variety of procedures for determining total rare earths have been based upon combinations of these separation procedures. In most of them there is a significant loss of rare earths, amounting to as little as 3 per cent or as much as 25 per cent or more. These losses can be observed and corrected by adding an active isotope of one or more of the rare earths before the first separation. Cerium-144 and yttrium-90 have been suggested, although active isotopes of other rare earths may also be used.

The procedure described below is based upon that given by Varshal and Ryabchikov.[16] The silicate rock material is decomposed by evaporation with hydrofluoric and perchloric acids in the usual way. A separation stage involving precipitation with sodium hydroxide is then combined with a chlorination procedure prior to the precipitation of the rare earths as the mixed oxalates.

The determination of total rare earths may be combined with those of chromium, vanadium, chlorine, barium and zirconium by using an alkali carbonate fusion to decompose the sample material. The rare earths are recovered by hydroxide precipitation with ammonia after the recovery of zirconium by phosphate precipitation. The hydroxide precipitate containing phosphate is then dissolved in hydrofluoric acid and the fluoride residue recovered for subsequent conversion to oxalates and then to oxides.

If the determination of zirconium is not required, the ammonium phosphate solution should not be added, and the hydroxide precipitate obtained by adding ammonia to the sulphuric acid solution. This precipitate can be calcined and chlorinated as described below. If the determination of barium can also be omitted the residue obtained after extracting the original melt with water can be dissolved in dilute hydrochloric acid, the hydroxides precipitated with ammonia, and then calcined prior to chlorination.

Method

Apparatus. The apparatus required for the chlorination is shown in Fig. 83. It is based upon that described by Iordanov and Daiev[17] and consists of a silica-tube furnace maintained at a temperature of 600°, and a means of providing an atmosphere containing carbon tetrachloride vapour. Either dry carbon dioxide or dry nitrogen can be used as carrier gas.

Reagents: *Sodium hydroxide solution,* dissolve 40 g of the reagent in water
and dilute to 100 ml. Store in a polyethylene bottle.

Sodium hydroxide wash solution, dissolve 2·5 g of sodium hydroxide in 500 ml of water. Store in a polyethylene bottle.

Carbon tetrachloride.

Carbon dioxide or nitrogen, cylinder supply.

Hydrogen peroxide, 30 per cent.

Ammonium nitrate wash solution, dissolve 10 g of the reagent in
500 ml of water and make just alkaline to methyl red with
ammonia.

Ammonium oxalate solution, dissolve 4 g of ammonium oxalate in
100 ml of water.

Oxalic acid solution, saturated.

Hexamethylenetetramine (urotropine) solution, dissolve 2·5 g of
the reagent in 10 ml of water.

Oxalic acid wash solution, dissolve 4 g of oxalic acid in 100 ml of
water and add 2 drops of the hexamethylenetetramine solution.

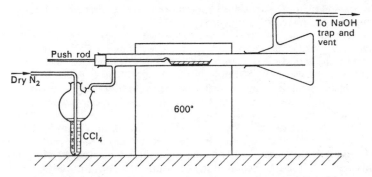

Fig. 83. Apparatus for the chlorination of rare earth residues.

Procedure. Accurately weigh 5 g of the finely powdered silicate rock material
into a platinum dish, moisten with water and add 10 ml of concentrated
perchloric acid and 40 ml of concentrated hydrofluoric acid. Transfer the dish
to a hot plate and evaporate to copious fumes of perchloric acid. Allow to
cool, add 10 ml of concentrated hydrofluoric acid and again evaporate to
fumes. Allow to cool, rinse down the sides of the dish with a little water and
evaporate to complete dryness. Allow to cool, moisten with perchloric acid
and again evaporate to complete dryness. Repeat the evaporation with a little
perchloric acid, but this time remove the dish from the hot plate before the
residue is completely dry. Allow to cool and dissolve the moist perchlorates
in about 400 ml of water.

If any residue remains it should be collected carefully on a small filter, washed with a little water, dried and ignited in a small platinum crucible. Fuse the residue with a little anhydrous sodium carbonate, extract with water, acidify with a little dilute perchloric acid and add to the main rock solution.

Heat the rock solution to a temperature of about 60°, add sodium hydroxide solution until no further precipitation is apparent and add 10 ml in excess. Allow to stand on a hot plate for about 15 minutes, allow to cool and collect the precipitate on a hardened, open textured filter paper. Wash the precipitate four or five times with the sodium hydroxide wash solution. Discard the filtrate and washings.

Rinse the residue back into the original beaker with water and dissolve by adding 10 ml of concentrated hydrochloric acid. Filter the solution through the paper used to collect the hydroxide precipitate and wash well with water. Dilute with water to a volume of about 250 ml, and precipitate the hydroxides by adding concentrated ammonia solution until no further precipitation is apparent, then adding 15 ml in excess. Allow the beaker to stand on a hot plate for 30 minutes, and collect the precipitate on a new open-textured filter paper and wash five times with the ammonium nitrate wash solution. Discard the filtrate and washings. Redissolve the residue in hydrochloric acid as before and reprecipitate the hydroxides with ammonia and again collect the precipitate on an open-textured filter paper.

Transfer the paper and precipitate to a silica crucible, dry and ignite the residue in an electric furnace set at a temperature of about 600°. Allow to cool and brush the oxide residue into a silica combustion boat. Assemble the apparatus as shown in Fig. 83, with the combustion boat outside the hot zone of the furnace which is maintained at a temperature of 600°. Pass dry nitrogen gas (or carbon dioxide) through the carbon tetrachloride and through the furnace tube. Slowly push the combustion boat into the hot zone of the furnace, where the process of chlorination takes place, until a completely white residue is obtained. This can only be observed by temporarily removing the boat from the furnace using the iron push rod, which is terminated with a hook for this purpose. The time required may amount to an hour or more, and will depend largely upon the total weight of the oxide residue.

Remove the boat from the tube, allow to cool and moisten the residue with water. Rinse into a 50-ml beaker, add a few ml of concentrated nitric acid and a few drops of hydrogen peroxide, cover with a clock glass and digest on a hot plate for about 30 minutes. Remove the cover and evaporate the solution almost but not quite to dryness. Allow to cool, add a further few ml of concentrated nitric acid and again repeat the evaporation almost to dryness. Add about 4 ml of hot water to the residue, stir until all soluble material has dissolved and collect the residue on a very small, close-textured filter paper. Wash the residue with a little hot water containing a few drops of dilute nitric acid and discard it (note 1).

Collect the combined filtrate and washings, with a total volume of about 7 ml, in a 25-ml beaker. Heat on a steam bath and precipitate the hydroxides by adding concentrated ammonia solution drop by drop until precipitation appears to be complete, then add 1 ml in excess. Digest on a steam bath for

15 minutes and collect the precipitate on a small filter paper. Test the filtrate for calcium by adding ammonium oxalate solution. If calcium is present, dissolve the rare earth precipitate in a little dilute nitric acid and reprecipitate the hydroxides as described above. Collect the precipitate and wash with the ammonium nitrate wash solution.

Dissolve the residue in a small amount of dilute nitric acid and filter through the paper into a 10-ml beaker. Remove the excess acid by evaporation almost to dryness and dissolve the moist residue in 2 ml of water. Add ammonia to neutralise the last traces of nitric acid, and heat on a steam bath to remove the excess ammonia. Cool, add 3 ml of water, 1 drop of hexamethylenetetramine solution and 3 ml of oxalic acid solution. Stir and set aside overnight.

Collect the precipitate on a small close-textured filter paper and wash four times with the oxalic acid wash solution. Discard the filtrate and washings. Transfer the paper and precipitate to a weighed platinum crucible, dry, ignite and weigh as total rare earth oxides.

Notes: 1. This residue consists largely of silica but may contain an appreciable proportion of the total rare earths from the rock material. The rare earths may be recovered by evaporation with hydrofluoric and nitric acids, and the nitrate solution added to the main solution.

2. If a radioactive tracer has been added to the rock material, then the several filtrates and washings should be examined for radio-activity before being discarded. On completion of the gravimetric determination, the rare earth oxides should be fused with a little sodium carbonate, extracted with water, dissolved in dilute nitric acid and the activity compared with that added originally.

GRAVIMETRIC DETERMINATION IN CARBONATITE ROCKS

Certain carbonatite rocks contain large amounts of rare earth elements, largely in the form of the mineral monazite, but also as bastnaesite and other acid-soluble minerals. Some specimens have been encountered with as much as 15 per cent rare earths, and 5–10 per cent is by no means uncommon in rocks from these particular occurrences. The total rare earth content can be determined gravimetrically using the separation procedures outlined previously.

The carbonate fraction of the carbonatite rock material is decomposed by warming with dilute hydrochloric acid. Monazite and other minerals in the insoluble residue are decomposed by prolonged fuming with concentrated sulphuric acid, after an evaporation with a mixture of hydrofluoric and sulphuric acids. Ammonia solution is used to precipitate the hydroxides which are then dissolved in hydrofluoric acid leaving a fluoride residue containing the rare earths. This is decomposed by evaporation with perchloric acid, the perchlorates dissolved in dilute hydrochloric acid and the mixed rare earths precipitated as oxalates in the usual way. For rocks containing 2–15 per cent of rare earths, isotope dilution analysis has indicated a recovery of about 97 per cent.

PHOTOMETRIC DETERMINATION OF RARE EARTHS IN SILICATE ROCKS

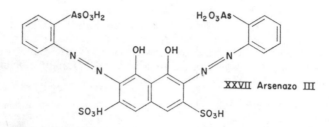

XXVII Arsenazo III

Reagents suggested for the photometric determination of total rare earths include alizarin red-S, aluminon, xylenol orange, arsenazo I and III, PAN (1-(2-pyridylazo)-2-naphthol) and PAR (4-(2-pyridylazo) resorcinol). None of these is specific for the rare earths, and the selectivity of all of them is poor. The best combination of high sensitivity with some degree of selectivity is given by arsenazo III (XXVII). This reagent forms complexes with a large number of other elements, including thorium, uranium and zirconium at low pH, and iron, yttrium, the rare earths and other elements at higher pH. In the procedure described below a pH of 1·8–2·0 is used for the determination of the rare earths.

The complexes formed by yttrium and the rare earths with arsenazo III, all have similar absorption spectra with maximum absorption occurring at a wavelength of about 660 nm. The arsenazo III complex of scandium has a maximum absorption at a slightly higher wavelength and interferes somewhat with the rare earths determination. A broadband red filter has been recommended for the determination.[18]

Before this photometric procedure can be applied it is necessary to separate the rare earth elements from all other elements that react with arsenazo III. This can be accomplished by precipitation as hydroxides with ammonia, followed by precipitation as oxalates with calcium added as carrier. The procedure described in detail below is based upon that given by Goryushina, Savvin and Romanova.[16]

Method

Reagents: *Hydrogen peroxide solution*, 30 per cent.

Ammonium nitrate wash solution, dissolve 10 g of the reagent in 500 ml of water and add a few drops of ammonia.

Calcium carrier solution, dissolve 0·50 g of pure calcium carbonate in the minimum amount of dilute hydrochloric acid and dilute to 100 ml with water. This solution contains 2·0 mg calcium per ml.

Oxalic acid.

Oxalic acid wash solution, dissolve 5 g of oxalic acid in 500 ml of water.

Potassium chloride buffer solution, mix together 80 ml of 0·2 N hydrochloric acid, 250 ml of 0·2 M potassium chloride solution and 670 ml of water.

Arsenazo III solution, dissolve 0·10 g of the reagent in 100 ml of 0·01 N hydrochloric acid.

Procedure. Accurately weigh approximately 1 g (note 1) of the finely powdered silicate rock material into a platinum dish, moisten with water and add 15 ml of concentrated hydrofluoric acid, 1 ml of concentrated nitric acid and 5 ml of 20 N sulphuric acid. Transfer the dish to a hot plate and evaporate to fumes of sulphuric acid. Allow to cool, rinse down the sides of the dish with a little water, add 5 ml of concentrated hydrofluoric acid and again evaporate to fumes of sulphuric acid. Allow to cool, again rinse down the sides of the dish with a little water and evaporate, this time to remove most of the excess sulphuric acid and to leave a moist residue of sulphates. Allow the dish to cool, add about 10 ml of water, break up the residue with a glass rod and rinse the contents of the dish into a 250-ml beaker with about 100 ml of water. Add 3 drops of hydrogen peroxide solution and 10 ml of concentrated hydrochloric acid and heat on a hot plate until a clear solution is obtained.

If any residue remains it should be collected, washed with water, dried and ignited in a small platinum crucible and fused with a little anhydrous sodium carbonate. Extract the melt with water, acidify with a little dilute hydrochloric acid and add to the main rock solution.

Heat the solution to near boiling and add concentrated ammonia solution until the precipitation of hydroxides is apparently complete and then add 15 ml in excess. Allow the precipitate to settle, then filter whilst still hot onto a hardened, open-textured filter paper and wash well with the ammonium nitrate wash solution. Discard the filtrate and washings.

Rinse·the residue back into the original beaker with about 100 ml of water and dissolve by warming with 10 ml of concentrated hydrochloric acid. Filter the warm solution through the paper previously used and wash well with water. Add 10 ml of the calcium carrier solution to the filtrate, and 1·5 g of oxalic acid. Stir, and heat the solution almost to boiling. Now add concentrated ammonia to bring the pH of the solution to a value of 5, as shown by close-range pH indicator papers. Allow to digest on a hot plate for 1 hour and then set the beaker aside overnight.

Collect the oxalate precipitate on a close-textured filter paper and wash with the oxalic acid wash solution. Discard the filtrate and washings. Transfer the paper and residue to a small platinum or silica crucible and dry and ignite in an electric furnace at a temperature of 500° in order to convert oxalates to carbonates. Allow to cool and dissolve the residue in a small volume of hydrochloric acid containing a few drops of hydrogen peroxide. Transfer the solution to a 50-ml beaker and evaporate to dryness on a steam bath. Dissolve the chloride residue in $0·01$ N hydrochloric acid and dilute to volume in a 100-ml volumetric flask with $0·01$ N hydrochloric acid.

Transfer an aliquot of this solution (note 2) containing 5–30 μg of rare earths to a 50-ml volumetric flask, add 5 ml of the potassium chloride buffer solution and 2 ml of the arsenazo III solution. Dilute to volume with $0·01$ N hydrochloric acid and mix well. Measure the optical density of this solution using an absorptiometer fitted with a red filter. The reference solution should be prepared from $0·01$ N hydrochloric acid to which 5 ml of buffer solution and 2 ml of arsenazo III solution have been added.

Calibration. The calibration graph can be obtained by plotting the optical densities of rare earth solutions containing 0–30 μg rare earths, to which buffer solution and arsenazo III solution have been added as described above for the reference solution. Ideally a mixture of rare earths from the silicate rock under examination should be used for the calibration, but the use of a mixed rare earth precipitate from other silicate rocks or from many other sources will not introduce significant errors.

Notes: 1. This sample weight is suggested for rock materials containing not more than about 500 ppm total rare earths. It should be reduced to $0·2$ g for rock materials containing higher concentrations of rare earths.

2. For basic and other rocks containing considerably less than 100 ppm rare earths, it may be necessary to take the whole of the sample solution for photometric determination.

References

1. BORISENKOV L. F., *Geokhimiya* (1959) (7), 623.
2. TAYLOR S. R., *Geochim. Cosmochim. Acta* (1962) **26**, 81.
3. FLANAGAN F. J., *Geochim. Cosmochim. Acta* (1967) **31**, 289.
4. FRYKLUND V. C. Jr. and FLEISCHER M., *Geochim. Cosmochim. Acta* (1963) **27**, 643.
5. FLEISCHER M., *Geochim. Cosmochim. Acta* (1965) **29**, 755.
6. AHRENS L. H., *Progress Sci. Tech. Rare Earths* (1964) **1**, 1 (Pergamon).
7. KEMP D. M. and SMALES A. A., *Anal. Chim. Acta* (1960) **23**, 410.
8. DESAI H. B., KRISHNAMOORTHY IYER R. and SANKAR DAS M., *Talanta* (1964) **11**, 1249.
9. HAMAGUCHI H., WATANABE T., ONUMA N., TOMURA K. and KURODA R., *Anal. Chim. Acta* (1965) **33**, 13.

10. Brudz V. G., Titov V. I., Osiko E. P., Drapkina D. A. and Smirova K. A., *Zhur. Anal. Khim.* (1962) **17**, 568.
11. Belopol'skii M. P. and Popov N. P., *Zavod. Lab.* (1964) **30**, 1441.
12. Shimizu T., *Anal. Chim. Acta* (1967) **37**, 75.
13. Shimizu T. and Momo E., *Anal. Chim. Acta* (1970) **52**, 146.
14. Galkina L. L. and Strel'tsova S. A., *Zhur. Anal. Khim.* (1970) **25**, 889.
15. Edge R. A. and Ahrens L. H., *Anal. Chim. Acta* (1962) **26**, 355.
16. Varshal G. M. and Ryabchikov D. I., *Zhur. Anal. Khim.* (1964) **19**, 202.
17. Iordanov N. and Daiev Khr., *Zhur. Anal. Khim.* (1962) **17**, 429.
18. Goryushina V. G., Savvin S. B. and Romanova E. V., *Zhur. Anal. Khim.* (1963) **18**, 1340.

CHAPTER 38

SELENIUM AND TELLURIUM

Occurrence

These two elements strongly resemble sulphur, and in silicate rocks are largely associated with the accessory sulphide minerals. The ratio Se:S has been given by Goldschmidt[1] as 1:6000 for magmatic rocks, giving an average selenium content of $0 \cdot 09$ ppm. This is in broad agreement with recent values for selenium in a series of standard rocks obtained using a neutron activation procedure.[2] The average tellurium content is probably at least one order of magnitude less than that of selenium. Sulphide minerals contain somewhat greater amounts of both selenium and tellurium, up to a few ppm being by no means uncommon.

A number of rare selenides and tellurides are known, mostly of the heavier metals—lead, copper, bismuth, gold, silver and the platinum metals—but the commercial sources of both elements are the flue dusts and anode slimes arising from the winning or refining of metals from sulphide ores.

The Determination of Selenium

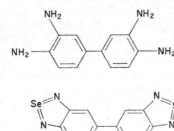

XXVIII a   Diaminobenzidine

XXVIII b   Piazselenol

Selenium reacts with 3,3'-diaminobenzidine (XXVIIIa) to give an intense yellow coloured compound known as piazselenol (XXVIIIb).

This reaction, reported by Hoste and Gillis,[3] was used as the basis of a determination of selenium by Cheng,[4] who used EDTA as a masking agent to prevent interference from a number of metals, and extracted the complex into toluene for photometric measurement. These solutions of piazselenol have absorption maxima at wavelengths of 340 and 420 nm—the latter wavelength being preferred as the reagent itself absorbs much less light at 420 nm than at 340 nm. the Beer–Lambert Law is followed over the concentration range 5–25 μg selenium in 10 ml of toluene.

Approximately 30 minutes are required for complete colour development, which is achieved at a pH of 2 to 3 in the presence of formic acid. For extraction into toluene, the pH of the solution is adjusted to a value between 6 and 7.

This colour reaction has been used to determine selenium in rocks and soils by Rapp, Willigman and Patraw[5] and in soils and sediments by Stanton and McDonald.[6] These two procedures were devised for sample material arising from mineralised areas, where the selenium content might be expected to be a great deal higher than that of normal silicate rocks. Belopol'skaya[7] has described a rapid method for selenium in sulphide and silicate minerals involving decomposition by heating with eschka mixture for $1 \cdot 5$–2 hours at a temperature of 750–800°. The aqueous extract is acidified with hydrochloric acid and heated with hydrobromic acid. Sulphosalylic acid and EDTA are added, followed by diaminobenzidine solution after adjustment of the pH to 2–3. After 30 minutes, the pH is again adjusted, this time to a value between 7 and 8, and the piazselenol extracted into toluene for photometric measurement.

Chau and Riley[8] have given procedures for selenium in sea water, silicates and marine organisms which are rather more rigorous than the earlier procedures, involving co-precipitation with ferric hydroxide, separation from iron by ion-exchange and finally extraction of the piazselenol into toluene. The amount of selenium recovered by this procedure, usually about 95 per cent, is determined by adding radioactive selenium-75 at the decomposition stage, and comparing the activity of the aqueous solution after completing the separation of selenium with that of a similar aliquot of the active solution. This method is described below.

A procedure for determining selenium in glass involving a fusion with sodium carbonate, volatilisation as tetrabromide and photometric measurement as piazselenol has been described by Blankley.[9] This is not directly applicable to the determination of those small amounts found in most silicate rocks and minerals, but could be used for rocks from mineralised areas containing several ppm of selenium.

Method

Apparatus. *Ion-exchange column*, this consists of a cation exchange resin such as Zeo-Carb 225, 52–100 mesh, in the form of a column 10 cm in length and 1·5 cm in diameter. This resin should be washed with N hydrochloric acid until the eluate no longer gives a reaction for iron, then with water until almost free from acid.

Spectrophotometer cells, for this determination 4-cm cells with a capacity of only a few ml are required ("micro cells").

Reagents: *Diaminobenzidine solution*, dissolve 50 mg of the hydrochloride form of the reagent in 10 ml of water. Store in a refrigerator and discard after 3 days or earlier if darkening occurs. The reagent is expensive and solutions deteriorate. They should therefore be prepared only in small volumes as required.

EDTA solution, dissolve 3·72 g of the disodium salt of EDTA in 100 ml of water.

Formic acid solution, 2·5 M.

Iron carrier solution, dissolve 2 g of anhydrous ferric chloride in 100 ml of water containing a few ml of concentrated hydrochloric acid to prevent hydrolysis.

Selenium-75 solution, carrier-free selenite solution, dilute with water as required to give a solution with approximately 10,000 counts per minute per ml.

Standard selenium stock solution, dissolve 0·25 g of pure selenium in 1–2 ml of concentrated nitric acid and dilute to 250 ml with water. This solution contains 1 mg selenium per ml.

Standard selenium working solution, containing 2 μg selenium per ml. Prepare by dilution of the stock solution with water as required.

Procedure. Accurately weigh 1–2 g of the finely ground silicate rock material (note 1) into a platinum dish, moisten with water and add 2-ml of the selenium-75 solution. At the same time transfer a similar 2-ml aliquot of the active solution to a counting tray, evaporate to dryness and set aside. To the sample material in the platinum dish add 10 ml of concentrated hydrofluoric acid and 10 ml of concentrated nitric acid, cover the dish and set aside overnight.

Rinse and remove the cover and evaporate to dryness on a steam bath. Add 10 ml of concentrated hydrofluoric acid and 10 ml of concentrated nitric acid and repeat the evaporation to dryness. Add 5 ml of concentrated nitric acid to the dry residue and again evaporate to dryness. Repeat this evaporation with two further 5 ml portions of nitric acid to decompose fluorides and remove fluorine as hydrogen fluoride.

Moisten the residue with nitric acid, rinse into a 100-ml beaker and evaporate to dryness. Add 25 ml of 4 N hydrochloric acid and boil gently for 5 minutes to convert any selenate to selenite. If any residue remains at this stage collect on a small filter, wash with water and combine the filtrate and washings Discard the residue (note 2).

Dilute the solution to about 5 litres with water and add dilute sodium hydroxide solution to bring the pH to a value between 3·5 and 4. Add 60 g of solid sodium chloride, stir to dissolve and then with stirring add 3 ml of the iron carrier solution. Now bring the pH to a value between 4·5 and 5·0 by adding dilute aqueous ammonia, stir, and allow to stand for 2 hours. Add a further 3 ml of the iron carrier solution, check the pH and adjust to 4·5–5·0 if necessary, stir and allow to stand for at least 2 days. Decant or siphon off the supernatant liquid and collect the precipitate on a small paper. Wash the ferric hydroxide precipitate with dilute ammonium nitrate solution and discard the filtrate and washings.

Rinse the precipitate into a small beaker, dissolve in 1–2 ml of concentrated nitric acid (note 3), and dissolve any remaining traces of residue from the paper. Dilute the solution to give an acid concentration of 0·2 N and pass through the cation exchange column. Elute with 350 ml of 0·2 N nitric acid. Combine the percolate and eluate, add 1 ml of 2 N sodium hydroxide and evaporate almost to dryness. Transfer to a counting tray and complete the evaporation to dryness. Compare the activity with that of the active residue previously set aside, and hence calculate the selenium recovery.

Rinse the residue from the counting tray into a small beaker, add 2 ml of 2·5 M formic acid and 5 ml of EDTA solution and dilute to about 25 ml with water. Adjust the pH of the solution to a value in the range 2–3 by adding dilute nitric acid or aqueous ammonia as necessary. Now add 2 ml of the diaminobenzidine reagent, allow to stand for 30 minutes then add aqueous ammonia to bring the pH to 6–7, and rinse the solution into a separating funnel.

Add 5 ml of toluene and shake for 3 minutes. Discard the lower, aqueous layer and measure the optical density of the toluene extract in 4-cm micro cells, using the spectrophotometer set at a wavelength of 429 nm, with pure toluene as the reference solution.

Calibration. Transfer aliquots of 0–5 ml of the standard selenium solution containing 0–10 μg selenium to separate beakers, add 2 ml of 2·5 M formic acid to each and continue as described above. Plot the relation of optical density to selenium concentration.

Notes: 1. Marine sediments and other rock samples containing water-soluble chlorine should be washed with water until no further chloride ion can be detected in the washings, and then dried before analysis.

2. If the residue contains silicate minerals, it should be retreated by evaporation with small quantities of hydrofluoric and nitric acids as described, the residue dissolved in hydrochloric acid and added to the main rock solution.

3. If any difficulty is experienced in obtaining complete solution, add 0·1 ml of concentrated hydrochloric acid to the nitric acid extract.

The Determination of Tellurium

Small quantities of tellurium are difficult to determine. Hanson[10] has described a spectrophotometric procedure based upon the reduction of tellurium to a colloidal form by adding stannous chloride solution. Anderson and Peterson[11] have used this for natural materials, but give a lower limit of 0·5 ppm, for which 10 g sample portions are used. This concentration is greatly in excess of the tellurium content of most silicate rocks.

Lovering, Lakin and McCarthy[12] determined down to 0·1 ppm tellurium in jasperoid samples by a method based upon the induced precipitation of elemental gold from a 6 N hydrochloric acid solution containing gold chloride, cupric chloride and hypophosphorous acid. The amount of gold reduced is proportional to the amount of tellurium present.[13]

Concentrations of down to 0·005 ppm (5 ppb) of tellurium were determined by Hubert[14] measuring the catalytic effect on the reduction of gold by hypophosphorous acid. The method is suitable only for those materials from which the tellurium can be released by digestion with bromine and hydrobromic acid. An extraction into methyl isobutyl ketone is used to isolate and concentrate the tellurium.

The Determination of Selenium and Tellurium by Atomic Absorption Spectroscopy

Although this technique has been suggested for the determination of both selenium and tellurium in geological materials,[15, 16, 17] such procedures do not have sufficient sensitivity for direct application to silicate rocks. A separation stage is essential, and even then the sensitivity is barely adequate.

References

1. GOLDSCHMIDT V. M., *Geochemistry*, Oxford, 1954, p. 532.
2. BRUNFELT A. O. and STEINNES E., *Geochim. Cosmochim. Acta* (1967) **31**, 283.
3. HOSTE J. and GILLIS J., *Anal. Chim. Acta* (1955) **12**, 158.
4. CHENG K. L., *Analyt. Chem.* (1956) **28**, 1738.
5. RAPP G. Jr., WILLIGMAN M. G. and PATRAW J., *Proc. S. Dakota Acad. Sci.* (1964) **43**, 57.
6. STANTON R. E. and McDONALD A. J., *Analyst* (1965) **90**, 497.
7. BELOPOL'SKAYA T. L., *Tr. Vses. Nauchn.-Issled, Geol. Inst.* (1964) **117**, 85.
8. CHAU Y. K. and RILEY J. P., *Anal. Chim. Acta* (1965) **33**, 36.
9. BLANKLEY M., Tech. Note 131, Determination of Selenium in Glass, Brit. Glass Res. Assoc. (1970).
10. HANSON C. K., *Analyt. Chem.* (1957) **29**, 1204.
11. ANDERSON W. L. and PETERSON H. E., *U.S. Bur. Mines Rept. Invest.* 6201, 1963.
12. LOVERING T. G., LAKIN H. W. and McCARTHY J. H., *U.S. Geol. Surv. Prof. Paper* 550-B, p. B138, 1966.
13. LAKIN H. W. and THOMPSON C. E., *Science* (1963) **141**, 42.
14. HUBERT A. E., *U.S. Geol. Surv. Prof. Paper* 750-B, p. B188 (1971).
15. SEVERNE B. C. and BROOKS R. R., *Talanta* (1972) **19**, 1467.
16. NAKAGAWA A. M. and THOMPSON C. E., *U.S. Geol. Surv. Prof. Paper* 600-B, p. B138 (1968).
17. SEVERNE B. C. and BROOKS R. R., *Anal. Chim. Acta* (1972) **58**, 216.

CHAPTER 39

SILICON

Occurrence

After oxygen, silicon is the most abundant of all the elements, amounting to almost 28 per cent of the rocks of the lithosphere.[1] Even when present in trace amounts, silicon is almost always reported as silica, SiO_2, which is the form used here. Silicate rocks have been classified in a number of different ways, based upon rock texture, crystallite size, mode of emplacement, etc., but the simplest subdivision following the chemical composition, is that based upon silica content (Fig. 84).

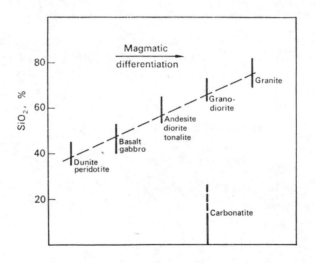

84. Silica content of igneous rocks.

The distribution of silica in silicate rocks was investigated by Richardson and Sneesby[2] who showed that maximum values occurred at 52·5 per cent and 73 per cent silica, corresponding to the two most common igneous rocks, basalt and granite. Details of this distribution pattern have been criticised by several workers, including Ahrens,[3] who attributed errors in the distribution relating to the basaltic mode to the inclusion of an undue proportion of unusual and rare rocks in the total of analyses used. It is also clear that the systematic error in the early gravimetric determinations of silica will have been included in all or almost all of the analyses.

Most silicate rocks contain between 35 and 80 per cent of silica, although the range of values that an analyst can expect to meet is extended by sandstones and quartzites at one end, and by siliceous and argillaceous carbonates at the other.

Quartzites and sandstones may contain up to about 99 per cent silica, although 90–95 per cent is more frequently encountered. Whilst the silica content of these rocks is rarely of much interest to the petrologist or geochemist, it is of importance to the economic mineralogists, as these materials are used in large quantities in the glass, refractory, chemical and building industries.

The carbonate rocks are a large and varied group of igneous (carbonatite), sedimentary and metamorphic rocks, with silica contents ranging from less than 1 per cent (as, for example, in certain limestones such as chalk and some forms of marble) to 30 per cent or more in some sediments and carbonatites.

Silica occurs in several crystalline forms (quartz, tridymite, crystobalite, coesite) and in the amorphous state (opal, onyx, chalcedony). But the commonest occurrence in silicate rocks is combined with magnesium, iron, aluminium, alkali and alkaline earth elements in the form of complex silicate minerals. The commonest of these are the rock-forming minerals—namely the minerals of the olivine, pyroxene, amphibole, mica and felspar series. Other silicate minerals include the felspathoids (nepheline, nosean, leucite, sodalite, etc.) aluminosilicates (kyanite, kaolin, sillimanite) and a variety of accessory minerals (zircon, tourmaline, sphene).

Gravimetric Determination of Silica in Quartzites and Sandstones

In the classical procedure for determining silica, a silica residue, obtained by chemical processing of the rock material, is subjected to evaporation with a mixture of hydrofluoric and sulphuric acids and the loss in weight caused by volatilisation of silicon tetrafluoride taken as a measure of the silica content of the residue. With rocks rich in silica, such as quartzites, the silica residue obtained by the usual chemical processing is likely to contain elements, particularly sodium, in amounts greater than in the quartzite itself. Furthermore, not all the silica will have been collected in the silica residue. For these reasons the preliminary separation stage is best omitted, and the loss in weight that occurs on volatilisation of the silica determined directly from the ground rock material.

This method is applicable only to those samples with high silica content, where the remaining constituents amount to no more than 1 or 2 per cent of the total. This method is, however, often used for sands and sandstones containing much less silica, where appreciable errors can arise. Carbonate minerals are rare components of these rocks, but ferruginous carbonate material has been noted cementing the silica grains in some sandstones. The loss of carbon dioxide that occurs during the initial heating stage does not give rise to error in the silica determination, but any ferrous iron will be oxidised to ferric during the decomposition of the sulphates and hence gives a negative error. A similar error has been observed to occur when the sandstone contains grains of magnetite or ilmenite, from which the ferrous iron is converted to the higher valency state. Another source of error arises from the volatilisation of alkali metals if, as frequently happens, the sandstone rock contains grains of felspar.

Method

Reagents: *Hydrofluoric acid*, concentrated.
Sulphuric acid, 20 N.

Procedure. Ignite a clean platinum crucible of about 30-ml capacity over the full flame of a Meker burner, allow to cool and weigh empty. Transfer to it approximately 1 g of the finely ground quartzitic material and reweigh to give the weight of the sample (note 1). Again ignite the crucible over the Meker burner, gently at first and then finally over the full flame of the burner for a period of 1 hour. Allow to cool and reweigh the crucible and contents. Repeat the ignition until no further loss in weight occurs. Report the change in weight as the "loss on ignition". This loss is due to the oxidation of ferrous iron and organic matter, and to the volatilisation of water, carbon dioxide and other gases.

Add 4 drops of 20 N sulphuric acid to the residue, followed by 20 ml of concentrated hydrofluoric acid. Transfer the crucible to a hot plate or sand bath and remove the silica and excess hydrofluoric acid by volatilisation in the usual way. Gently heat the crucible on the hot plate to remove the excess sulphuric acid and then ignite over the full flame of the Meker burner for 10 minutes. Allow to cool and reweigh the crucible. Repeat the ignition until no further loss in weight occurs. Determine a reagent blank by volatilisation in the same way, but omitting the sample material. The blank value obtained with analytical grade hydrofluoric acid will probably amount to between $0\cdot3$ and $1\cdot0$ mg. This should be *added* to the loss on volatilisation to give the total silica content of the sample.

In a very few samples some silicate mineral grains may remain undecomposed. These can usually be attacked by a second evaporation with smaller amounts of hydrofluoric and sulphuric acid.

Note: 1. Tsubaki[4] describes a modification to this procedure in which 3 g of boric acid are added to a $0\cdot5$ g sample portion prior to igniting at 1000°. The determination is then completed as described above.

Gravimetric Determination in Silicate and Carbonate Rocks

The classical procedure for the determination of silica in silicate rocks has already been described in detail (p. 37). It consists of an alkali carbonate fusion, extraction with water, dissolution and dehydration with hydrochloric acid and finally determination of the loss on volatilisation with hydrofluoric acid. Two successive dehydrations are usually employed and any silica remaining in the filtrate from the second is recovered from the ammonia precipitate.

This procedure is very time consuming, partly in itself and partly because the subsequent stages of the rock analysis by this classical method cannot begin until the silica determination is complete. This may take up to 3 days. It is therefore not surprising that considerable efforts have been devoted to finding a better and more rapid way of recovering the silica from the hydrochloric acid solution of the rock melt.

Sulphuric and perchloric acids have been used to effect the dehydration of silica. Sulphuric acid is more effective than hydrochloric acid for this purpose, whilst perchloric acid is more effective still, although care must be taken to avoid explosions when perchloric acid is used. These can arise either through a dramatic oxidation of organic matter present in the solution, or more likely through incomplete removal of perchloric acid from the filter paper used to collect the dehydrated silica.

DETERMINATION IN CARBONATE ROCKS

Perchloric acid has been found to be particularly useful for the determination of major amounts (5 per cent and upwards) of silica in carbonate rocks. The following procedure has been used.

Procedure. Accurately weigh approximately 1 g of the finely powdered carbonate rock material into a platinum crucible, partly cover with a platinum lid and ignite strongly over a Meker burner for 30 minutes (note 1). Allow to cool, moisten the cake with water and rinse the residue into a platinum dish (note 2). Add 50 ml of water, 5 ml of concentrated nitric acid and 10 ml of concentrated perchloric acid. Rinse the crucible and lid with a little dilute nitric acid and add to the solution in the platinum dish. Transfer the dish to a steam bath and evaporate until most of the water has been removed. Place the dish on a hot plate or sand bath and continue the evaporation to fumes of perchloric acid. Allow to fume gently for 10 minutes, then allow to cool and add 50 ml of water. Warm to allow all soluble salts to pass into solution and then collect the residue on a close-textured filter paper. Wash the residue twice with 0·1 N hydrochloric acid and then several times with hot water. A few milligrams of silica can be recovered from the filtrate by a second dehydration if required, otherwise the combined filtrate and washings should be reserved for the determination of iron, calcium, magnesium and other constituents of the limestone.

The residue, or combined residues if a second dehydration has been made, is transferred to a weighed platinum crucible and the silica content determined by volatilisation with hydrofluoric acid as described in the classical method for the analysis of silicates (p. 38).

Notes: 1. If required, the "loss on ignition" can be determined at this stage. Carbonate minerals are very largely converted to oxides during the ignition, which also converts acid-insoluble silicate minerals to acid-soluble calcium silicate.

2. Silica dishes and both silica and borosilicate glass beakers have also been described for this dehydration of silica with perchloric acid.

GRAVIMETRIC DETERMINATION WITHOUT DEHYDRATION

The substitution of perchloric acid for hydrochloric acid in the dehydration of silica saves only a little time. A much bigger saving can be obtained by avoiding the dehydration step altogether. This can be done by adding gelatin to the acid solution obtained after the addition of hydrochloric acid to the aqueous extract of the melt. The gelatin serves to coagulate the silica, enabling it to be collected and determined in the usual way. This technique has long been used in industrial practice for silicate slag analysis, where the emphasis is more on speed than accuracy. Some 2–4 per cent of the silica will remain in solution, the exact amount depending upon the conditions used and the composition of the slag. This small amount of silica is not recoverable except by dehydration in the usual way, when some part of it will be collected.

As an alternative to gelatin, Bennett and Reed[5] have suggested the addition of industrial coagulating agents, such as polymers of ethylene oxide. A number of "Polyox" polymers were tried and all were effective at coagulating the silica and obviating the need for dehydration. This procedure has the advantages of removing all but about 0·5 per cent of the silica from the solution (this trace can be determined photometrically), the introduced polyethyleneoxide does not interfere with the subsequent photometric determination of iron, titanium or aluminium, and very little boron will accompany the silica even when borate has been added to the alkali carbonate used for sample decomposition. The following detailed procedure is essentially that described by Bennett and Reed.[5]

Method

Reagents: *Fusion mixture*, an equimolecular mixture of potassium and sodium carbonates.

Boric acid.

Polyethylene oxide solution, dissolve 0·25 g of a polyethylene oxide coagulant in 100 ml of water. (Material supplied by BDH as polyethylene oxide polymer is suitable.)

Procedure. Accurately weigh approximately 1 g of the finely powdered silicate rock material into a platinum dish, add 3 g of fusion mixture and 0·4 g of boric acid (note 1). Mix using a small spatula, and fuse gently over a burner until the vigorous stage of the decomposition is complete, then transfer to an electric furnace at a temperature of 1200° for 10 minutes. Allow to cool, add to the dish 10 ml of water and 15 ml of concentrated hydrochloric acid. Cover, and warm on a steam bath until the melt has completely disintegrated and no further evolution of carbon dioxide can be detected. Remove the lid and rinse any liquid adhering to it back into the dish with a small amount of water.

Replace the dish on the steam bath and heat for a period of 30 minutes, then stir into the solution, which now contains gelatinous silica, a little macerated filter paper, 5 ml of the polyethylene oxide solution and 10 ml of water. Allow to stand for about 5 minutes and then collect the residue on a fine-textured filter paper (note 2). Wash the residue three times with 0·5 N hydrochloric acid and then with water until the filtrate is free from chloride ion. Combine the filtrate and washings and reserve for the determination of silica traces (see p. 42), iron, titanium and aluminium. Transfer the paper and residue to a weighed platinum crucible, dry and ignite, and determine the silica by evaporation with hydrofluoric acid in the usual way (p. 38).

Notes: 1. This flux is recommended for clays and for high-silica materials such as silicate rocks. For refractories and other high-alumina materials, 2 g of the fusion mixture and $0·4$ g of boric acid should be used, together with a longer period for the final stage of the rock fusion. This should be increased to up to 30 minutes.

2. Difficulties may be encountered at the filtration stage if the polyethylene oxide solution is added before the gelling of the silica has taken place. With materials containing less than about 30 per cent silica, this gelling will not be observed, but provided that the sample solution is digested on the steam bath as suggested, polymerisation will occur with these samples and difficulties with filtration will be avoided.

A Titrimetric Method for Silicate Rocks

The only titrimetric procedure of general application to the determination of silica is that based upon the fluoride reaction:

$$Si(OH)_4 + 6F^- + 4H^+ = SiF_6{}^{2-} + 4H_2O$$

which occurs in acid solution to which excess of alkali fluoride has been added. In order to apply this reaction to silicate rocks Khalizova *et al.*[6] decomposed the rock material by fusion with potassium hydroxide and sodium peroxide, dissolved the melt in water, acidified with hydrochloric acid and removed aluminium, iron, titanium, calcium, magnesium and other elements by passing the solution through a column containing a cation exchange resin. The solution is at a sufficient dilution to prevent polymerisation of the silica.

The eluate is neutralised to methyl orange and a known excess of standard hydrochloric acid added. Potassium fluoride and potassium chloride are then added, and the excess hydrochloric acid titrated with standard sodium hydroxide solution. Some experience is required to judge the exact end point, although this could probably be improved by the use of a screened indicator.

Method

Apparatus. *Ion exchange column*, 20 cm length, $1·2–1·3$ cm diameter, packed with a strongly acidic cation exchange resin (KU-2).

Reagents: *Potassium hydroxide solution*, approximately $2·5$ M.

Sodium hydroxide solution, $0·25$ M standard.

Hydrochloric acid, 0·3 M standard.

Potassium fluoride solution, dissolve 25 g of the normal fluoride in
water and dilute to 100 ml.

Potassium chloride.

Procedure. Fuse approximately 5 g of potassium hydroxide in a nickel
crucible and allow to cool. Accurately weigh approximately 0·5 g of the finely
powdered silicate rock material onto the melt and add approximately 0·5 g
of sodium peroxide. Cover the crucible with a nickel lid and fuse the contents
at a temperature of 500–600° for a period of 3 minutes. Chill the crucible by
immersing the lower part in cold water, and then place the crucible on its
side with its lid in a 250-ml beaker. Add 100 ml of hot water, and warm if
necessary to give complete disintegration of the melt. Rinse and remove the
crucible and lid. Dilute to about 150 ml, cover with a clock glass and carefully add 15 ml of concentrated hydrochloric acid. At this stage the solution
may be turbid, but this turbidity will disappear on warming.

Cool the solution, transfer to a 250-ml volumetric flask, dilute to volume
with water, and mix well. Transfer this solution in small volumes at a time
(about 10 ml) to the ion exchange column and collect the eluate. Reject the
first portion of 50 ml of the eluate and collect subsequent portions of about
60 ml in separate 100-ml beakers.

Using a pipette, transfer 50 ml of the eluate to a polyethylene or polypropylene beaker, and add 3 drops of methyl orange indicator solution. Neutralise
by adding first 2·5 M potassium hydroxide solution to a yellow-pink end-point,
then 0·25 M sodium hydroxide solution to a pure yellow end point. Now add
by pipette 25 ml of the 0·3 M hydrochloric acid solution followed by 5 ml of
potassium fluoride solution and 35 g of solid potassium chloride. Stir to
dissolve, allow to stand in an ice bath or refrigerator for 1 hour, and then
titrate the excess hydrochloric acid with 0·25 M sodium hydroxide solution.
A blank titration should also be made.

Photometric Determination of Silica

There are a very large number of papers describing the photometric
determination of silica, but almost all of these are variations of two
basic methods. These are the determination as yellow silicomolybdate
and as the molybdenum blue to which this yellow complex can be
reduced.

Phosphorus, arsenic and germanium form similar yellow molybdate
complexes that can be reduced to molybdenum blues. Fortunately
neither arsenic nor germanium are likely to be present in silicate rocks
in amounts sufficient to interfere, whilst the yellow phosphomolybdate
can be decomposed by the addition of tartaric, citric or oxalic acid. This
addition serves also to prevent interference from iron.

Reagents suggested for the reduction of the yellow silicomolybdate to molybdenum blue include stannous chloride, hydroxylamine, hydro-quinone, ascorbic acid, ferrous ammonium sulphate and 1-amino-2-naphthol-4-sulphonic acid. Stannous chloride is a very powerful reducing agent and has been favoured because of its extremely rapid action. The blue colours obtained, however, are not as stable as those produced with 1-amino-2-naphthol-4-sulphonic acid, which, in combination with alkali sulphite and bisulphite, has been recommended by a number of authors.

Solutions of the yellow silicomolybdate have maximum absorption at a wavelength of about 350 nm; some authors have recommended higher wavelengths with some loss of sensitivity. This is not of impor-tance in the analysis of silicate rocks, as advantage can seldom be made of the full sensitivity that is available. Molybdenum blue methods are even more sensitive than those based on the yellow complex. Solutions containing molybdenum blue have a maximum absorption at a wave-length of 810 nm (Fig. 85), although a somewhat lower wavelength of 650 nm has been recommended[7] as giving more reproducible values of optical density in the presence of iron.

The procedure given in detail below is based upon the molybdenum blue procedure given by Shapiro and Brannock[8] for the determination of silica in their compilation of methods for the rapid analysis of sili-cate rocks.

Method

The combination of good sensitivity to silica with its presence as the major component—in many rocks exceeding the total for the remaining oxides—results in the need to take only a very small sized sample portion for the determination. Care in weighing this small sample must be combined with attention to detail at all the subsequent stages of the analysis if accurate results are to be obtained.

Some care must also be exercised in the choice of silicate material to be used as standard. Previously analysed silicate rock samples have been recom-mended for this, but any error in the earlier analyses will be repeated in the subsequent photometric determinations. Rock standards such as granite G-1, diabase W-1 or the NBS felspar, No. 99, provide a better standard material. Other good standards include suitably prepared quartzite or glass sand (99+ per cent silica) or pure crystal quartz (99·5+ per cent silica), for which accur-ate silica determinations can be made by evaporation with hydrofluoric acid as described on p. 406.

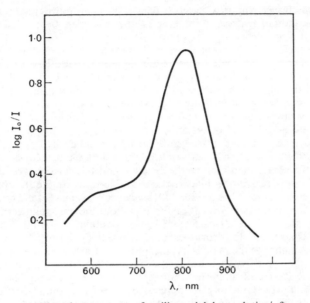

Fig. 85. Absorption spectrum of a silicomolybdate solution after reduction to molybdenum blue (2-cm cells, 125 μg SiO$_2$).

Reagents: *Sodium hydroxide solution,* dissolve 30 g of sodium hydroxide pellets in water and dilute to 100 ml. Store in a polyethylene bottle.

Ammonium molybdate solution, dissolve 7·5 g of ammonium molybdate in 80 ml of water and add 20 ml of 9 N sulphuric acid. Store in a polyethylene bottle.

Tartaric acid solution, dissolve 25 g of the reagent in water and dilute to 250 ml.

Reducing solution, dissolve 0·7 g of sodium sulphite and 0·15 g of 1-amino-2-naphthol-4-sulphonic acid in 100 ml of water, add 9 g of sodium metabisulphite and stir until solution is complete. Store in a cool, dark cupboard and prepare freshly every few days.

Procedure. Transfer 5-ml aliquots of the sodium hydroxide solution to a series of six nickel crucibles using either a polypropylene pipette or more simply by direct weighing on a laboratory rough balance. Transfer all six crucibles to a hot plate and evaporate the solutions to dryness. Care should be taken to avoid any great loss of reagent, although a small amount of spattering can be ignored. Allow to cool in a desiccator.

Accurately weigh approximately $0 \cdot 05$ g of the finely powdered rock material onto the sodium hydroxide in three of the crucibles, $0 \cdot 05$ g of the finely powdered silicate rock standard into two more of the crucibles, and reserve the last crucible for the reagent blank. Cover all six crucibles with nickel lids and transfer to a hot plate set at its highest temperature to melt the sodium hydroxide and allow the fusion process to begin. Take each crucible in turn and heat to dull redness over a gas burner for a period of about 5 minutes, swirling to give a good melt (note 1). Allow to cool.

Place each crucible on its side together with its lid, in a small polypropylene beaker and add 100 ml of water. Cover the beaker with a polyethylene or polypropylene cover and warm gently on a hot plate until all soluble material has passed into solution. Treat each of the solutions as follows. Rinse and remove the cover, then remove the crucible and lid, rinsing carefully. Using a rubber-tipped rod remove all particles of residue adhering to the crucible and lid and return these to the beaker with a fine jet of water. Set the crucible and lid aside for the next determination (note 2). Pour the rock solution containing the hydroxide residue into a 600-ml beaker containing 400 ml of water and 10 ml of concentrated hydrochloric acid. Rinse the polypropylene beaker and add the washings to the main rock solution. If a clear solution is not obtained, warm gently for a few minutes until it clears, then cool, transfer to a 1-litre volumetric flask and dilute to volume with water. Mix well. The acid rock solution is fairly stable in glass apparatus, but if the colour development is likely to be delayed, the solution should be transferred to a polyethylene bottle for storage.

Transfer 10 ml of each of the three sample solutions, the two standard solutions and the reagent blank solution to separate 100-ml volumetric flasks, dilute each to 60 ml with water, add by pipette 2 ml of the ammonium molybdate solution, swirl to mix, and set aside for 10 minutes. After exactly 10 minutes add by pipette 4 ml of the tartaric acid solution to each flask, swirling again to mix. Now add 1 ml of the reducing solution, dilute to volume with water and mix well. Set all six solutions aside for at least 30 minutes and preferably for 1 hour. Measure the optical density of each solution relative to water in 1-cm cells, with the spectrophotometer set at a wavelength of 650 nm.

Calibration. As the Beer–Lambert Law is valid, the average optical density of the two standard rock solutions, after deduction of the reagent blank value, can be used to establish the position of the calibration line. Alternatively, the silica content of each sample portion can be calculated directly:

$$\text{Per cent SiO}_2 = \frac{S_1 \times (\log I_0/I)_2 \times w_1}{(\log I_0/I)_1 \times w_2}$$

where w_1 and w_2 are the weights of sample and standard and $(\log I_0/I)_1$ and $(\log I_0/I)_2$ the respective optical densities. S is the silica content in per cent, of the standard. An average value for the silica content of the sample can then be calculated.

Notes: 1. On remelting the sodium hydroxide, some of the sample material may float on the surface of the melt and remain unattacked by it. When this occurs, gently swirl the melt to disperse the powder as much as possible, then allow the melt to solidify. On further remelting the material complete fusion should occur.

2. The crucible and lid should be rinsed with dilute hydrochloric acid before being used for the next determination.

3. The method described here for the determination of silica in silicate rocks can be applied directly to carbonate rocks. A sample weight of up to 0·2 g can be taken for the analysis, which is otherwise carried out as for silicate rocks. A silicate rock is used as the reference standard.

A Combined Gravimetric and Photometric Determination of Silica

The precision of the photometric determination of silica is controlled to a large extent by the limitations inherent in spectrophotometry. Such determinations cannot by their nature be as precise as those obtained by good gravimetric methods, although of course the possibility of systematic errors exists in both methods. In order to combine the advantages and avoid some of the disadvantages of both methods Jeffery and Wilson[7] have suggested a new procedure based upon a single dehydration with hydrochloric acid, giving a major silica fraction as an insoluble residue, and a minor fraction in the filtrate that can be determined photometrically. As only one dehydration is made, this procedure is more rapid than the classical method, and as all the silica not collected in the residue is determined photometrically, it is also more accurate. Moreover, since the photometric determination is now of a minor constituent (2–8 mg, equivalent to less than 1 per cent of the rock composition), the limiting accuracy of photometric methods is no longer restrictive.

A disadvantage that has been introduced is that in the filtrate from the collection of the major silica fraction, all the remaining elements have been increased in concentration relative to silica. Under these circumstances elements that are not usually present in amounts sufficient to interfere may be concentrated up to and beyond this point. Interference has been noted from titanium and phosphorus, although up to 5 per cent TiO_2 in the rock material does not have a significant effect, and up to 10 per cent P_2O_5 can also be tolerated.

Fluorine in trace amounts does not interfere, although if present in major amounts, some of the silica may be lost by volatilisation during the dehydration stage, as in the classical method for determining silica.

Method

Reagent: *Ammonium molybdate solution*, dissolve 10 g of ammonium molybdate in N aqueous ammonia and dilute to 100 ml with N ammonia.

Oxalic acid solution, dissolve 10 g in water and dilute to 100 ml.

Reducing solution, dissolve 0·15 g of 1-amino-2-naphthol-4-sulphonic acid, 0·7 g of anhydrous sodium sulphite and 9 g of sodium metabisulphite in 100 ml of water. Prepare freshly each month.

Procedure. Full details for the fusion of silicate rock material, dissolution of the melt, dehydration of the silica and determination by volatilisation with hydrofluoric acid are given on p. 37. This part of the procedure is given below in outline only.

Fuse 1 g of the rock material with 5 g of anhydrous sodium carbonate in a platinum crucible and digest the melt with water. Acidify with concentrated hydrochloric acid, adding approximately 10 ml in excess and evaporate to dryness in a large platinum dish. When completely dry add a further 10 ml of concentrated hydrochloric acid and sufficient water to dissolve all soluble material. Collect the silica residue on a small filter, wash well with hot water and determine the major silica fraction in the usual way. Collect the filtrate and washings from the silica residue in a 200-ml volumetric flask and dilute to volume with water.

Using a pipette, transfer 5 ml of this solution to a 100-ml volumetric flask and add 10 ml of water. Now add 1 ml of ammonium molybdate solution and set aside for 10 minutes to complete the formation of the silicomolybdate complex. After exactly 10 minutes, add 5 ml of the oxalic acid solution, gently swirl the flask to mix the contents, and then add 2 ml of the reducing solution. The addition of the reducing solution should not be delayed. Dilute the solution to volume with water and set aside for at least 30 minutes, but preferably for 1 hour, and then measure the optical density of the solution in 2-cm cells with the spectrophotometer set at a wavelength of 650 nm.

Calibration. Transfer aliquots of the standard silicate solution containing 0–200 μg of silica to separate 100-ml volumetric flasks, dilute to 15 ml with water and add 1 ml of 3 N hydrochloric acid to each. Now add 1 ml of ammonium molybdate solution to each, allow to stand for 10 minutes and then reduce the silicomolybdate to molybdenum blue as described above. Dilute each solution to volume with water and measure the optical density in 2-cm cells with the spectrophotometer set at a wavelength 650 nm. Plot the relation of optical density to silica concentration. As the calibration graph is a straight line passing through the origin, a single point method can be used for subsequent calibrations.

The Determination of Silica by Atomic Absorption Spectroscopy

Although attempts were made, soon after the general introduction of atomic absorption spectrophotometers into rock analysis laboratories, to determine silica by this technique, the results obtained tended to be poor in both accuracy and precision. Some improvement was obtained with the introduction of the higher temperature nitrous oxide burner, and the relative freedom from interference from other elements at the concentrations found in silicate rocks was noted. However, both consistently high[9] and consistently low[10] results were reported— both possibly arising from a failure to appreciate the chemistry of silicon in the rock solutions used.

Satisfactory silica results were reported by Van Loon and Parissis[9] using a decomposition technique similar to that recommended by Ingamells.[11]

Procedure. Accurately weigh 0·2 g of the finely powdered silicate rock material into a small platinum crucible and mix with 1 g of pure lithium metaborate. Fuse at a temperature of 1000° for 30 minutes, swirl the contents and quench by placing the crucible upright in a 100-ml beaker containing 25 ml of diluted (1+24) nitric acid. Keeping the crucible in an upright position, add 50 ml of the diluted nitric acid and stir without heating until dissolution is complete. Transfer the solution to a 250-ml volumetric flask and dilute to volume with the diluted nitric acid.

Measure the absorption, spraying the solution into an atomic absorption spectrophotometer using a nitrous oxide flame, at a wavelength of 251·6 nm. Measure also the absorption of standard silica solutions similarly prepared from pure silica to give the calibration.

References

1. GOLDSCHMIDT V. M., *Geochemistry*, Oxford, 1954.
2. RICHARDSON W. A. and SNEESBY G., *Min. Mag.* (1922) **19**, 303.
3. AHRENS L. H., *Geochim. Cosmochim. Acta* (1964) **28**, 271.
4. TSUBAKI I., *Bunseki Kagaku* (1967) **16**, 610.
5. BENNETT H. and REED R. A., *Analyst* (1967) **92**, 466.
6. KHALIZOVA V. A., ALEKSEEVA A. YA. and SMIRNÓVA E. P., *Zavod. Lab.* (1964) **30**, 530.
7. JEFFERY P. G. and WILSON A. D., *Analyst* (1960) **85**, 478.
8. SHAPIRO L. and BRANNOCK W. M., *U.S. Geol. Surv. Circ.* 165, 1952, *Bull.* 1036-C, 1956 and *Bull.* 1144-A, 1962.
9. VAN LOON, J. C. and PARISSIS C. M., *Anal. Lett.* (1968) **1**, 519.
10. KATZ A., *Amer. Mineral.* (1968) **53**, 283.
11. INGAMELLS C. O., *Analyt. Chem.* (1966) **38**, 1228.

CHAPTER 40

SILVER, GOLD AND
THE PLATINUM METALS

Occurrence

Basic rocks have been shown by Hamaguchi and Kuroda[1] to contain slightly more silver than acidic rocks, Table 34. This behaviour is contrary to that predicted from considerations of electronegativity and ionisation potential, which indicate that silver should accumulate in residual melts.[2] This apparent anomaly arises from the ease with which silver is removed from silicate melts by incorporation in a sulphide phase. Galena and chalcopyrite are particularly indicated as silver carriers.

TABLE 34. AVERAGE SILVER
CONTENT OF SILICATE
ROCKS[1]

Rock type	Ag, ppb*
Ultrabasic	60
Basic	110
Intermediate	73
Acidic	46

(Mean crustal abundance 80 ppb.)
* Parts per billion, i.e. 1 in 10^{-9}.

These sulphides provide the major industrial source of silver, in preference to silver minerals, which are available in only small quantities. Such minerals include native silver which usually contains small amounts of other elements, particularly gold and copper, halides such as cerargyrite AgCl, bromyrite AgBr, iodyrite AgI and embolite

418

Ag(Cl,Br), sulphides such as argentite Ag_2S, stephanite Ag_5SbS_4, proustite Ag_3AsS_3 and even rarer tellurides such as hessite Ag_2Te.

Studies by Vincent and Crocket[3] on the rocks of the Skaergaard intrusion suggest that 1–10 ppb was a likely magnitude for the occurrence of gold in silicate rocks. A somewhat similar range of values was reported by DeGrazia and Haskin,[4] for a larger range of silicates and some other rock types, Table 35.

TABLE 35. AVERAGE GOLD CONTENT OF SOME
ROCK TYPES[4]

Rock type	Au, ppb
Acid igneous and metamorphic rocks	2·4
Basic igneous and metamorphic rocks	2·6
Mid-atlantic ridge basalts	10
Carbonates	2·5
Shales	4·7
Sandstones	6·0
Pelagic clays	12

(Mean crustal abundance 2·5 ppb.)

In view of the apparent uniformity of distribution of gold in silicate rocks, DeGrazia and Haskin[4] have suggested that the gold occurs in finely divided particles, possibly colloidal, that remain chemically inert during most geochemical processes and tend to remain suspended in melts or solutions. Rather higher gold values were reported for some silicate rocks by Shcherbakov and Perezhogin[5] (Table 36). Although there may be regional differences in the distribution of gold, further work is clearly necessary to establish reliable abundance values.

Unlike silver, the main source of gold is from deposits containing free or native metal. Under favourable conditions gold-bearing rocks containing as little as 3–4 ppm of gold can be worked economically. Other gold minerals occurring largely as mineral curiosities include calverite $AuTe_2$, petzite $(Ag,Au)_2Te$ and sylvanite $AuAgTe_4$. Certain sulphide and arsenide minerals also contain appreciable amounts of gold.

The platinum metals resemble gold in occurring largely in the metallic state; they are almost always associated with each other and often with

TABLE 36. THE AVERAGE GOLD CONTENT OF SOME ROCKS AND MINERALS[5]

	Au, ppb		Au, ppb
Granite	3·2	Limestones and marbles	3·2
Granodiorite	4·0	Muscovite	3·8
Quartz porphyry	5·4	Sphene	3·9
Syenites	4·4	Biotite	4·0
Diorites	3·5	Felspar	4·0
Altai-Sayan gabbro	6·4	Amphiboles	5·9
Gabbro in traps	10·0	Quartz	11
Neutral gabbro	8·7	Tourmaline	12
Diabase and porphyrites	6·5	Olivine	14
Ultrabasic rocks	9·4	Pyroxenes	16
Shales and sandstones	3·6	Magnetite	48

gold, copper and other metals. A small number of other minerals, mostly sulphides and tellurides are known.

The platinum metals do not substitute for other metals in the lattices of rock-forming silicates, but tend to crystallise in the metallic form. This gives rise to a non-uniform distribution[6] and leads to difficulties in obtaining representative samples for analysis. This is evident from the work of analysts who have attempted to assess the platinum metal content of a single rock exposure, and noted a wide range of values differing by as much as an order of magnitude for adjacent samples. The determination of the metals in this group is difficult, and analytical errors may have contributed towards the wide spread of reported values.

Very little is known of the distribution of these six metals in silicate rocks and, for the reasons noted above, many of the results are clearly unreliable or atypical. Crocket and Skippen[7] found $1·9 \pm 1·2$ ppb palladium in nonorogenic oceanic basalts and $8·2$ ppb in orogenic and continental basalts. For W-1 a value of 16 ppb was given, and for G-1 a value of $1·6$ ppb. Other similar values for palladium have been tabulated by Wright and Fleischer.[6] Vinogradov[8] has reported a few determinations by earlier workers including a value of 40 ppb iridium in basic rocks. For platinum individual values of $0·5$ and $0·3$ ppm were reported for ultrabasic rocks, $0·2$ and $0·1$ ppm for basic rocks and $0·02$ for acidic rocks. These values appear to be unduly high and may need considerable revision as more sensitive techniques become available. Thus

Baedecker and Ehmann[9] have reported less than $0 \cdot 07$ and $0 \cdot 05$ ppb iridium in G-1 and W-1 respectively, whilst Sarma, Sen and Chowd-hury[10] obtained $8 \cdot 2$ and $9 \cdot 2$ ppb platinum in these two rocks. Values for osmium include $0 \cdot 05$ ppb in G-1 and $0 \cdot 026$ ppb in W-1, by Morgan.[11] Somewhat higher values of $0 \cdot 2$ ppb and $0 \cdot 46$ ppb were suggested by Bate and Huizenga.[12] No reliable values appear to be available for rhodium or ruthenium in silicate or other rocks.

The Determination of Silver, Gold and the Platinum Metals

At the concentration levels encountered in silicate rocks, spectro-photometric and spectrochemical techniques cannot usually be employed and recourse must be made to more sensitive procedures such as neutron activation.[13, 14] Even this technique is not sufficiently sensitive for determining rhodium and ruthenium in normal silicate rocks, and barely adequate for the remaining metals. Fire-assay procedures[15] are custo-marily used for the analysis of ores containing silver, gold and the plati-num metals. Using fire-assay procedures to collect the noble metals from gangue material, some improvement in sensitivity can be obtained by completing the determination by spectrochemical or spectrophoto-metric methods. This technique has been reviewed by Chow et al.[16] Acid-soluble palladium in the parts per billion range has been deter-mined by Grimaldi and Schnepfe[17] using a 10-g sample, extraction with aqua regia, co-precipitation with added platinum and tellurium and finally spectrophotometric determination with p-nitrosodimethyl-aniline.

A comprehensive review of methods for determining silver, gold and the platinum metals by atomic absorption spectroscopy has been given by Sen Gupta.[18] Methods for the determination of gold at the ppm and ppb levels in gold ores,[19] and of silver in soils, sediments and rocks,[20] make use of an acid digestion and an extraction into organic solution containing methylisobutyl ketone.

Of some interest is the fluorimetric method for gold in rocks using rhodamine B, described by Marinenko and May.[21] A cyanide extrac-tion is used to remove gold from a 10- to 30-g rock portion and the gold precipitated with tellurium using hydrazine as reductant. Rhodamine B chloraurate is extracted into 10 ml of isopropyl ether and the fluorescence measured at 575 nm, with excitation at 550 nm.

References

1. HAMAGUCHI H. and KURODA R., *Geochim. Cosmochim. Acta* (1959) **17**, 44.
2. TAYLOR S. R., *Phys. Chem. Earth* (1965) **6**, 177.
3. VINCENT E. A. and CROCKET J. H., *Geochim. Cosmochim. Acta* (1960) **18**, 130.
4. DEGRAZIA A. R. and HASKIN L., *Geochim. Cosmochim. Acta* (1964) **28**, 559.
5. SHCHERBAKOV YU. G. and PEREZHOGIN G. A., *Geokhimiya* (1964) 518.
6. WRIGHT T. L. and FLEISCHER M., *U.S. Geol. Surv. Bull.* 1214-A, 1965.
7. CROCKET J. H. and SKIPPEN G. B., *Geochim. Cosmochim. Acta* (1966) **30**, 129.
8. VINOGRADOV A. P., *Geochemistry* (Eng. transl.) (1956) **1**, 1.
9. BAEDECKER P. A. and EHMANN W. D., *Geochim. Cosmochim. Acta* (1965) **29**, 329.
10. SARMA B. D., SEN B. N. and CHOWDHURY A. N., *Econ. Geol.* (1965) **60**, 373.
11. MORGAN J. W., *Anal. Chim. Acta* (1965) **32**, 8.
12. BATE G. L. and HUISZENGA J. R., *Geochim. Cosmochim. Acta* (1963) **27**, 345.
13. BEAMISH F. E., CHUNG K. S. and CHOW A., *Talanta* (1967) **14**, 1.
14. MALESZEWSKA H., *Chem. Anal.*, Warsaw (1967) **12**, 281.
15. SMITH E. A., *The Sampling and Assay of the Precious Metals*, Griffin, London, 2nd ed., 1947.
16. CHOW A., LEWIS C. L., MODDLE D. A. and BEAMISH F. E., *Talanta* (1965) **12**, 277.
17. GRIMALDI F. S. and SCHNEPFE M. M., *U.S. Geol. Surv. Prof. Paper*, 575-C, p. C141, 1967.
18. SEN GUPTA J. G. *Minerals Science Engineering* (1973) **5**, 207.
19. HILDON M. A. and SULLY G. R., *Anal. Chim. Acta* (1971) **54**, 245.
20. CHAO T. T., BALL J. W. and NAKAGAWA H. M., *Anal. Chim. Acta* (1971), **54**, 77.
21. MARINENKO J. and MAY I., *Analyt. Chem.* (1968) **40**, 1137.

CHAPTER 41

STRONTIUM

Occurrence

With the possible exception of ultrabasic rocks, some of which contain only a few parts per million, all silicate rocks contain appreciable quantities of strontium, amounting to several hundred parts per million. In view of this and the relative ease with which strontium can be determined by emission spectrography, it is not surprising that considerable attention has been paid to the geochemistry of this element. Values for the occurrence in silicate and other rocks have been given by a number of workers[1-4] and some typical values from these and other sources are given in Table 37. Most analysts have reported a considerable spread of individual values about the mean for each particular rock group.

TABLE 37. TYPICAL VALUES FOR THE
STRONTIUM CONTENT OF ROCKS

Rock type	Sr, ppm
Granite	100
Granodiorite	400
Rhyolite	200
Diorite	300
Basaltic rocks	450
Ultrabasic rocks	25
Limestone	600
Shales	300
Calcareous deep-sea sediments	2000

The relation of strontium to calcium in silicate rocks is a complex one.[4] In certain rocks, granites and granodiorites, for example, the strontium content of individual samples taken from the same rock body

423

increases linearly with calcium content, whilst the inverse relation has been shown to hold for rocks belonging to a single basaltic rock body.

Although strontium is present in amounts of up to several thousand parts per million in certain silicate rocks, it is seldom reported as occurring in the form of discrete strontium minerals. It tends to be incorporated into the lattice of silicate minerals, particularly felspars and felspathoids such as orthoclase, microcline, leucite and nepheline.

The principal ore of strontium is celestite, $SrSO_4$, an evaporite mineral often associated with gypsum rock, limestones and dolomites. Strontianite $SrCO_3$ also exists in some evaporite deposits, is known to occur also as a major component of certain carbonatites, and has been reported from certain late stage granite pegmatites.

An accurate knowledge of the strontium concentration in certain silicate minerals is necessary for the calculation of geological age by the rubidium–strontium method. This is based upon the accumulation of strontium-87 throughout geological time by the decay of radioactive rubidium-87, and requires an accurate knowledge not only of the total strontium content, but also of the total rubidium content and the strontium isotopic composition.

The Determination of Strontium

Many of the strontium values that have been reported were obtained by emission spectrographic methods; these are probably a great deal more accurate than the gravimetric methods previously available. The most popular chemical method appears to have been that based upon the precipitation with calcium as oxalate in the filtrate from the removal of iron, aluminium, titanium and other elements by precipitation with ammonia. The mixed oxalates were converted to the anhydrous nitrates, and the strontium nitrate separated from calcium nitrate by dissolution of the latter in concentrated nitric acid. Groves[5] has suggested that the error involved in making this separation is not as great as that arising from the failure to collect the strontium from the rock material in the oxalate precipitate.

An alternative gravimetric procedure is based upon the precipitation of barium and strontium together with calcium in the form of their sulphates, from the rock solution to which dilute sulphuric acid and ethanol have been added. The sulphates are converted to carbonates by alkali carbonate fusion, and the calcium removed by dissolution of the nitrate in concentrated nitric acid as described above. Barium is then precipitated as chromate and the strontium determined in the filtrate by precipitation as sulphate. This procedure is long and tedious. It has been reported by Groves in his book on *Silicate Analysis*,[5] but appears to have been little used.

Spectrophotometric methods have not found wide application to the determination of strontium, possibly because of interference from other alkaline earth elements. Reagents that have been used include *o*-cresolphthalein complexone,[6] murexide[7] and chlorophosphonazo III.[8] For the determination of strontium in silicate and other rocks, flame photometric and atomic absorption techniques now appear to have displaced the gravimetric procedures. As both of these techniques can be used to determine a whole range of elements in silicate rocks, they are now used where previously emission spectrography might have been considered.

Determination by Flame Photometry

In common with calcium and barium, strontium has a very characteristic flame emission. The strongest emission of strontium is at the resonance line of $460 \cdot 7$ nm. Other strontium lines are at $407 \cdot 8$ nm and $421 \cdot 6$ nm, and bands occur in the orange and red parts of the spectrum. As with other alkaline earth elements, the flame emission depends not only upon the flame conditions, but also upon the nature of the solution being sprayed and the concentration of other elements in it. The calibration curve for strontium, based upon the resonance line, approximates to a straight line. Where accurate results are required the procedure known as the "method of additions" should be used.

Sodium and potassium in high concentrations tend to enhance the strontium emission at $460 \cdot 7$ nm, as does appreciable amounts of calcium. Magnesium, iron, silica, phosphate, sulphate and more particularly aluminium depress the strontium emission.

To overcome most of these interferences Fabrikova and Isaeva[9] have suggested making a prior separation of the calcium, strontium and magnesium from iron, aluminium and phosphorus by a double precipitation of the latter with aqueous ammonia. The combined filtrates from the two precipitations are evaporated to small volume and the emission at $460 \cdot 7$ nm measured. The method is described as giving only approximate results, with a deviation from the mean of up to 28·5 per cent in parallel determinations.

In the procedure described below all separations, other than that of silica, are avoided and the determination is made by adding a standard solution of strontium to separate aliquots of the rock solution. The depressing effect of aluminium and other elements is avoided by adding a considerable excess of calcium to the rock solution as described by Fornaseri and Grandi[10] and more recently by Kirillov and Alkhimankova.[11] Silica is removed by evaporation with hydrofluoric and perchloric acids.

Method

Reagents: *Calcium perchlorate solution*, dissolve 25 g of pure calcium carbonate in dilute perchloric acid, avoiding an excess, and dilute to 500 ml with water. This solution contains 20 mg calcium per ml.

               Standard strontium stock solution, dissolve 0·1685 g of pure strontium carbonate in dilute perchloric acid, also avoiding excess acid, and dilute to 500 ml with water. This solution contains 200 µg Sr per ml.

               Standard strontium working solution, prepare the working standard solution by dilution with water as required.

Procedure. Accurately weigh approximately 2 g of the finely powdered silicate rock material into a small platinum basin, moisten with water, add 20 ml of concentrated perchloric acid and evaporate to dryness in the usual way. Cool, add 3 ml of perchloric acid and again evaporate to dryness. Cool, add 1 ml of perchloric acid and evaporate again, this time to give a residue of the moist perchlorates. Allow to cool, dissolve the residue in about 10 ml of water, add 10 ml of the calcium perchlorate solution and dilute to volume in a 100-ml volumetric flask (note 1). Pipette 25 ml of this solution into a clean 50-ml volumetric flask and dilute to volume with water. Measure the flame emission of this solution at a wavelength of 460·7 nm using a flame photometer set according to the maker's instructions. If a recording attachment is available, trace the emission from about 480 nm to 450 nm. From the emission obtained and a strontium calibration graph calculate the approximate strontium content of the dilute solution.

Prepare two new solutions of the rock sample by transferring two 25-ml portions of the rock solution into separate 50-ml volumetric flasks, adding aliquots of the standard strontium solution, and diluting to volume with water. The strontium addition should be chosen to give new concentrations of strontium approximately two and three times that of the original dilute rock solution.

Measure the flame emission of all three dilute solutions of the rock together with that of a reagent blank solution (note 2), and hence determine the strontium content of the sample material.

Notes: 1. With normal silicate rocks and minerals, the precipitation of potassium perchlorate is not likely to occur at this dilution. For rocks and minerals rich in potassium where this does occur, allow the crystalline precipitate to settle before removing the 25-ml aliquots.

2. Even so-called "pure" grades of calcium carbonate may contain small amounts of strontium, and a blank determination is therefore essential.

In a recent study of the development of analytical methods for barium and strontium Ingamells *et al.*[12] reported that they had investigated numerous procedures for the flame photometric determination of strontium, but only that using a cool flame and a neutral nitrate solution of calcium and strontium after an ammonium carbonate precipitation was successful.

Determination by Atomic Absorption Spectroscopy

The resonance line at $460 \cdot 7$ nm can also be used to determine strontium by atomic absorption spectroscopy. Elements that interfere include phosphorus as phosphate[13] and aluminium,[14] although it has been shown[15] that both of these effects can be suppressed by adding lanthanum to the solution. The method of addition used in the previous section can also be used for this determination, although probably to less advantage as the atomic absorption technique is not so sensitive to interference from the presence of other elements.

The determination of strontium by atomic absorption spectroscopy using an air–acetylene flame is highly influenced by the matrix composition.[16] This matrix effect is reduced considerably using a nitrous oxide–acetylene flame, which has been proposed for this determination.[17]

Method

Reagents: *Lanthanum solution,* dissolve $23 \cdot 4$ g of lanthanum oxide by heating with 100 ml of 9 M hydrochloric acid. Add perchloric acid and evaporate first to fumes and then to leave a moist perchlorate residue. Dissolve in water and dilute to 500 ml with water. This solution contains 40 mg lanthanum per ml.

Procedure. Prepare a perchlorate solution of the rock material as described above but adding 10 ml of the lanthanum solution to the rock in place of calcium, before diluting the aqueous perchlorates to 100 ml. Dilute 25 ml of this solution to 50 ml in a volumetric flask and using a fuel-rich, air–acetylene flame, determine the absorption at 460·7 nm in the usual way. Prepare new dilutions of the rock solution with added strontium as described in the previous section, measure the flame absorption of each solution and hence calculate the strontium content of the rock material.

References

1. NOLL W., *Chemie der Erde* (1934) **8**, 507.
2. AHRENS L. H., *Min. Mag.* (1948) **28**, 277.
3. WAGER L. R. and MITCHELL R. L., *Geochim. Cosmochim. Acta* (1951) **1**, 131.
4. TUREKIAN K. K. and KULP J. L., *Geochim. Cosmochim. Acta* (1956) **10**, 245.
5. GROVES A. W., *Silicate Analysis*, Allen & Unwin, London, 2nd ed., 1951.
6. POLLARD F. H. and MARTIN J. V., *Analyst* (1956) **81**, 348.
7. RUSSELL D. S., CAMPBELL J. B. and BERMAN S. S., *Anal. Chim. Acta* (1961) **25**, 81.
8. LUKIN A. M., ZELICHENOK S. L. and CHERNYSHEVA T. V., *Zhur. Anal. Khim.* (1964) **19**, 1513.
9. FABRIKOVA E. A. and ISAEVA A. G., *Trudy Inst. Mineral. Geokhim. i Kristallokhim. Redk. Elementov, Akad. Nauk SSSR* (1963) (18), 175.
10. FORNASERI M. and GRANDI L., *Geochim. Cosmochim. Acta* (1960) **19**, 218.
11. KIRILLOV A. I. and ALKHIMENKOVA G. I., *Zavod. Lab.* (1965) **31**, 57.
12. INGAMELLS C. O., SUHR N. H., TAN F. C. and ANDERSON D. H., *Anal. Chim. Acta* (1971) **53**, 345.
13. DAVID D. J., *Analyst.* (1962) **87**, 585.
14. BELCHER C. B. and BROOKS K. A., *Anal. Chim. Acta* (1963) **29**, 202.
15. DINNIN J. I., *U.S. Geol. Surv. Prof. Paper* 424-D, 1961.
16. BECCALUVA L. and VENTURELLI G., *Atom. Absorpt. Newsl.* (1971) **10**, 50.
17. ABBEY S., *Geol. Surv. Canada Paper* 71-50, 1972.

CHAPTER 42

SULPHUR

Occurrence

In the great majority of silicate rocks the sulphur content is very small, amounting to no more than a few hundred parts per million. This is in contrast to the meteoric abundance where, in the order of frequency, sulphur is the fifth element by weight, coming before aluminium.[1] It is conjectured that the cause of this apparent discrepancy lies in the ease with which the sulphides of iron, nickel and cobalt crystallise from parental magma and segregate from it by sinking downwards through the liquid silicate melt of lower specific gravity. The differentiated sulphides often appear associated with intrusive gabbros and other basic rocks, sometimes in the form of hydrothermal veins, forming important sulphide ore bodies. High-level intrusives and the more acidic rocks appearing later in the magmatic sequence contain very little sulphur. Sedimentary rocks, particularly marine muds and shales may contain appreciable amounts of sulphur, which can collect in natural gas and petroleum. A variety of sulphide (and some sulphate) minerals have been reported from carbonatite rocks, but otherwise carbonate rocks and sandstones contain very little sulphur.

Sulphur minerals noted in silicate rocks include the sulphides pyrite, marcasite, pyrrhotite, pentlandite, sphalerite, chalcopyrite, molybdenite and many others in smaller amounts. The sulphates include barite, gypsum, celestite, kainite, polyhalite and a number of silicate minerals such as lazulite, hauyne and nosean. Sulphur also occurs rarely in the uncombined or free sulphur form.

Free Sulphur

Methods for the determination of free sulphur are based upon extraction from the ground rock material with an organic solvent. In the pro-

429

cedure described by Volkov,[2] a soxhlett-type apparatus is used and the sample refluxed for 16 hours with acetone. The elemental sulphur, plus any soluble organic compounds containing sulphur, are recovered by evaporation of the acetone and are converted to sulphuric acid by oxidation with a solution of bromine in carbon tetrachloride. The sulphate is then reduced with hydriodic acid to hydrogen sulphide, which is distilled from the solution and collected in a solution of cadmium acetate. An excess of iodine is added and the amount of the excess determined by titration with standard sodium thiosulphate solution.

Extraction times of only 10 minutes were suggested by Ozawa[3] using carbon disulphide or benzene as solvent. The organic solutions of sulphur obey the Beer–Lambert Law, and the optical densities can be measured directly at 390 or 395 nm (CS_2 solution) and 330 or 360 nm (benzene solution).

Sulphide–Sulphur

For most purposes the determination of nitric acid–soluble sulphur can be regarded as a measure of the sulphide content of silicate rocks. This determination will, however, include sulphur from those sulphate minerals that are soluble in nitric acid, and traces of sulphur from those that are not—barite, for example. An alternative approach to the determination of sulphide–sulphur that does not include sulphate–sulphur is to reduce the sulphides to hydrogen sulphide with hydriodic acid.[4] The gases evolved are bubbled through a suspension of cadmium hydroxide, and the cadmium sulphide determined with iodine in the usual way. Most sulphides are decomposed, but the presence of mercury is necessary[5] to complete the decomposition of pyrite and chalcopyrite.

Method

Reagents: *Hydriodic acid,* mix equal volumes of concentrated hydrochloric acid and an aqueous solution containing 50 g of potassium iodide in 100 ml. Add a few crystals of sodium hypophosphite to reduce any liberated iodine (but avoid an excess) and decant from the precipitated potassium chloride.

Cadmium acetate solution, dissolve 2 g of cadmium acetate in water and dilute to 100 ml.

Iodine, standard $0 \cdot 1$ N *solution.*
Acetic acid, N.
Sodium thiosulphate standard $0 \cdot 1$ N *solution.*

Procedure. Accurately weigh 2 g or more of the finely powdered rock material into the reaction flask and add a few mg of mercury. Transfer 25 ml of cadmium acetate solution to two gas wash bottles and add 5 ml of N sodium hydroxide solution to each. Assemble the apparatus as shown in Fig. 86 and displace the air from the flask with nitrogen or hydrogen. Add 10 ml of hydriodic acid to the flask and warm gently to assist the decomposition of the sulphide minerals. Continue passing gas through the apparatus for about 1 hour.

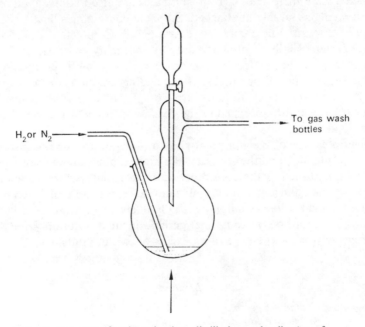

FIG. 86. Apparatus for the reduction, distillation and collection of sulphur in sulphide minerals.

Allow to cool, remove the two gas wash bottles and combine their contents. Add an excess of standard iodine solution and sufficient acetic acid to give a final acid concentration of about $0 \cdot 5$ N and titrate the excess iodine with standard thiosulphate solution.

Water-soluble Sulphur

The extraction of more than a trace of sulphate ions from silicate rocks into aqueous solution indicates the presence of evaporite minerals such as kieserite $MgSO_4.H_2O$, kainite $KCl.MgSO_4.3H_2O$, etc. The sulphate ion is easily determined gravimetrically by precipitation with barium chloride as described below.

Acid-soluble Sulphur

As noted above, the decomposition of metallic sulphides can be effected with nitric acid. Certain sulphates are also dissolved, although barite remains largely unattacked. Sulphate-containing silicates such as lazulite appear to be completely decomposed. Complete dissolution of silicates can usually be obtained only by adding hydrofluoric acid. This procedure has been used by Wilson et al.[6] for the determination of total sulphur in silicate rocks. By including perchloric acid, organic matter and resistant sulphides are completely dissolved. Vanadium pentoxide is added to expedite the oxidation, for which a PTFE dish can be used.

The presence of barium gives rise to precipitation of the insoluble barium sulphate, and the recovery of a high sulphur content was significantly diminished by the occurrence of $0 \cdot 2$ per cent barium oxide in the rock sample. Barium oxide equivalent to 1 per cent of the sample caused a serious loss of sulphur. It is unlikely that sulphur will be lost by evaporation on heating the perchloric acid solution in PTFE, particularly in the presence of an excess of calcium salts.

Method

Reagents: *Vanadium pentoxide.*
Calcium chloride.
Cupferron solution, dissolve $0 \cdot 5$ g of the reagent in 50 ml of chloroform. Prepare as required and discard if not used.
Barium chloride solution, dissolve 3 g of the dihydrate in 100 ml of water.

Procedure. Accurately weigh 1 g of the finely powdered rock material into a PTFE dish, moisten with water and add in succession, 15 ml of concentrated nitric acid, 2 ml of concentrated hydrochloric acid and 10 ml of concentrated hydrofluoric acid. Cover the dish and set aside overnight. Remove the cover and evaporate to dryness on an air-bath. If the rock powder is completely decomposed, add a few ml of water and 10 ml of concentrated nitric acid and again evaporate to dryness. Repeat this last operation once more. If the decomposition is not complete after the first evaporation, add 5 ml of concentrated perchloric acid (60 per cent w/v) and, if organic matter is present, 100 mg of vanadium pentoxide. In addition if the material is low in calcium, add not less than 100 mg of calcium chloride. (There should be present at least two equivalents of calcium for each equivalent of sulphur.) Evaporate to dryness, and, if any organic or black mineral particles remain, evaporate with further portions of perchloric acid until the decomposition is complete.

After the final evaporation, add 3 ml of concentrated hydrochloric acid and about 20 ml of water. Digest on the air bath for 15 minutes with occasional stirring, then cool and transfer the solution to a 250-ml separating funnel. Dilute to a volume of about 100 ml, and extract iron and titanium with successive 50-ml portions of cupferron solution until the extract is no longer brown. Wash the solution twice with 50-ml portions of chloroform, and finally with 50 ml of light petroleum. Run off the aqueous layer into a 400-ml beaker and wash the light petroleum twice with 10-ml portions of water. Dilute the combined aqueous layer and washings to about 200 ml, and filter if necessary.

Heat the solution to boiling and add a slight excess of a hot barium chloride solution. Allow the solution to stand for an hour on a steam bath and then set the beaker aside overnight. Collect the precipitated barium sulphate on a small close-textured filter paper and wash with successive small quantities of cold water. Transfer the paper to a weighed platinum crucible, dry, ignite and weigh as $BaSO_4$.

After ignition, the barium sulphate residue should be perfectly white. Blank values are usually insignificant.

Total Sulphur

All sulphur in the sample material, whether present as free sulphur, organic sulphur, sulphide or sulphate–sulphur, is converted to alkali sulphate by fusion with sodium carbonate containing a little potassium nitrate. The melt is extracted with water and the sulphate in the aqueous filtrate recovered by adding acid and precipitating with barium chloride. High and somewhat variable blanks can sometimes be obtained, particularly if gas burners are used for the fusion—the use of an electric furnace is recommended. Barium does not interfere.

A combustion tube method in which the sample is heated with vanadium pentoxide at a temperature of 900–950° in pure nitrogen has also been described.[7] Any SO_3 formed is reduced to SO_2 with copper turnings, and the sulphur dioxide determined volumetrically or spectro-photometrically. Gaseous oxygen has also been used as oxidant and carrier for this determination. Sen Gupta[8] has published a comparative study of the usefulness of resistance type and induction furnaces for determining sulphur in rocks, ores and stony meteorites.

The procedure given in detail below is based upon fusion with sodium carbonate.

Method

Procedure. Accurately weigh approximately 2 g of the finely powdered rock material into a large platinum crucible and mix with 10 g of anhydrous sodium carbonate and 0·25 g of potassium nitrate (note 1). Transfer the crucible to a small electric furnace, cover with a platinum lid, and slowly raise the temperature to about 1000°. Maintain at this temperature for about 30 minutes and then allow to cool. Extract the melt with hot water and rinse the solution and residue into a 250-ml beaker, add a few drops of ethanol to reduce any manganate formed in the fusion, and dilute to about 150 ml with water. Cover the beaker and digest upon a steam bath, using a glass rod to reduce the insoluble material to a fine state of subdivision.

Filter the solution through an open-textured filter paper and wash the residue with hot sodium carbonate solution. Discard the residue (note 2). Collect the filtrate and washings in a 800-ml beaker and dilute to a volume of approximately 500 ml. Add a few drops of methyl red indicator solution and concentrated hydrochloric acid until the neutral point is reached, followed by 10 ml in excess. Boil the solution to expel carbon dioxide and precipitate the sulphur from the boiling solution with a slight excess of barium chloride solution as described in the previous section. Collect, ignite and weigh the precipitate. If the rock contains an appreciable amount of sulphur, it may be necessary to purify the barium sulphate precipitate (p. 133).

Notes: 1. The amount of potassium nitrate should be increased up to 1 g for rocks containing much ferrous iron or organic matter.

2. This residue may be used for the determination of barium, zirconium, etc., if these are required.

Vlisidis[6] has described a procedure for determining sulphate–sulphur in the presence of sulphide–sulphur, which can then be determined by difference. The sample is digested with acidified barium chloride solution in an inert atmosphere, converting all sulphates to barium sulphate. Cadmium chloride is added to precipitate any sulphide ion that may be liberated. Sulphate present as barium sulphate in the precipitate is then determined indirectly by weighing the barium as barium sulphate. Sulphate-containing silicates such as lazulite, helvite and danalite are soluble in hydrochloric acid and present no problems.

References

1. GOLDSCHMIDT V. M., *Geochemistry*, Oxford, 1954, p. 524.
2. VOLKOV I. I., *Zhur. Anal. Khim.* (1959) **14**, 592.
3. OZAWA T., *J. Chem. Soc. Japan, Pure Chem. Sect.* (1966) **87**, 587.
4. MURTHY A. R. V., NARAYAN V. A. and RAO M. R. A., *Analyst* (1956) **81**, 373.
5. MURTHY A. R. V. and SHARADA K., *Analyst* (1960) **85**, 299.
6. WILSON A. D., SERGEANT G. A. and LIONNEL L. J., *Analyst* (1963) **88**, 138.
7. GUPTA J. G. Sen., *Analyt. Chem.* (1963) **35**, 1971.
8. SEN GUPTA J. G., *Anal. Chim. Acta* (1970) **49**, 519.
9. VLISIDIS A. C., *U.S. Geol. Surv. Bull.* 1214-D, 1966.

CHAPTER 43

THALLIUM

Occurrence

Normal spectrographic methods are not sufficiently sensitive to record the presence of thallium in silicate rocks, although a special double arc technique has been devised for this.[1] Using this procedure Shaw has reported that thallium is present in parental mafic magmas to the extent of 0·13 ppm. Early ultrabasic rocks contain very little thallium, which appears to an increasing extent in the later rocks, Table 38. These figures are based upon the analysis of a very few samples, but are in rough agreement with later values of Brooks and Ahrens,[2] Table 39, who used an ion-exchange procedure to concentrate the thallium from a large sample and completed the determination by normal spectrographic methods.

As a result of the tendency to accumulate in residual magmas the later minerals also contain an increasing amount of thallium. This has been particularly noted in potassium minerals (especially micas and to a

TABLE 38. THALLIUM CONTENT OF SOME
SILICATE ROCKS[1]

Rock type	Tl, ppm
Anorthosite	0·015
Peridotite and pyroxenite	0·06
Basalt	0·124
Gabbro	0·134
Diorite and andesite	0·15
Granodiorite, etc.	0·43
Syenite and monzonite	1·4
Granite	3·1
Rhyolite and obsidian	3·5
Nepheline rocks	3·6

Table 39. Thallium Content of Some
Silicate Rocks[2]

Rock type	Tl, ppm
Granites and related rocks	0·73
Basalt, diabase and gabbro	0·11
Sedimentary rocks	0·34

lesser extent felspars) where Tl$^+$, can substitute for K$^+$, in the same way as rubidium Rb$^+$—all ions of similar size. The greatest concentration of thallium occurs in pegmatite minerals such as lepidolite, where 120 ppm Tl has been reported.[2]

The Photometric Determination of Thallium in Silicate Rocks

There is no lack of photometric reagents for thallium, although none of them is sufficiently selective to be used without a prior separation. Dithizone,[3] brilliant green,[4] crystal violet,[5] methyl violet[6] and rhodamine B[7] have all been suggested for particular applications. The procedure given in detail below has been adapted from that described by Voskresenskaya[4, 8] using brilliant green. It is first necessary to separate thallium from interfering elements, particularly antimony, tin, mercury, cadmium, chromium and tungsten. For this a solvent extraction of thallic bromide into diethyl ether is described.

The reaction between the anion TlBr$_4^-$ and the brilliant green cation gives an organic-soluble green coloured complex with maximum absorption at a wavelength of 627 nm (Fig. 87). The Beer–Lambert Law is obeyed in the range 0·1–5μg Tl per ml of extract, part of this range is shown in Fig. 88.

Silicate rock material is decomposed by evaporation with hydrofluoric acid or a mixture of hydrofluoric and sulphuric acids in the usual way. The dry residue is converted to bromides by evaporation with hydrobromic acid. Fusion with alkali carbonates must be avoided as this leads to volatilisation of any thallium present.[9]

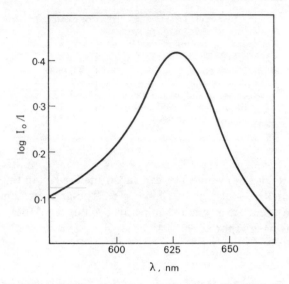

FIG. 87. Absorption spectrum of the thallium complex with brilliant
green (1-cm cells, 5 μg Tl/5 ml).

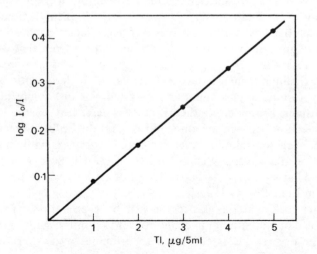

FIG. 88. Calibration graph for thallium using brilliant green (1-cm
cells, 627 nm).

Method

Reagents: *Hydrobromic acid*, concentrated and N.

Hydrobromic acid–bromine reagents, both concentrated and N hydrobromic acid saturated with bromine.

Potassium bromide.

Diethyl ether reagent, saturated with N hydrobromic acid.

Brilliant green solution, 10 mg of reagent dissolved in 100 ml of water.

Amyl acetate.

Standard thallium stock solution, dissolve 0·0618 g of thallous sulphate in water and dilute to 500 ml to give a solution containing 100 μg Tl/ml.

Standard thallium working solution, dilute 5 ml of the stock solution to 500 ml with water to give a working standard containing 1 μg Tl/ml.

Procedure. Accurately weigh approximately 1 g of the finely powdered rock material into a platinum basin or crucible, and evaporate to dryness with 5 ml of concentrated hydrofluoric acid. Cool, add a further 5 ml of hydrofluoric acid together with 2 ml of 20 N sulphuric acid, and repeat the evaporation, first to fumes of sulphuric acid then to complete dryness. Add a few ml of water to the residue, warm gently and break up the residue. Rinse the residue with water into a 100-ml beaker, add 5 ml of concentrated hydrobromic acid and evaporate to dryness. Moisten the dry residue with concentrated hydrobromic acid saturated with bromine and again evaporate to dryness. Repeat this evaporation with hydrobromic acid and bromine, but this time do not allow the residue to dry out. Dissolve this residue in 25 ml of N hydrobromic acid saturated with bromine, and transfer to a 100-ml separating funnel.

Add to the solution 25 ml of diethyl ether saturated with N hydrobromic acid, and shake for 1 minute to extract the thallium. Allow the layers to separate and remove the ether extract. Add a further 25 ml of ether saturated with N hydrobromic acid and repeat the extraction. Discard the aqueous layer. Combine the ether extracts, wash with 2–3 ml of N hydrobromic acid and transfer the ether layer to a 100-ml beaker. Remove the ether by evaporation on a water bath.

Add 2 ml of concentrated hydrochloric acid to the small residue followed by 2 ml of bromine water, and evaporate to dryness. Repeat this sequence of additions and evaporations twice more, finally dissolving the dry residue in 3 ml of N hydrochloric acid. Now add 2 ml of bromine water and heat on a hot plate to remove all free bromine. Cool the solution, transfer to a 100-ml separating funnel, dilute to 25 ml with water and add 1 ml of brilliant green solution. Shake for 1 minute, then add exactly 5 ml of amyl acetate and shake again to extract the green thallium complex. Allow to stand for 20 to 30 minutes then separate the organic layer and measure the optical density in

1-cm cells using the spectrophotometer set at a wavelength of 627 nm. Measure also the optical density of a reagent blank solution prepared in the same way as the sample solution but omitting the rock powder.

Calibration. Transfer aliquots of 1–5 ml of the standard thallium solution containing 1–5 μg Tl, to separate 100-ml beakers. Add to each 2 ml of concentrated hydrochloric acid and 2 ml of bromine water. Evaporate each solution to dryness on a hot plate, and then repeat the sequence of additions and evaporations twice more. Add 3 ml of N hydrochloric acid, 10 mg of potassium bromide and 2 ml of bromine water. Heat the solution to remove free bromine, and continue as described above for the sample solution. Plot the relation of optical density to thallium concentration (Fig. 88).

The Fluorimetric Determination of Thallium in Silicate Rocks

The complex of thallium(III) with rhodamine B exhibits a violet fluorescence in benzene solution that can be used for the fluorimetric determination of thallium. Other elements behave similarly including gold, gallium and antimony(V). A procedure for the determination of thallium in silicate rocks based upon the use of this reaction was described by Matthews and Riley.[10] The rock material was decomposed by evaporation with nitric and hydrofluoric acids, thallium oxidised to the trivalent state with bromine and separated by ion exchange chromatography.

Method

Reagents: *Diethyl ether*, redistilled.

Sulphur dioxide, saturated aqueous solution.

Bromine water, saturated.

Rhodamine B solution, dissolve 0·1 g of the solid reagent in 100 ml of 3·5 M hydrochloric acid.

Ion exchange column, remove the fine material from 50–100-mesh Deacidite FF or similar resin by decantation in water. Digest twice with 2 M hydrochloric acid, wash well with water and transfer to a column 6 mm diameter with a bed depth of about 75 mm.

Standard thallium stock solution, dissolve 0·065 g of thallous nitrate in 500 ml of water. This solution contains 100 μg thallium per ml.

Standard thallium working solution, prepare from the stock solution by dilution as required, to give 0·5 μg thallium per ml and to be 2·7 M in hydrochloric acid.

Procedure. Accurately weigh approximately 1 g of the finely powdered silicate rock material (note 1) into a 50-ml PTFE beaker and add 10 ml of concentrated hydrofluoric acid and 5 ml of concentrated nitric acid. Cover the beaker and warm on a steam bath overnight. Rinse and remove the cover and evaporate the contents of the beaker to dryness on the water bath. Moisten the dry residue with a little concentrated nitric acid and evaporate to dryness. Repeat the evaporation to dryness with a little nitric acid. Repeat the evaporation twice more but with 6·5 M hydrochloric acid to remove the nitrate ion.

Add to the dry residue 7·7 ml of 6·5 M hydrochloric acid and 20–25 ml of water. Warm until all soluble salts have dissolved then dilute to 500 ml with water and add 5 ml of bromine water. Allow this solution to percolate through the ion exchange column at a rate of about 3 ml per minute. Wash the column with 20 ml of water followed by 350 ml of 0·5 M nitric acid and 250 ml of 0·5 M hydrochloric acid to elute the interfering elements. Rinse with 25 ml of water and remove the thallium from the column by elution with 35 ml of the sulphur dioxide saturate solution. Collect the eluate in a small silica beaker, add 1 ml of 6·5 M hydrochloric acid and evaporate on a water bath to a volume of about 15 ml. Remove the last traces of sulphur dioxide by adding 4 ml of bromine water and warming.

Cool, add a further 3 ml of bromine water, transfer to a 50-ml separating funnel using no more than 5 ml of 0·3 M hydrochloric acid to effect the transfer. Add 15 ml of redistilled ether and shake to extract the thallium. Remove the organic layer and repeat the extraction with a further 15 ml of ether. Combine the extracts and wash with 5 ml of 0·3 M hydrochloric acid. Discard the acid layer and transfer the ether extract to a small beaker containing 5 ml of 2·7 M hydrochloric acid. Allow the ether to evaporate and oxidise the thallium with 1 ml of bromine water. Warm gently to remove the excess bromine water (note 2), and transfer to a 50-ml separating funnel using 5 ml of 2·7 M hydrochloric acid. Add 2 ml of rhodamine B solution and 5 ml of benzene and shake for 2 minutes to extract the complex. Measure the fluorescence of the extract at 600 nm, using excitation at about 546 nm.

Measure also the fluorescence of a reagent blank extract prepared in the same way but omitting the sample material. Prepare also a calibration graph from aliquots of the thallium solution containing 0·05 to 1·5 μg of thallium.

Notes: 1. Use a sufficient quantity of the powdered rock material to give 0·05 to 1·5 μg of thallium.
2. Do not overheat at this stage. A temperature of 70° should not be exceeded.

References

1. Shaw D. M., *Geochim. Cosmochim. Acta* (1952) **2**, 118.
2. Brooks R. R. and Ahrens L. H., *Geochim. Cosmochim. Acta* (1961) **23**, 100.
3. Clarke R. S. Jr. and Cuttitta F., *Anal. Chim. Acta* (1958) **19**, 555.
4. Voskresenskaya N. T., *Zavod. Lab.* (1958) **24**, 395.

5. PATROVSKY V., *Chem. Listy* (1963) **57**, 961.
6. OSHMAN V. A., *Trudy Ural'sk. Nauch.-Issled. i Proekt. Inst. Medn. Prom.* (1963) (7), 417.
7. MINCZEWSKI J., WIETESKA E. and MARCZENKO Z., *Chem. Anal., Warsaw* (1961) **6**, 515.
8. VOSKRESENSKAYA N. T., *Zhur. Anal. Khim.* (1956) **11**, 623.
9. KUZNETZOV V. I. and MYASOEDOVA G. V., *Zhur. Prikl. Khim.* (1956) **29**, 1875.
10. MATTHEWS A. D. and RILEY J. P., *Anal. Chim. Acta* (1969) **48**, 25.

CHAPTER 44

THORIUM

Occurrence

The geochemistry of thorium has been reviewed together with that of uranium by Adams *et al.*[1] Thorium does not appear to any great extent in the early magmatic rocks, but is concentrated in the residual magmas and hence in granitic rocks and more particularly in granitic pegmatites. Average values for the thorium content of a number of silicate rock types are given in Table 40. This table has been compiled from data given by Turekian and Wedepohl,[2] Whitfield *et al.*[3] and others.

TABLE 40. THE THORIUM CONTENT OF SOME
SILICATE ROCK TYPES

Rock type	Th, ppm
Ultrabasic rocks	0·003
Basalt and diabase	2
Syenite	10
Granodiorite	10
Granite	12
Shale	12
Limestone	1

Most of the thorium in silicate rocks is present in the accessory minerals, especially in the rare earth minerals monazite and allanite, and in smaller amounts in zircon, sphene and epidote. Two independent thorium minerals are known, thorianite $(Th,U)O_2$ and thorite $ThSiO_4$, both of which occur as minor constituents of granite pegmatites.

The Determination of Thorium

GRAVIMETRIC DETERMINATION

Beach sand deposits and pegmatites containing appreciable amounts of thorium can be decomposed by fusion with potassium hydroxide, and the thorium recovered by precipitation as the insoluble fluoride. After separation from rare earths and other elements, thorium is precipitated as oxalate, ignited and weighed as oxide. This gravimetric procedure can be applied to normal silicate rocks only by taking a very large sample, and has been replaced by photometric methods.

PHOTOMETRIC METHODS

Only a few reagents are known that form coloured complexes with thorium suitable for photometric determination. Of these the four most commonly used are thoron, and arsenazo I, II and III (XXIX–XXXII). These compounds all contain arsenic and have related structures.

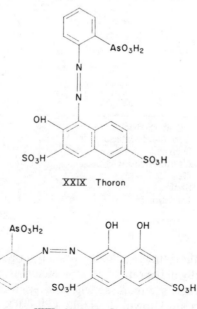

XXIX Thoron

XXX Arsenazo I

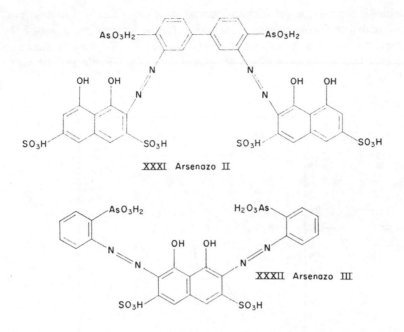

XXXI Arsenazo II

XXXII Arsenazo III

Thoron is known by several other names including thorone, thorin, APANS and naphtharsen. This material reacts also with a number of other elements, although the selectivity can be improved by the masking action of mesotartaric acid.[4] Arsenazo I and II are more sensitive to thorium than thoron, but are generally similar in that they form stable complexes with thorium and other elements in weakly acid solution. Arsenazo III is appreciably more sensitive, and in addition can be used in fairly strongly acid solution. This gives increased selectivity to thorium, and only zirconium, hafnium and uranium(IV) interfere. Within limits zirconium and hafnium can be masked with oxalic acid, whilst uranium can be oxidised to uranium(VI), which does not react with arsenazo III.

The reagent arsenazo III is a dark-red powder that dissolves in water to give a rose-red solution. The thorium complex is emerald green in colour, although solutions are usually coloured purple from the presence of excess reagent. The absorption spectra of the thorium complex is shown in Fig. 89. There is a small absorption of the reagent at 665 nm, which is the wavelength of maximum absorption of the thorium complex. As with other complexes of arsenazo III, the absorption band of the thorium complex is very narrow, and care must therefore be taken to ensure that all measurements of optical density are made at their maximum value.

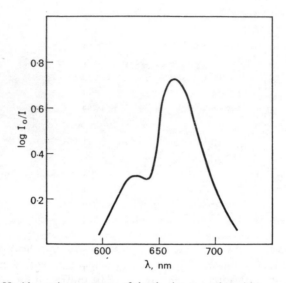

FIG. 89. Absorption spectrum of the thorium complex with arsenazo III.

The calibration graphs given by Savvin[5] and by Savvin and Bareev[7] show a curious change of slope occurring at a concentration of about 0·4 ppm thorium. This feature has been confirmed by Abbey[6] who noted that in a mixed hydrochloric–perchloric acid solution, this break in the slope of the graph occurred at a concentration of about 0·6 ppm. May and Jenkins[8] obtained linear calibration graphs up to 2 ppm thorium, but suggested using a calibration over the range 0–0·4 ppm.

SEPARATION METHODS

Ion exchange, column chromatography, solvent extraction and precipitation methods have all been advocated for the separation of thorium.

None of these methods gives a complete separation in a single operation, and a number of authors have suggested or recommended combinations of two or more procedures. Anion exchange resins have been used to remove those elements that form anionic complexes in chloride solution from thorium and other elements that do not.[8] Uranium and zirconium—elements that may interfere with the photometric determination of thorium with arsenazo III—can be separated in this way. Rare earths, which also interfere with the determination of thorium, accompany thorium in the eluate. Culkin and Riley[9] use a solvent extraction procedure with tributyl phosphate to recover thorium, zirconium (+ hafnium) and cerium from silicate rocks, and separate these elements on a column of cation exchange resin. Oxalic acid solutions are used to elute zirconium (+ hafnium) and thorium, and hydrochloric acid to elute cerium.

Column chromatography using cellulose pulp with and without alumina[10, 11] has been used for the analysis of thorium ores and minerals, but does not appear to have been applied to any extent to the analysis of silicate rocks. Solvent extraction procedures also have limited application in this branch of analysis, although extraction with mesityl oxide has been used,[12] and as noted above extraction with tributyl phosphate can be used to recover thorium with zirconium and cerium.[9]

One of the oldest procedures used to recover thorium is precipitation as oxalate. This was used by Abbey,[6] with calcium as carrier prior to a photometric determination. Rare earths accompany thorium in the oxalate precipitate. Rare earths and thorium can be precipitated together as insoluble fluorides, for which calcium can also be used as carrier.[13] A separation of thorium from the rare earths can be achieved by precipitation of the former as iodate from a nitric acid solution.[14, 15]

Extraction with a long-chain amine was used by Pakalns[16] to remove zirconium and large amounts of iron, but with many silicate rocks the titanium remaining with the thorium is likely to interfere with the determination.

The Determination in Silicate Rocks

The procedure described in detail below is based upon that described by May and Jenkins.[13] The silicate material is decomposed by evaporation with hydrofluoric and nitric acids and the thorium, rare earths and calcium precipitated together as fluorides after the addition of a calcium carrier solution. Precipitations with potassium hydroxide and potassium iodate are used to complete the separation of thorium from other elements before photometric determination with arsenazo III.

Method

Reagents: *Calcium carrier solution*, dissolve 5 g of calcium carbonate in a mixture of 100 ml of concentrated nitric acid and 200 ml of water. Boil to expel carbon dioxide, cool and dilute to 500 ml with water.

Hydrofluoric acid wash solution, dilute 5 ml of concentrated hydrofluoric acid with 95 ml of water.

Iron carrier solution, dissolve 0·875 g of ferric nitrate hexahydrate in 100 ml of water containing 1 ml of concentrated nitric acid.

Mercury carrier solution, dissolve 0·158 g of mercuric nitrate monohydrate in 2 ml of 8 N nitric acid and dilute to 100 ml with water.

Oxalic acid solution, dissolve 50 g of the dihydrate in a hot mixture of 500 ml of water and 500 ml of concentrated hydrochloric acid.

Potassium hydroxide concentrated solution, dissolve 50 g of potassium hydroxide in 50 ml of water.

Potassium hydroxide wash solution, dilute 5 ml of the concentrated solution to 1 litre with water.

Potassium iodate solution, dissolve 15 g of potassium iodate in water and dilute to 250 ml.

Potassium iodate wash solution, mix together 30 ml of concentrated nitric acid, 3 ml of 30 per cent hydrogen peroxide and 100 ml of potassium iodate solution. Dilute to 500 ml with water.

8-hydroxyquinoline solution, dissolve 0·5 g of the solid reagent in 100 ml of water containing 1 ml of concentrated nitric acid.

Hydrogen peroxide solution, 3 and 30 per cent solutions.

Arsenazo III solution, grind 50 mg of the reagent with 1 ml of water and dilute with water to 100 ml. Prepare freshly each day.

Standard thorium stock solution, fuse 0·114 g of pure thorium oxide with 2 g of potassium hydroxide. Extract the melt with water, acidify with hydrochloric acid, adding 25 ml in excess. Heat until the solution is complete, cool and then dilute to volume with water in a 1-litre flask. This solution contains 100 μg thorium per ml.

Standard thorium working solution, transfer 5 ml of the stock solution to a 500-ml volumetric flask, add 25 ml of concentrated hydrochloric acid and dilute to volume with water. This solution contains 1 μg thorium per ml.

Procedure. Accurately weigh approximately 1 g of the finely powdered sample material into a platinum dish, moisten with water, add 10 ml of concentrated nitric acid and 10 ml of concentrated hydrofluoric acid and evaporate to dryness on a steam bath. Allow to cool, add 5 ml of concentrated nitric acid and 5 ml of concentrated hydrofluoric acid and again evaporate to dryness on a steam bath. Repeat the evaporation to dryness four or five times with small quantities of concentrated nitric acid and finally dissolve the nitrate residue by warming with 5 ml of concentrated nitric acid and 30 ml of water.

If any residue remains at this stage, collect on a small filter, wash with hot water, dry, ignite and fuse in a platinum crucible with a small amount of anhydrous sodium carbonate. Exract the melt with water, acidify with nitric acid, heat to complete the dissolution and add to the main rock solution (note 1).

Transfer the rock solution, or an aliquot of it (note 2) to a platinum dish, add 5 ml of the calcium carrier solution and evaporate to dryness. Moisten the dry residue with water, add 20 ml of concentrated hydrofluoric acid and 20 ml of water and allow to stand on a steam bath for 2 hours. Allow to cool. Stir into the solution a macerated 9-cm filter paper and collect the paper pulp together with the fluoride precipitate on a small close-textured filter paper previously washed with the hydrofluoric acid wash solution. Wash the residue thoroughly with the wash solution and discard the filtrate and washings.

Transfer the paper and residue to a platinum crucible, dry, ignite and fuse with 2 g of potassium pyrosulphate. Allow to cool and dissolve the melt in about 10 ml of water containing 2 ml of nitric acid. Rinse the solution into a 400-ml beaker and dilute with water to a volume of about 150 ml. Add 1 ml of the iron carrier solution, 2 drops of concentrated hydrogen peroxide solution, concentrated potassium hydroxide solution to neutralise the acid present and then 15 ml in excess. Transfer the beaker to a steam bath, digest for about 15 minutes and then collect the precipitate on an open-textured filter paper. Rinse the beaker and wash the precipitate several times with the potassium hydroxide wash solution.

Rinse the residue into a 100-ml beaker with a little water and add 1 ml of the 8-hydroxyquinoline solution. Moisten the paper with 1 ml of the dilute hydrogen peroxide solution followed by 2 ml of hot 8 N nitric acid, added slowly drop by drop. When the paper has drained, add 5 ml of hot water and again allow the paper to drain. Repeat this sequence of additions to the paper twice more to ensure complete removal of the residue from the paper.

Using pipettes, add 10 ml of potassium iodate solution to the beaker, followed by 5 ml of the mercury carrier solution, stirring continuously. Allow the beaker to stand in an ice-bath for 45 minutes. Mix in a small quantity of macerated filter-paper pulp and collect the pulp and precipitate on a small close-textured filter paper. Rinse the beaker and wash the precipitate eight or ten times with ice-cold potassium iodate wash solution. Discard the filtrate and washings. Dissolve the precipitate from the filter by adding 5 ml of hot 6 N hydrochloric acid and rinsing with 5 ml of hot water. Repeat this addition of hydrochloric acid and water twice more, collecting the solution and washings in a 50-ml beaker. Transfer the beaker to a steam bath and evaporate to dryness. Moisten the residue with 2 ml of concentrated hydrochloric acid and again evaporate to dryness.

Add 1·6 ml of 6 N hydrochloric acid to the beaker, moistening the residue and the walls of the beaker. Add 3 ml of water, cover with a watch glass and warm for 5 minutes on a steam bath. Allow to cool, and rinse the solution into a 25-ml volumetric flask using several small portions of water with a total volume of not more than 8 ml. Add 10 ml of the oxalic acid solution followed by 2 ml of arsenazo III solution. Dilute to volume with water and mix well.

Measure the optical density of the solution in 4-cm cells with the spectrophotometer set at a wavelength of 665 nm, using a reference solution without added thorium, prepared as described below for the calibration graph.

Calibration. Transfer aliquots of 0–10 ml of the standard solution containing 0–10 μg of thorium to separate 25-ml beakers. Evaporate each to dryness and then add 1·6 ml of 6 N hydrochloric acid. Transfer each solution to a separate 25-ml volumetric flask and dilute to volume with water after adding 10 ml of the oxalic acid solution and 2 ml of the arsenazo III solution. Measure the optical density of each relative to the solution containing no added thorium, as described above for the sample solution. Plot the relation of optical density to thorium concentration.

Notes: 1. Zircon and other refractory silicate minerals are best decomposed by fusion with sodium carbonate as described. If the residue is composed largely of oxide minerals, it should be fused with potassium pyrosulphate. Extract the melt with water, dissolve by adding dilute nitric acid, add 1 ml of the iron carrier solution and precipitate iron, thorium and other elements with ammonia. Collect the precipitate, dissolve in dilute nitric acid and add to the main rock solution.

2. This procedure is designed for silicate materials containing 0–10 ppm of thorium. Many silicate rocks contain higher concentrations of thorium, and for these a smaller sample portion can be taken. Alternatively the nitric acid solution of the rock material can be diluted to volume, and an aliquot of the solution containing 10 μg or less of thorium taken for the subsequent stages of the analysis.

References

1. ADAMS J. A. S., OSGOOD J. K. and ROGERS J. J. W., *Phys. Chem. Earth* (1959) **3** 298.
2. TUREKIAN K. K. and WEDEPHOL K. H., *Bull. Geol. Soc. Amer.* (1961) **72**, 175.
3. WHITFIELD J. M., ROGERS J. J. W. and ADAMS J. A. S., *Geochim. Cosmochim. Acta* (1959) **17**, 248.
4. FLETCHER M. H., GRIMALDI F. S. and JENKINS L. B., *Analyt. Chem.* (1957) **29**, 963.
5. SAVVIN S. B., *Dokl. Akad. Nauk SSSR* (1959) **127**, 1231.
6. ABBEY S., *Anal. Chim. Acta* (1964) **30**, 176.
7. SAVVIN S. B. and BAREEV V. V., *Zavod. Lab.* (1960) **26**, 412.

8. ARNFELT A-L. and EDMUNDSSON I., *Talanta* (1961) **8**, 473.
9. CULKIN F. and RILEY J. P., *Anal. Chim. Acta* (1965) **32**, 197.
10. WILLIAMS A. F., *Analyst* (1952) **77**, 297.
11. KEMBER N. F., *Analyst* (1952) **77**, 78.
12. LEVINE H. and GRIMALDI F. S., *Geochim. Cosmochim. Acta* (1958) **14**, 93.
13. MAY I. and JENKINS L. B., *U.S. Geol. Surv. Prof. Paper* 525-D, p. 192, 1965.
14. GRIMALDI F. S., JENKINS L. B. and FLETCHER M. H., *Analyt. Chem.* (1957) **29**, 848.
15. FURTOVA E. V., SADOVA G. F., IVANOVA V. N. and ZAIKOVSKII F. V., *Zhur. Anal. Khim.* (1964) **19**, 94.
16. PAKALNS P., *Anal. Chim. Acta* (1972) **58**, 463.

CHAPTER 45

TIN

Occurrence

Much of the early work on the occurrence of tin in silicate rocks was handicapped by the lack of a suitable method of analysis, i.e. one of sufficient sensitivity. One of the earliest systematic studies of the geochemistry of tin was that of Onishi and Sandell[1] based upon determinations by a colorimetric method.[2] Other studies by Barsukov[3] and Hamaguchi et al.[4] have given similar values for tin in silicate rocks. A summary of average values for some rock types is given in Table 41.

TABLE 41. THE AVERAGE TIN
CONTENT OF SOME IGNEOUS
ROCKS

Rock type	Sn, ppm
Granite	3·6
Intermediate	1·5
Basic	0·9
Ultrabasic	0·35
Shale	6

Figures of the same order but slightly higher were given by Durasova[5] who suggested that intermediate granitoids are of two types that she described as tin rich (mean value 8·5 ppm) and tin poor (mean value 3·3 ppm); the tin rich possibly being derived by assimilation of tin-bearing sediments.

Studies of the geochemistry of tin have included reference to a "regional factor", that is to the occurrence of granitic rocks with high tin values in those areas where tin deposits are found. A tin content of 15–30 ppm is by no means uncommon for these rocks, whilst values of up to 400 ppm and more have occasionally been noted.

Although Ahrens and Liebenberg[6] have suggested that tin does occur in part at least, in the lattice of mica minerals, it seems probable that the greater part of the tin content of certain silicate rocks is as the accessory mineral cassiterite SnO_2. This mineral is widely dispersed in certain types of granite as very small sized grains. It occurs in more massive form in pegmatites where it may be accompanied by minerals of other less common elements, such as tantalum and lithium.

The Determination of Tin

At the concentration level encountered in normal silicate rocks, atomic absorption, flame photometric and polarographic procedures are not sufficiently sensitive even to detect the presence of tin. Spectro-photometric methods can be used, but some of the procedures described are not applicable to those rocks that contain less than 1 ppm Sn. Neutron activation analysis[7] can be used where increased sensitivity is required.

SAMPLE DECOMPOSITION

Those rocks that contain tin within the silicate lattice present no problem in sample decomposition. Such samples may be evaporated with hydrofluoric acid and either sulphuric or perchloric acid in the usual way. The residue may be fused with potassium pyrosulphate and the melt obtained dissolved in hydrochloric or sulphuric acid.

Where the rock contains discrete particles of cassiterite, the dissolution is considerably more difficult. This is because of the unreactive nature of cassiterite, particularly towards acids and acidic fluxes, and also from the difficulty in transferring the very fine grains of unattacked cassiterite from one vessel to another following the decomposition of the silicate phase with hydrofluoric acid. These grains are easily decomposed by fusion with sodium peroxide, but because of the attack of platinum apparatus that occurs it is not possible to use this procedure directly after the volatilisation of silica with hydrofluoric acid. Attempts to mix the dry residue with sodium peroxide and sinter at 500° have not given complete decomposition of residual cassiterite.

Fusion of the rock material with a mixture of sodium carbonate and sulphur converts all the tin present into a thiosalt, from which it is recovered together with other heavy metals as sulphide. This procedure has been used by Popov[8] for determining tin in rocks and tin ores, but is unsuitable for those silicates containing only very small amounts of tin.

One of the simplest ways of decomposing cassiterite present in rock samples is to ignite with ammonium iodide. This procedure has been used in geochemical prospecting for tin[9] as well as for the determination of tin in silicate rocks.[10] In its simplest form, the sample material is intimately mixed with ammonium iodide, transferred to a borosilicate test tube and ignited over a gas burner. Attempts to obtain quantitative recovery of tin from the tonalite T-1 have not been successful and a more elaborate version of this decomposition has been devised.[11] The apparatus used is shown in Fig. 90, and the procedure used as follows.

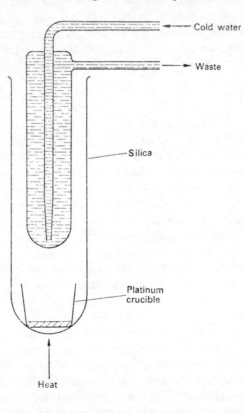

FIG. 90. Apparatus for the decomposition of silicate rocks containing cassiterite.

Five ml of concentrated hydrofluoric acid and 0·5 ml of concentrated sulphuric acid were added to 0·5 g of the sample material in a small platinum crucible, and the silica removed by evaporation to dryness. The residue was mixed with 0·25 g of ammonium iodide, and the crucible placed within the silica tube as shown in Fig. 90. The tube was slowly raised to dull red heat, and this temperature was maintained for 5 minutes. After cooling, the residue was extracted by heating with N hydrochloric acid. The sides of the tube and the cold finger were washed down with hot N hydrochloric acid and the washings combined with the acid extract. Free iodine was removed with ascorbic acid and the tin present in the solution determined with salicylideneamino-2-thiophenol (XXXIII).

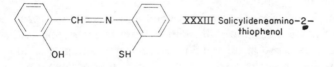

XXXIII Salicylideneamino–2–thiophenol

Although this way of decomposing cassiterite grains in silicate rocks was found to give better recoveries of tin than the simpler form of decomposition, occasionally very low recoveries were noted for which no explanation has yet been suggested.

Agterdenbos and Vlogtman[12] have drawn attention to the occurrence of tin in silicate lattices, as well as in the form of accessory cassiterite. When the silicate sample is ignited with ammonium iodide, only the tin in cassiterite is converted to tin(IV) iodide—lattice-bound tin remaining unattacked. If the silicate matrices are destroyed with hydrofluoric and mineral acids, the total tin content becomes available for conversion to iodide.

SEPARATION OF TIN

Ion-exchange separations have been described by Huffman and Bartel[13] and by De Laeter and Jeffery.[14] The former used an oxalate form of an anion exchange resin and absorbed the tin from a solution containing hydrochloric and oxalic acids. The tin was recovered by elution with dilute sulphuric acid. The latter workers used the chloride form of the resin and absorbed the tin from 6 M hydrochloric acid solution. Nickel and cobalt were eluted with 3 M acid, iron with M acid and tin with 0·1 M hydrochloric acid. A second anion exchange separation was recommended.

A similar anion exchange separation in 6 M medium was described by Smith[15] in a method for determining traces of tin in rocks, sediments and soils similar to that described below, but based upon the use of phenylfluorone.

A highly selective procedure for the recovery of tin from acidic solutions was described by Ross and White.[16] This involved the extraction of the tin complex with tris-(2-ethylhexyl)phosphine oxide into cyclohexane solution. The extraction of stannic iodide into nitrobenzene was reported by Gilbert and Sandell[17] and by Tanaka,[18] and into toluene by Newman and Jones.[19]

Stannic bromide is appreciably volatile, and a method for the separation of tin, based upon a distillation from hydrobromic acid solution has been used by Onishi and Sandell for determining small amounts of tin in silicate rocks. Arsenic, antimony and germanium were first removed by distillation from a hydrochloric acid solution. The only element to accompany tin in the bromide distillation that gave rise to interference in the subsequent determination was selenium.

PHOTOMETRIC DETERMINATION

Reagents advocated for the determination of tin include cacotheline, gallein, 8-hydroxyquinoline, haematoxylin, diethyldithiocarbamate, phenylfluorone and dithiol. None of these is ideal, although each has its own advocates. Dithiol has found more use than some of the other reagents on this list, principally because of the easily identified red complex with tin. As with some of the other reagents, the colour is present only in colloidal solution. Attempts to extract the tin complex into organic solution result in loss of sensitivity and the disappearance of the characteristic red colour. For photometric determinations the complex is dispersed in aqueous solution, when approximately 3 μg of tin can be detected. For visual comparisons, the detection limit is about 0·5 ug, permitting the determination of 0·5 ppm, when a 1-g sample portion has been taken.

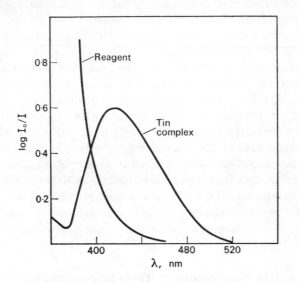

FIG. 91. Absorption spectrum of the tin complex with salicylidene-
amino-2-thiophenol.

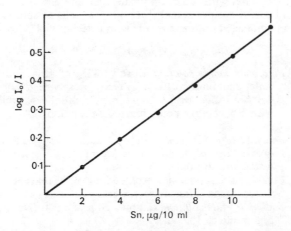

FIG. 92. Calibration graph for tin using salicylideneamino-2-thiophenol
(4-cm cells, 415 nm).

One of the most valuable reagents for tin is salicylideneamino-2-thiophenol.[20] This reagent is easily prepared and gives a yellow-coloured complex with tin that can be extracted into any of a number of organic solvents for photometric measurement (xylene is preferred). The maximum optical density occurs at a wavelength of 415 nm (Fig. 91) and the Beer–Lambert Law is valid up to about 45 μg Sn in 10 ml of xylene (Fig. 92). The molar extinction coefficient is about 15,000 which enables the reagent to be used for determining the small amounts of tin that occur in normal silicate rocks. A simple procedure for this, based upon a method used for tin ore analysis, gave erratic recoveries of tin. An initial separation and concentration stage should therefore be used.

A number of dyes form coloured complexes with tin. These include methyl violet and catechol violet. The complexes that are formed with tin are water soluble, allowing simpler manipulative techniques than are possible with such reagents as dithiol or phenylfluorone which give colloidal systems.

The Determination of Tin in Silicate Rocks

DETERMINATION WITH CATECHOL VIOLET

In the procedure described below, silica is removed by evaporation with hydrofluoric acid in the usual way. Any cassiterite in the acid-insoluble residue is recovered by fusion with sodium peroxide. Tin present in the solution is concentrated by extraction of stannic bromide into toluene solution, and the determination completed photometrically with catechol violet.

Method

Reagents: *Potassium iodide solution,* dissolve 83 g of potassium iodide in water and dilute to 100 ml. Prepare freshly each day.

Potassium iodide wash solution, mix 25 ml of 9 N sulphuric acid with 2·5 ml of the potassium iodide solution.

Toluene.

Sodium hydroxide solution, approximately 5 M and 0·1 M.

Ascorbic acid solution, dissolve 5 g of the reagent in 100 ml of water. Prepare only as required.

Catechol violet solution, dissolve 50 mg of the reagent in 100 ml of water. Prepare freshly each week.

Sodium acetate solution, dissolve 20 g of the trihydrate in water and dilute to 100 ml.

Standard tin stock solution, dissolve 0·100 g of pure tin by evaporation to fumes with 20 ml of concentrated sulphuric acid. Cool, cautiously dilute with water, add 65 ml of concentrated sulphuric acid, cool again and dilute with water to 500 ml in a volumetric flask. This solution contains 200 μg tin per ml.

Standard tin working solution, dilute 5 ml of the stock tin solution to 500 ml with water as required. This solution contains 2 μg tin per ml.

Procedure. Accurately weigh approximately 1 g (note 1) of the finely powdered silicate rock material into a platinum crucible, moisten with water and add 1 ml of 20 N sulphuric acid, 1 ml of concentrated nitric acid and 10 ml of concentrated hydrofluoric acid. Transfer the crucible to a hot plate and evaporate first to fumes of sulphuric acid, then to dryness. Allow the crucible to cool, add 2 g of potassium pyrosulphate and fuse over a burner, taking care to avoid loss by spitting. Allow the crucible to cool, dissolve the melt by warming with dilute sulphuric acid and rinse into a separating funnel to give a solution of about 25 ml in volume in 9 N sulphuric acid (note 2).

Add 2·5 ml of the potassium iodide solution and mix. Now add 10 ml of toluene, stopper the funnel and shake vigorously for 2–3 minutes. Allow the phases to separate. The toluene layer now contains stannic iodide together with a small quantity of iodine which gives a pink colour to the solution. Separate and discard the aqueous layer. Without shaking rinse the toluene layer with 5 ml of the potassium iodide wash solution and discard the washings.

Add to the toluene solution 5 ml of water and 5 M sodium hydroxide solution drop by drop, with shaking, until the pink colour of the toluene layer is discharged, then add 2 drops in excess. Stopper the funnel, shake for 30 seconds, allow the phases to separate and run the aqueous layer into a small beaker. Rinse the organic layer by shaking with 3 ml of 0·1 M sodium hydroxide solution and add the wash solution to the aqueous solution in the small beaker. Retain the toluene layer in the separating funnel.

Add 2·5 ml of 5 N hydrochloric acid to the solution in the beaker and add ascorbic acid solution drop by drop to decolorise any iodine that may have been liberated. Using a pipette, add 2 ml of the catechol violet solution and mix. Without shaking wash the toluene in the separating funnel with 5 ml of the sodium acetate solution and add the washings to the beaker. Add dilute ammonia drop by drop to the solution in the beaker to bring the pH to a value of 3·8 ± 0·1 (use a pH meter). Transfer the solution to a 25-ml volumetric flask, dilute to volume with water, mix well and allow to stand for 30 minutes. Measure the optical density of the solution in a 4-cm cell (note 3) using a spectrophotometer set at a wavelength of 552 nm. Prepare also a reagent blank solution in the same way as the sample solution but omitting the sample material (note 4).

Calibration (4-cm cells). Transfer aliquots of the standard tin solution containing 0–8 μg tin to a series of 50-ml beakers and dilute to 10 ml with water. Add 1 ml of 5 M sodium hydroxide solution, mix and add 2·5 ml of 5 N hydrochloric acid and by pipette, 2 ml of catechol violet solution. Adjust the pH to 3·8 ± 0·1, and transfer each solution to a separate 25-ml volumetric flask. Add 5 ml of the sodium acetate solution, dilute to volume with water, mix well and set aside for 30 minutes. Measure the optical densities relative to the solution containing no tin, in 4-cm cells at a wavelength of 552 nm, as described above.

Notes: 1. Use a 1-g sample portion for rocks containing up to 30 ppm tin, and smaller weights for samples of higher tin content.

2. This procedure is used only for those rocks that are completely decomposed by evaporation with hydrofluoric and sulphuric acids followed by fusion of the residue with potassium pyrosulphate. If the sample material contains cassiterite, the residue remaining after evaporation to dryness should be digested with dilute sulphuric acid and any remaining residue collected, ignited and fused in a nickel crucible with a small quantity of sodium hydroxide. The melt is extracted with water, acidified and added to the main rock solution. Although some loss may occur, the greater part of the tin present as cassiterite can be recovered in this way.[2] For rocks containing several hundred ppm or more of tin, it may be necessary to fuse the residue with sodium peroxide.

3. Use a 4-cm cell for up to 8 μg of tin, and a 1-cm cell for up to 30 μg.

4. Measure both the rock and the reagent blank solutions against a reference solution prepared without added tin as described in the calibration section.

DETERMINATION USING SALICYLIDENEAMINO-2-THIOPHENOL

This method also incorporates an evaporation with hydrofluoric acid for silica removal, and an extraction of stannic iodide into toluene solution.

Method

Reagents: *Salicylideneamino-2-thiophenol solution,* dissolve 1 g of ascorbic acid in 100 ml of warm ethanol. Add 0·1 g of the reagent and stir until dissolved. Store in a brown glass bottle away from direct sunlight and prepare freshly each day.

Lactic acid solution, mix 20 ml of lactic acid with 80 ml of water.

Xylene.

Procedure. Decompose and recover the tin in a 1-g portion of the rock material as described above by evaporation with hydrofluoric, nitric and sulphuric acids and fusion of the residue with potassium pyrosulphate (or sodium hydroxide if necessary). Extract stannic iodide into toluene solution and back extract into very dilute sodium hydroxide solution.

Dilute to about 20 ml with water and add 2 ml of lactic acid solution. Check the pH and adjust to a value of 2 $\pm$ 0·1. Using a pipette add 5 ml of the salicylideneamino-2-thiophenol reagent solution, mix and allow to stand for 20 minutes. Add exactly 10 ml of xylene, shake vigorously for 20 seconds and allow the phases to separate. After 5 minutes, measure the optical density of the organic extract in 4-cm cells, with the spectrophotometer set at a wavelength of 415 nm. Prepare a reagent blank extract and use this as the reference solution. For the calibration use aliquots of the standard solution containing 0–10 μg tin.

The Determination of Tin by Atomic Absorption Spectroscopy

The use of atomic absorption spectroscopy for the determination of tin in geological materials, including silicate rocks and sulphide minerals was explored by Moldan et al.[21] Interferences including complex interelement effects were noted, and even with a long path-(45 cm) absorption tube, the sensitivity was scarcely adequate for even average tin contents.

References

1. ONISHI H. and SANDELL E. B., Geochim. Cosmochim. Acta (1957) **12**, 262.
2. ONISHI H. and SANDELL E. B., Anal. Chim. Acta (1956) **14**, 153.
3. BARSUKOV V. L., Geokhimiya (1957) (1), 36.
4. HAMAGUCHI H., KURODA R., ONUMA N., KAWABUCHI K., MITSUBAYASHI T. and HOSOHARA K., Geochim. Cosmochim. Acta (1964) **28**, 1039.
5. DURASOVA N. A., Geochem. Internat. (1967) **4**, 671.
6. AHRENS L. H. and LIEBENBERG W. R., Amer. Mineral. (1950) **35**, 571.
7. HAMAGUCHI H., KAWABUCHI K., ONUMA N. and KURODA R., Geochim. Cosmochim. Acta (1964) **30**, 335.
8. POPOV M. A., Byull. Nauch.-Tekh. Inform. Tsentr. Nauch.-Issled. Inst. Olova, Sur'my i Rtuti (1962) (3), 38.
9. STANTON R. E. and MCDONALD A. J., Trans. Inst. Min. Metall. (1961) **71**, 27.
10. MARTINET B., Chim. Anal. (1961) **43**, 483.
11. KERR G. O., unpublished work.
12. AGTERDENBOS J. and VLOGTMAN J., Talanta (1972) **19**, 1295.
13. HUFFMAN C. Jr. and BARTEL A. J., U.S. Geol. Surv. Prof. Paper 501-D, 1964.
14. DE LAETER J. R. and JEFFERY P. M., J. Geophys. Res. (1965) **70**, 2895.
15. SMITH J. D., Anal. Chim. Acta (1971) **57**, 371.
16. ROSS W. J. and WHITE J. C., Analyt. Chem. (1961) **33**, 421.
17. GILBERT D. D. and SANDELL E. B., Microchem. J. (1960) **4**, 491.
18. TANAKA K., Japan Analyst (1962) **11**, 332.
19. NEWMAN E. J. and JONES P. D., Analyst (1966) **91**, 406.
20. GREGORY G. R. E. C. and JEFFERY P. G., Analyst (1967) **97**, 293.
21. MOLDAN B., RUBESKA I., MIKSOVSKY M. and HUKA M., Anal. Chim. Acta (1970) **52**, 91.

CHAPTER 46

TITANIUM

Occurrence

The overall abundance of titanium in terrestial silicate rocks is given by Clarke and Washington[1] as $0 \cdot 64$ per cent Ti. This figure is in agreement with the $0 \cdot 61$ per cent given by Hevesy[2] for the analysis of composite silicate samples, but is probably too high. This has arisen partly as a result of the inclusion of too great a proportion of basic and rare rock types rich in titanium, and partly from a bias in the older analyses towards high titanium results.[3]

The amount of titanium present in individual silicate rocks varies from less than $0 \cdot 1$ per cent in certain granites and other acidic rocks, to as much as 10 per cent or more in certain basic rocks—some norites for example. In general, basic rocks contain considerably more titanium than acidic rocks. This is shown in Table 42, compiled from published analyses of igneous silicate rocks of the United Kingdom.[4]

TABLE 42. THE TITANIUM CONTENT OF SOME
SILICATE ROCKS

Rock type	Ti, per cent
Granite	0·11
Picrite, peridotite and serpentine	0·27
Diorite and granodiorite	0·67
Andesite and trachyandesite	0·79
Dolerite, gabbro and other basic rocks	1·23

The titanium ion is not readily accommodated in a silicate lattice, and any titanium present in the parental magma will be largely concentrated in the accessory minerals in the course of magmatic evolution.

Although rare, a number of titantium-containing silicate minerals are known from their occasional occurrence in basic-alkaline and calc-alkaline rocks, but the titanium minerals most frequently encountered are rutile, ilmenite, titaniferous magnetite, sphene and perovskite.

Rutile and ilmenite are particularly resistant to weathering and appear as such in residual sediments. Other titanium minerals, such as sphene and perovskite, can be accumulated in this way, but may also be decomposed. Any titanium that passes into solution is likely to be removed by hydrolysis and subsequently be incorporated into the sediment. The titanium content of residual sediments is therefore of a similar order to that of igneous rocks, $0 \cdot 2$–1 per cent Ti, but with higher concentrations where preferential collection or enrichment has occurred. Carbonate rocks contain smaller amounts of titanium, generally not exceeding $0 \cdot 1$ per cent Ti, and more frequently $0 \cdot 01$ per cent or less.

The Photometric Determination of Titanium in Silicate Rocks

In the course of chemical analysis, titanium minerals are considerably more resistant to decomposition than the silicate matrix in which they occur, and care must be taken to ensure that all the mineral grains are completely attacked. Most of the accessory minerals can be decomposed by evaporation to fumes with a mixture of sulphuric and hydrofluoric acids. Some grains of perovskite are likely to remain unattacked if this mineral is present in quantity, although a second evaporation is usually sufficient to remove these. Mixtures of hydrofluoric acid with either nitric or perchloric acid are less effective for this decomposition. The most effective decomposition is probably a single evaporation to dryness with sulphuric and hydrofluoric acids followed by a fusion with potassium pyrosulphate. This can be done in a single platinum crucible and serves also to remove fluorine which can otherwise interfere with the determination of titanium.

Gravimetric and occasionally titrimetric methods were formerly used for determining titanium in those rocks containing 2 or 3 per cent or more, but even at the higher level spectrophotometric methods are now commonly employed. There is no lack of reagents for titanium, although few of these are very specific. They are all based upon the formation of coloured complexes with titanium in acid solution. Hydrogen peroxide is probably the most widely used, and although the method is capable of high precision, it is not very sensitive, Table 42. Two more sensitive reagents are chromotropic acid (1,8-dihydroxynaphthalene-3,6-disulphonic acid) and tiron (1,2-dihydroxybenzene-3,5-disulphonic acid). Less subject,to interference from other metallic ions is the reaction with diantipyrylmethane.[5] These three reagents are some 20 times as sensitive as hydrogen peroxide, Table 43.

TABLE 43. MOLAR ABSORPTIVITY
OF SOME REAGENTS FOR TITANIUM

Reagent	
Hydrogen peroxide	740
Chromotropic acid	17,000
Tiron	15,000
Diantipyrylmethane	13,000*

* Polyak[5] gives 18,000.

METHOD USING HYDROGEN PEROXIDE

This method is based upon the yellow-coloured complex formed by titanium with hydrogen peroxide in acid solution. The maximum value for the optical density occurs at a wavelength in the range 400–410 nm (Fig. 93). There is evidence that the position of the peak depends to some small extent upon the conditions used, but as the absorption band is fairly wide any wavelength within this range can be used for measurement without appreciable loss of sensitivity.

A number of other metals form coloured complexes with hydrogen peroxide including vanadium, uranium, niobium, molybdenum and under certain circumstances chromium. Of these only vanadium is likely to give rise to any interference with the determination of titanium in rocks and minerals, and this only rarely. The maximum absorption of solutions containing the vanadium complex occurs at a wavelength of 460 nm, and it is therefore possible to determine both titanium and vanadium in the same solution by measuring optical densities at both 400 and 460 nm. The concentrations of the two elements can then be calculated from the simultaneous equations:

$$\log I_0/I_{(400 \text{ nm})} = a\,[\text{Ti}] + b\,[\text{V}]$$
$$\text{and } \log I_0/I_{(460 \text{ nm})} = a'[\text{Ti}] + b'[\text{V}]$$

where a, a', b, and b' are the slopes of the calibration graphs for titanium and vanadium at the two wavelengths.

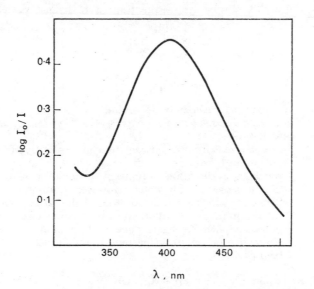

FIG. 93. Absorption spectrum of the titanium complex with hydrogen peroxide (4-cm cells).

Iron, chromium and nickel, elements that form coloured ions in solution, can also interfere. In silicate rocks, however, only iron is likely to be present in amounts sufficient to give rise to serious error, and this can be avoided by adding measured amounts of phosphoric acid to the rock solution and to the standards used for the calibration.

Fluoride ions interfere with the titanium determination by bleaching the yellow colour. A slight bleaching effect has also been observed in the presence of alkali salts, citric acid and phosphoric acid, but these do not interfere with the determination in silicate rocks. High acid concentrations also cause some reduction in colour, and for most purposes a concentration of between 1·5 and 3 N sulphuric acid should be used. Perchloric acid concentrations of up to 3·5 M can be used, but hydrochloric acid should be avoided because of the strong colour given by ferric iron in this medium.

Even the very small quantities of platinum removed from old crucibles in the course of a pyrosulphate fusion can catalytically decompose hydrogen peroxide, and result in a slow fading of the yellow titanium–peroxide colour. A similar decomposition of hydrogen peroxide has been observed with a number of samples containing appreciable amounts of cerium.

Method

Reagents: *Hydrogen peroxide*, 20 or 30 volume.

Potassium pyrosulphate solution, dissolve 20 g in water and dilute to 100 ml.

Standard titanium stock solution, this can be prepared from either potassium titanyl oxalate or potassium fluotitanate by direct weighing, evaporation to fumes with sulphuric acid and dilution to volume with water. However, both these salts are hydrated, and the solutions obtained should be standardised by precipitation of titanium with ammonia, cupferron or *N*-benzoylphenylhydroxylamine and ignition to oxide. An alternative procedure using pure titanium metal foil is given below (note 1).

Procedure. Accurately weigh approximately 0·5 g (note 2) of the finely powdered rock material into a 25-ml platinum crucible, add 0·5 ml of concentrated nitric acid (note 3), 1 ml of 20 N sulphuric acid and 10 ml of concentrated hydrofluoric acid. Place the crucible on a hot plate and evaporate to fumes of sulphuric acid. Allow to cool, rinse down the sides of the crucible with a little water and again evaporate, this time to dryness. Add 2 g of potassium pyrosulphate to the dry residue and fuse in the covered crucible over a low flame for the minimum time required to give a clear melt. Allow to cool.

Extract the melt with water containing 10 ml of 20 N sulphuric acid and transfer the solution to a 100-ml volumetric flask. Add 5 ml of phosphoric acid, mix the solution, dilute to volume with water and again mix well (note 4). Fill two matched 1-cm spectrophotometer cells with the solution. To one cell add 1 drop only of hydrogen peroxide solution and mix with a small glass rod. Measure the optical density of this coloured solution relative to the solution in the other cell to which hydrogen peroxide has not been added. A wavelength in the range 400–410 nm should be used. If the recorded optical density at this wavelength is much less than 0·1, the measurement should be repeated in 4-cm cells.

Calibration. Although this method is not very sensitive it is very precise, and as there is a linear relation between optical density and titanium concentration, it is possible to use a calibration factor to convert measured densities to titanium concentrations (an optical density of 1·000 was obtained from a solution containing 11·56 mg TiO₂ in 100 ml). Alternatively a calibration graph can be used. This can be produced as follows.

Transfer aliquots of 4–20 ml of the standard titanium solution containing 2–20 mg TiO₂, to separate 100-ml volumetric flasks and add to each 5 ml of phosphoric acid, 10 ml of 20 N sulphuric acid and 10 ml of the potassium pyrosulphate solution. Dilute to volume, mix well and determine the optical densities of each solution by filling two 1-cm spectrophotometer cells with each solution, adding 1 drop of hydrogen peroxide to one of each pair and measuring the coloured solution against the solution to which hydrogen peroxide has not been added, as described for the sample solution above. Plot the relation of optical density to titanium concentration (Fig. 94).

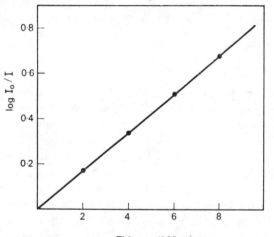

FIG. 94. Calibration graph for titanium using hydrogen peroxide (1-cm cells, 400 nm).

Notes: 1. For the preparation of the standard titanium solution from metal foil proceed as follows. Weigh 0·149 g into a 150-ml beaker, add 25 ml of 20 N sulphuric acid and 50 ml of water and gently boil on a hot plate until all the titanium has passed into solution, this usually takes 3 to 4 hours. Now add dilute hydrogen peroxide solution drop by drop until the violet colour of the solution just disappears and a very slight yellow colour is apparent, but avoiding an excess. Add a piece of platinum (a crucible lid will do) and gently boil the solution to decompose the titanium–peroxide complex, giving a completely colourless solution. Allow to cool, transfer to a 500-ml volumetric flask, add a further 25 ml of 20 N sulphuric acid and dilute to volume with water. Mix well. This solution contains 0·5 mg TiO₂ per ml.

2. A 0·5-g sample is adequate for most silicate rocks. If material is limited a much smaller portion can be taken and the quantities of reagents and final volumes reduced accordingly.

3. The volume of nitric acid added should be increased for those samples containing much ferrous iron or organic matter.

4. If the rock material contains more than a trace of barium, a white residue will be obtained on dissolution of the melt. Once the solution has been diluted to volume, this precipitate can be allowed to settle and may then be ignored.

METHOD USING DIANTIPYRYLMETHANE

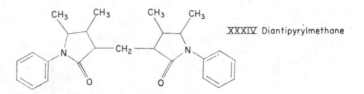

XXXIV Diantipyrylmethane

The reaction of the titanium ion Ti^{4+} with diantipyrylmethane (XXXIV) to give an intensely yellow-coloured solution was originally described by Minin[6] for the photometric determination of titanium in the presence of iron, vanadium, fluorides and phosphates. Polyak[5] made a more extensive study of the use of this reagent, whilst Jeffery and Gregory[7] used it for the analysis of ores, rocks and minerals.

Solutions of diantipyrylmethane deteriorate slowly, particularly when exposed to direct sunlight. This deterioration is evidenced by gradual yellowing of the solution, and can be considerably reduced by adding ascorbic acid and by storing in the dark. The colour given with titanium forms in dilute hydrochloric or sulphuric acid solution, some reduction in intensity being noted at acid concentrations greater than 4 N hydrochloric. Perchloric acid precipitates the reagent. The colour develops rapidly, reaching a maximum after 3 hours (Fig. 95), and is stable for several months. The maximum absorption occurs at a wavelength of 380 nm (Fig. 96), and the Beer–Lambert Law is valid up to 400μ g 100 ml (Fig. 97).

Very few metallic ions interfere with the reaction. Niobium and tantalum precipitate with the reagent giving low recoveries of titanium, but this can be avoided by adding tartaric acid. Iron and vanadium do not interfere in their lower valency states. Slight interference is encountered

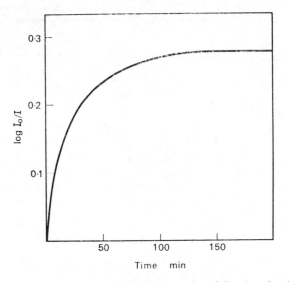

FIG. 95. The formation of the titanium complex of diantipyrylmethane
(4-cm cells, 40 μg Ti/100 ml).

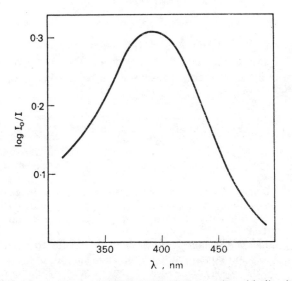

FIG. 96. Absorption spectrum of the titanium complex with diantipyrylmethane (4-cm cells, 40 μg Ti/100 ml).

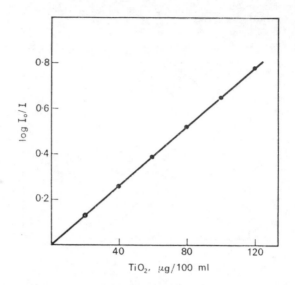

FIG. 97. Calibration graph for titanium using diantipyrylmethane (4-cm cells, 380 nm).

from uranium, molybdenum and tin when present in the solution, although 3–4 mg of each can be tolerated in 100 ml. There is no interference from other elements present in normal silicate or carbonate rocks.

Method

Reagents: *Ascorbic acid solution*, dissolve 10 g of the reagent in 100 ml of water.

    Diantipyrylmethane solution, dissolve 5 g of the reagent and 5 g of ascorbic acid in 150 ml of 2 N sulphuric acid and dilute to 500 ml with water. Transfer to a dark glass bottle and store in the dark.

    Standard titanium working solution, pipette 10 ml of the stock solution containing 0.5 mg TiO_2 per ml into a 500-ml volumetric flask, add 83 ml of concentrated hydrochloric acid and dilute to volume with water. This solution contains 10 μg TiO_2 per ml.

Procedure. For granitic and other acidic rocks containing little titanium a sample weight of $0 \cdot 25$ g should be taken, for most other silicate rocks a $0 \cdot 1$ g portion will be sufficient.

Accurately weigh approximately $0 \cdot 1$ g of the finely powdered silicate rock into a 10-ml platinum crucible and evaporate to fumes of sulphuric acid with $0 \cdot 5$ ml of concentrated nitric acid, 1 ml of 20 N sulphuric acid and 4 ml of hydrofluoric acid. Allow to cool, rinse down the sides of the crucible and again evaporate, this time to dryness. Fuse the residue with a small quantity of potassium pyrosulphate—not more than 2 g should be required—to give a completely clear melt. Dissolve this melt in 2 N hydrochloric acid, transfer to a 100-ml volumetric flask and dilute to volume with 2 N hydrochloric acid.

Pipette an aliquot of this solution containing not more than 500 μg TiO_2 to a 100-ml volumetric flask and add sufficient 2 N hydrochloric acid to bring the total volume of this acid up to 50 ml. Add 5 ml of ascorbic acid solution, mix by gently swirling the flask, and allow to stand for 30 minutes. Add 25 ml of the diantipyrylmethane solution, dilute to volume with water, mix well and allow to stand for 3 hours or overnight. Measure the optical density relative to a reference solution containing the same quantity of rock solution, 50 ml of 2 N hydrochloric acid, 5 ml of ascorbic acid and diluted to volume in a 100-ml volumetric flask without the addition of diantipyrylmethane reagent. The spectrophotometer should be set at a wavelength of 380 nm, and either 1-cm or 4-cm cells used.

Calibration. Prepare the calibration curve by taking aliquots of the standard titanium-solution containing 10 μg/ml TiO_2 and diluting to volume with hydrochloric acid, ascorbic acid and diantipyrylmethane solution as described above. Aliquots of 2–8 ml containing 20–80 μg TiO_2 can be used for the calibration of the 4-cm cells, and 10–50 ml containing 100–500 μg TiO_2 for the 1-cm cells.

An optical density of $1 \cdot 000$ was obtained for a solution containing 153 μg TiO_2, in 100 ml, measured in 4-cm cells.

METHOD USING TIRON

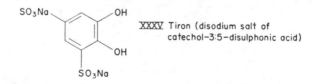

XXXV Tiron (disodium salt of catechol-3:5-disulphonic acid)

Tiron, the trivial name for the disodium salt of 1,2-dihydroxybenzene-3,5-disulphonic acid (XXXV), was used by Yoe and Jones[8] for the determination of iron, and by Yoe and Armstrong[9] for the simultaneous determination of iron and titanium. Rigg and Wagenbauer[10] extended the study of this reagent, and used it for the determination of titanium in silicate rocks.

The complex formed with titanium is yellow in colour and has a maximum absorption at a wavelength of 380 nm. The Beer–Lambert Law is obeyed for concentrations of titanium up to at least 4 mg per litre. The optical density of coloured solutions is, however, dependent upon the length of time they have been standing, and also upon the pH of the solution. Close control of the conditions used for the determination is essential.

Interference from iron can be avoided by reduction to a lower valency state with thioglycollic acid. Sodium dithionite has been used but it is not recommended because of its tendency to deposit sulphur at low pH, giving rise to turbid solutions. No interference is encountered from other elements likely to be present in silicate rocks, with the possible exception of vanadium, chromium, copper, tungsten, molybdenum and uranium when these are present in unusually large amounts as in certain rare rocks and minerals. The tolerance limits are given in Table 44.

TABLE 44. TOLERANCE LIMITS
FOR 0·01 PER CENT ERROR IN
ANALYSIS OF ROCK CONTAINING
0·50 PER CENT TiO_2[10]

Constituent	Per cent in rock
Cr_2O_3	0·84
V_2O_3	0·13
CuO	0·06
WO_3	0·19
MoO_3	0·01
U_3O_8	>0·6

Method

Reagents: *Tiron solution*, dissolve 5 g of the reagent in water and dilute to 100 ml with water. Discard at the first sign of yellow discoloration.
Thioglycollic acid solution, dilute 2 ml to 100 ml with water.
Acetic acid, N.

Buffer solution, dissolve 136 g of sodium acetate trihydrate in 1 litre of water and add 390 ml of glacial acetic acid.

Procedure. The procedure as described below is applicable to those rocks containing up to $1 \cdot 0$ per cent TiO_2. For rocks containing more titanium than this, the method should be followed as described but a smaller aliquot of the sample solution taken for colour development.

Accurately weigh approximately $0 \cdot 1$ g of the finely powdered sample material into a 10-ml platinum crucible and decompose by evaporation with hydrofluoric, sulphuric and nitric acids as described in the previous section. Remove the excess sulphuric acid by evaporation to dryness on a hot plate, and fuse the residue with a small quantity of potassium pyrosulphate. Allow the melt to cool and extract with 50 ml of N acetic acid. Any residue remaining at this stage is probably barium sulphate which may be collected on a small close-textured filter, washed with a little water and discarded. No unattacked mineral grains should be present. Combine the filtrate and washings and dilute to volume with water in a 100-ml volumetric flask. The acetic acid concentration should be $0 \cdot 5$ N.

Transfer 10 ml of this rock solution to a 50-ml volumetric flask and add in the following order: 25 ml of buffer solution, 5 ml of the aqueous tiron solution and 2 ml of the thioglycollic acid solution. At this stage the pH of the solution should be $3 \cdot 8$. Mix the contents of the flask by shaking, dilute with water to volume and again mix well. Allow the flasks to stand for at least 1 hour and preferably overnight, before measuring the optical densities with the spectrophotometer set at a wavelength of 380 nm.

Calibration. Calibration factors should not be used for this determination as the optical density measurements will vary with the length of time taken for colour development. The standard titanium working solution required for the calibration can be prepared from the stock solution by adding 30 ml of glacial acetic acid to a 10-ml aliquot and diluting to 500 ml with water in a volumetric flask. This solution contains 10 μg TiO_2 in N acetic acid. To prepare a calibration graph transfer 2–10 ml aliquots of this solution containing 20–100 μg TiO_2, to separate 50-ml volumetric flasks and continue as described above for the sample solution. At least one point on the calibration should be checked at the same time as the sample solution is prepared and measured.

The Determination of Titanium by Atomic
Absorption Spectroscopy

The ease with which titanium absorption can be measured is deceptive. Several authors have recorded interference by other elements in the determination of titanium by this technique, and great care must be taken if accurate titanium figures are to be obtained. In particular, the enhancement due to the presence of iron, aluminium, manganese, calcium, sodium and potassium in the rock solution must be taken into account. In one such procedure Van Loon and Parissis[11] add aluminium to the rock solution in 4 N hydrochloric acid to bring the total aluminium level to 750–1000 ppm. This addition was shown to stabilise the titanium absorption and to mask the interference from the other constituents. The absorption is measured at a wavelength of 364 nm, and a nitrous oxide–acetylene burner used.

References

1. CLARKE F. W. and WASHINGTON H. S., *U.S. Geol. Surv. Prof. Paper* 124, 1924.
2. HEVESY G. V., ALEXANDER E. and WURSTLIN K., *Zeit. Anorg. Chem.* (1930) (194), p. 316.
3. TATLOCK D. B., *Geochim. Cosmochim. Acta* (1966) 30, 123.
4. GUPPY E. M., *Chemical Analyses of Igneous Rocks, Metamorphic Rocks and Minerals 1931–54*, Mem. Geol. Surv. Gt. Brit. 1956.
5. POLYAK L. YA., *Zhur. Anal. Khim.* (1962) 17, 206.
6. MININ A. A., *Uch. Zap. Permsk. Univ.* (1955) 9, 177.
7. JEFFERY P. G. and GREGORY G. R. E. C., *Analyst* (1965) 90, 177.
8. YOE J. H. and JONES A. L., *Ind. Eng. Chem., Anal. Ed.* (1944) 16, 111.
9. YOE J. H. and ARMSTRONG A. R., *Analyt. Chem.* (1947) 19, 100.
10. RIGG T. and WAGENBAUER H. A., *Analyt. Chem.* (1961) 33, 1347.
11. VAN LOON J. C. and PARISSIS C., *Anal. Lett.* (1968), 1, 249.

CHAPTER 47

URANIUM

Occurrence

The geochemistry of uranium has been reviewed, together with that of thorium, by Adams *et al.*[1] and by Turovskii[2] and Whitfield *et al.*[3] Table 45 summarises the abundance data from these and other sources.

TABLE 45. THE URANIUM CONTENT
OF SOME ROCK TYPES

Rock type	U, ppm
Granitic rocks	4
Intermediate rocks	3
Basic rocks	1
Ultrabasic rocks	0·01
Shales	4
Sandstones and quartzites	0·4
Carbonate rocks	2

As with certain other elements, the concentration in a particular rock type shows evidence of regional variation. Thus, for example, Ingerson[4] has reported granites from the southern Californian batholith to contain 3–6 ppm uranium, whereas those from the White Mountain series in New Hampshire contains 12–14 ppm.

The concentration of uranium in marine shales containing sulphides and carbonaceous material has been noted by a number of workers including Swanson,[5] who recorded that normal marine shales contain 1–4 ppm uranium, mostly in such minerals as zircon and sphene, whereas the black shales with 2 per cent or more of organic carbon may contain as much as 250 ppm, associated partly with phosphatic nodules and partly with organic material.

The average crustal ratio of thorium to uranium is about 3·5, and with small variation this ratio holds for the average values of most rock types. Individual values, however, vary considerably. Both elements tend to increase with increasing silica content of the rock. Some thorium and uranium are contained in the silicate minerals, but most appears in the accessory minerals, particularly zircon, thorite, apatite, monazite, sphene and xenotime.

The recent considerable interest in uranium has led to an extensive investigation of the mineralogy of uranium deposits, and to the characterisation of a large number of new minerals. Primary uranium minerals include pitchblende, coffinite, uraninite and allanite. In the oxidised zone U(IV) is converted to U(VI), giving rise to a large number of minerals containing the uranyl group UO_2^{2+}, including complex sulphates, carbonates, arsenates, vanadates and phosphates.

The Determination of Uranium

One of the most sensitive of methods in general use is based upon measurement of the fluorescence produced by uranium compounds in fused media.[6] The fluorescence intensity depends upon the medium used, as well as on the conditions used for the fusion. A number of elements have enhancing or quenching effects and for this reason prior separation of the uranium is necessary before the method can be applied to silicate rocks. Polarographic methods have also been widely applied for the determination of uranium, but these are generally less sensitive than recent photometric methods.

PHOTOMETRIC METHODS

There is no lack of general photometric reagents for uranium. Not all of these are sufficiently sensitive for application to the analysis of silicate rocks, although some of them were formerly used for this purpose. These include thiocyanate,[7] which gives a yellow colour with uranium, and peroxide which gives a somewhat similar yellow colour in alkaline solution.[8]

More sensitive reagents that have been suggested for the determination of uranium include dibenzoylmethane,[9, 10] arsenazo I,[11] 1-(2-pyridylazo)-2-naphthol (PAN)[12] and 4-(2-pyridylazo)-resorcinol

(PAR).[13] The molar absorptivities of the uranium complexes with these reagents range up to about 35,000 but interference occurs from the presence of many other elements, making prior separation essential.

Arsenazo III[14, 15] has been used without extensive separation from other elements, but interference does occur and a separation procedure is recommended. The molar absorptivity of the uranium complex has a value of about 100,000. The absorption spectrum (Fig. 98) shows maximum absorption at a wavelength of 662·5 nm but this is a very

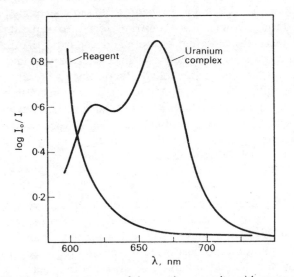

FIG. 98. Absorption spectrum of the uranium complex with arsenazo III.

sharp peak and when a spectrophotometer is being used, care must be taken to ensure that the optical densities are measured at this peak value. It is advisable to prepare a calibration solution and measure this at the same time as the sample solution, as slight variations in the slope of the calibration graph have been found to occur from day to day. The reagent itself has considerable absorption at wavelengths below 600 nm, and for this reason all measurements of optical density are made against a "blank solution" containing the same quantity of arsenazo III reagent. The complex formation is between the reagent and uranium (IV).

SEPARATION OF URANIUM

Uranyl nitrate is appreciably soluble in a number of organic solvents, and solvent extraction procedures have often been used to recover uranium from aqueous solutions. The efficiency of the separation depends very much upon the nature and amount of other constituents present in the solution, and small quantities of other metals are usually extracted with the uranium into the organic phase. A number of organic solvents can be used, but diethyl ether is the one most frequently reported.

Ion exchange methods have been applied to the recovery of uranium from solutions. Korkisch and Arrhenius[16] used a strongly basic anion exchange resin (Dowex 1 × 8, nitrate form) to separate uranium, thorium, rare earths, cadmium, bismuth and lead from all other elements present in a deep-sea sediment, and in a later paper Hazan et al.[17] separated uranium and lead from all other elements present in a silicate rock by absorption from a hydrochloric acid solution containing methyl glycol. Ion exchange procedures have not, however, been widely applied for the determination of uranium in silicate rocks, although the separation of microgram quantities of uranyl ions by sorption on a column of silica gel was reported by Sulcek and Sixta.[18] Separation on a column of activated charcoal was used by Malyasova[19] to determine uranium in rocks and minerals.

One of the procedures commonly used for the recovery of uranium from rocks, ores and minerals involves elution with an ether–nitric acid mixture from a column packed with cellulose and alumina.[20-22] This method of separation has been widely used and reported; it is given in detail below for the determination of uranium in silicate rocks. Small amounts of both thorium and zirconium tend to bleed through the cellulose–alumina column and give rise to high values for uranium where the determination is completed photometrically with arsenazo III. This error can be avoided by dividing the solution into two parts in only one of which is the unreactive uranium(VI) reduced to the reactive uranium(IV). After the addition of arsenazo III reagent, the unreduced solution is used as the reagent blank for the reduced sample solution.

The Determination in Silicate Rocks

In the procedure described here, silica is removed by evaporation with concentrated hydrofluoric and nitric acids, leaving a nitrate residue for the separation of uranium on a column packed with cellulose and alumina. After elution of the uranium with an ether–nitric acid mixture, the solvent is removed and the solution diluted to volume. Metallic zinc is added to an aliquot of the solution to reduce uranium(VI) to uranium(IV), and the optical density of the uranium complex with arsenazo III measured relative to a second, unreduced aliquot of the solution.

Method

Reagents: *Diethyl ether.*
Ferric nitrate.
Ascorbic acid.
Cellulose powder (Whatman Column Chromedia CF 11 is suitable).
Activated alumina, 100–200 mesh.
Tartaric acid.
Zinc pellets, approximately 0·5 g each.
Arsenazo III solution, grind together 50 mg of the solid reagent with 1 ml of water and 2 drops of 2·5 M sodium hydroxide solution. Transfer to a 100-ml volumetric flask, add 50 ml of concentrated hydrochloric acid and dilute to volume with water. Prepare freshly each day.
Standard uranium stock solution, dissolve 0·118 g of uranium oxide in a few ml of concentrated nitric acid and evaporate just to dryness. Moisten with water, add hydrochloric acid and again evaporate to dryness. Dissolve the chloride residue in water and dilute to 1 litre. This solution contains 100 μg uranium per ml.
Standard uranium working solution, dilute 5 ml of the stock solution with water to 250 ml in a volumetric flask. This solution contains 2 μg uranium per ml.

Apparatus. A glass column and plunger (Fig. 99) is required for each sample portion. The tube is made from borosilicate glass tubing of 2 cm internal diameter.

Procedure. Accurately weigh approximately 2·5 g of the finely powdered silicate rock material into a platinum dish, moisten with water and add 20 ml of concentrated hydrofluoric acid and 20 ml of concentrated nitric acid. Transfer the dish to a steam bath and evaporate to complete dryness. Allow to cool, add 5 ml of concentrated hydrofluoric acid and 5 ml of concentrated nitric acid and again evaporate to dryness. Repeat the evaporation to dryness

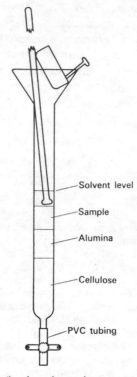

Fig. 99. Apparatus for the column chromatographic separation of
uranium (Crown copyright).

with small volumes of concentrated nitric acid to decompose any complex
fluorides and to remove all but trace amounts of hydrofluoric acid. Allow to
cool, add 10 ml of 4 N nitric acid and warm to dissolve all soluble matter
(note 1). Add 3 g of tartaric acid and 3 g of ferric nitrate to the warm solution
and stir until dissolution is complete. Add sufficient cellulose powder to give
a friable mixture, stir with a glass rod and set aside.

   Column separation. Set up the glass column as shown in Fig. 99. Prepare
also a quantity of the ether-nitric acid mixture by carefully stirring into diethyl
ether a sufficient quantity of fresh concentrated nitric acid to give an ether–
nitric acid ratio of 97:3. A total volume of about 300 ml is required for each
column used.

   Fill the glass tube with solvent mixture, add cellulose and gently work the
plunger up and down to give an uniform air-free column. When the cellulose
has settled, press the plug down with the plunger and add more cellulose.

Continue in this manner until a total length of about 8 cm of compacted cellulose is obtained. Allow the solvent to drain, at the same time adding sufficient alumina to give a 2 cm band at the top of the column. Do not allow the solvent level to fall below the level of the top of the alumina. Add further solvent to the glass tube and transfer the cellulose mixture containing the absorbed sample to the top of the column, again using the plunger to remove all trapped air and to compact the cellulose. Rinse the platinum dish thoroughly with the solvent solution and add the washings to the column. Remove the PVC tubing from the bottom of the column and collect the eluate in a 500-ml conical flask containing 50 ml of water. Use 250 ml of the ether–nitric acid mixture for the elution, and make sure that the solvent level is always above the level of the cellulose in the column.

When the elution is complete, remove the conical flask, place it on an asbestos plate across a steam bath and slowly evaporate the ether (note 2). When all the ether has evaporated transfer the aqueous solution to a 150-ml beaker and evaporate to dryness on a hot plate. Allow to cool, and add 5 ml of 4 N nitric acid and 2 ml of concentrated perchloric acid. Evaporate to dryness. Add 2 ml of perchloric acid and again evaporate to dryness.

Photometric determination. Add a few millilitres of 6 N hydrochloric acid and warm to ensure complete dissolution. Cool, transfer the solution to a 50-ml volumetric flask with 6 N hydrochloric acid, dilute to volume with acid and mix well. From this solution take two 20-ml aliquots, transfer one to a 150-ml beaker and the other to a dry 25-ml volumetric flask. To the solution in the volumetric flask add 5 ml of arsenazo III solution to bring the volume up to the mark, mix well and set aside.

Evaporate the solution in the beaker to dryness on a hot plate. Add 3 ml of 6 N hydrochloric acid and warm to ensure complete dissolution. Transfer the solution to a test tube and dilute with acid to give a total volume of about 10 ml; add 50 mg of ascorbic acid and 3 pellets of metallic zinc (about 1·5 g) and allow the reduction to proceed for 15 minutes. Pour the solution into a 25-ml volumetric flask containing exactly 5 ml of the arsenazo III solution, taking care to avoid transferring any metallic zinc. Rinse the test-tube, add the washings to the flask and dilute to volume with 6 N hydrochloric acid. Cool to room temperature, adjust the volume to the mark by adding 6 N hydrochloric acid and mix well.

Measure the optical density in 4-cm cells with the spectrophotometer set at a wavelength of 662·5 nm, using the unreduced solution as the reference. Measure also the optical density of a standard solution containing 10 μg of uranium, prepared as described below for the calibration. All measurements should be made within 30 minutes of preparation.

Calibration. Transfer aliquots of 0–5 ml of the standard solution containing 0–10 μg of uranium to 30-ml test tubes and dilute with water to a total of 5 ml as necessary. Add to each tube 5 ml of concentrated hydrochloric acid, about 50 mg of ascorbic acid and 3 pellets of zinc. Allow 15 minutes for the reduction, transfer each solution to a 25-ml volumetric flask containing 5 ml of arsenazo III solution, and dilute to volume with 6 N hydrochloric acid as described above. Measure the optical density of the solutions containing uranium in 4-cm cells with the spectrophotometer set at a wavelength of 662·5 nm, and using the solution containing no uranium as the reference. Plot the relation of optical density to uranium concentration (Fig. 100).

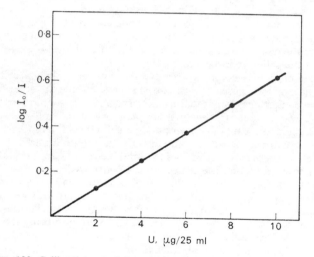

FIG. 100. Calibration graph for uranium using arsenazo III (4-cm cells, 662·5 nm).

Notes: 1. If an appreciable amount of the sample remains unattacked, evaporate the solution again to dryness. Add to the dry residue approximately five times its bulk of sodium peroxide, mix and place the dish in an electric furnace set at a temperature of 480° for 20 minutes. Remove the dish and allow to cool. Place the dish in a 400-ml beaker and add about 25 ml of water followed by concentrated nitric acid until the solution is just acid. Stand the beaker on a hot plate for 10 minutes, then rinse and remove the platinum dish. Evaporate the solution to dryness on a steam bath, add 10 ml of 4 N nitric acid, cover the beaker with a clock glass and return to the steam bath until a clear solution is obtained. Allow to cool, and then continue as described above.

2. A vigorous oxidation reaction may occur if insufficient water has been added.

References

1. ADAMS J. A. S., OSMOND J. K. and ROGERS J. J. W., *Phys. Chem. Earth* (1959) 3, 298.
2. TUROVSKII S. D., *Geokhimiya* (1957) (2), 199.

3. WHITFIELD J. M., ROGERS J. J. W. and ADAMS J. A. S., *Geochim. Cosmochim. Acta* (1959) **17**, 248.
4. INGERSON E., *Geochim. Cosmochim. Acta* (1954) **5**, 27.
5. SWANSON V. E., *U.S. Geol. Surv. Prof. Paper* 356-C, 1961.
6. STRASHEIM A., *Spectrochim. Acta* (1950) **4**, 200.
7. NEITZEL O. A. and DESESA M. A., *Analyt. Chem.* (1957) **29**, 756.
8. HARVEY C. O., *Bull. Geol. Surv. Gt. Brit.* (1951) (3), 43.
9. YOE J. H., WILL F. and BLACK R. A., *Analyt. Chem.* (1953) **25**, 1200.
10. UMEZAKI Y., *Bull. Chem. Soc. Japan* (1963) **36**, 769.
11. HOLCOMB H. P. and YOE J. H., *Analyt. Chem.* (1960) **32**, 616.
12. CHENG K. L., *Analyt. Chem.* (1958) **30**, 1027.
13. FLORENCE T. M. and FARRAR Y., *Analyt. Chem.* (1963) **35**, 1613.
14. SINGER E. and MATUCHA M., *Zeit. Anal. Chem.* (1962) **191**, 248.
15. NEMODRUK A. A. and GLUKHOVA L. P., *Zhur. Anal. Khim.* (1966) **21**, 688.
16. KORKISCH J. and ARRHENIUS G., *Analyt. Chem.* (1964) **36**, 850.
17. HAZAN I., KORKISCH J. and ARRHENIUS G., *Zeit. Anal. Chem.* (1965) **213**, 182.
18. ŠULCEK Z. and SIXTA U., *Anal. Chim. Acta* (1971) **53**, 335.
19. MALYASOVA Z. V., *Zhur. Anal. Khim.* (1972) **27**, 1275.
20. *The Determination of Uranium and Thorium*, Nat. Chem. Lab., D.S.I.R., H.M.S.O., 1963 (2nd ed.).
21. BURSTALL F. H. and WELLS R. A., *Analyst.* (1951) **76**, 396.
22. ADAMS J. A. S. and MAECK W. J., *Analyt. Chem.* (1954) **26**, 1635.

CHAPTER 48

VANADIUM

Occurrence

The ubiquitous distribution of vanadium in igneous and sedimentary rocks has been noted by many workers and summarised by Bertrand.[1] As early as 1930 Pondal[2] had reported that the vanadium content of basic silicate rocks was sometimes considerable, in contrast to that of acidic rocks which contained very little. Later analyses confirmed this pattern, as in a more recent summary by Taylor.[3] Some typical values from Turekian and Wedepohl,[4] Wager and Mitchell[5] and other sources are given in Table 46. The figures given for each group of silicate rocks should be considered only as rough average values, as the distribution of vanadium within each class can be considerable, Ahrens[6] for example, has reported values of between 5·5 and 630 ppm for a series of twenty-seven specimens of granite from the Ontario District of Canada, although most of these were in the range 5·5–50 ppm. The concentration of vanadium in basic rocks occurs largely in the iron ore fraction and is particularly associated with titaniferous magnetite. Certain

TABLE 46. THE VANADIUM
CONTENT OF SOME SILICATE
ROCKS

Rock type	V, ppm
Ultrabasic	130
Basic (basaltic)	250
Intermediate	130
Granitic	20
Shale	130
Sandstone	20
Carbonate	20

484

magnetite occurrences have been reported with nearly 2 per cent vanadium, forming an important commercial source of this element, although more commonly values of $0 \cdot 1 - 0 \cdot 4$ per cent vanadium are obtained.

Goldschmidt[7] has noted that the concentration of vanadium in ferric minerals is made possible by the similarity in ionic radius, V^{3+}, $R = 65$ pm; Fe^{3+}, $R = 67$ pm. The relative scarcity of ferric iron at the early stages of magmatic evolution may therefore account for the relative paucity of vanadium in these rocks.

Certain bituminous and carbonaceous shales of marine origin contain abnormally large amounts of vanadium, for which an organic origin is indicated. Sandstones and limestones contain very little, although varieties are known where the porous nature of these rocks has permitted the accumulation of vanadium-containing bitumen or hydrocarbon.

The Determination of Vanadium in Silicate Rocks

Volumetric methods were at one time used for rocks rich in vanadium. These methods have now been replaced by spectrographic and spectrophotometric procedures which are applicable not only to those silicates rich in vanadium, but also to granites and other rocks containing only a few ppm.

One of the simplest ways of determining vanadium is to measure the optical density of alkali vanadate solutions in the ultraviolet region of the spectrum. Many elements interfere and it is necessary to make a prior separation, as, for example, by extracting the complex of vanadium with 8-hydroxyquinoline into chloroform solution, in combination with a cation exchange separation as described by Yoshimura and Murakami.[8]

There is no shortage of colour-forming reagents for vanadium; these have given rise to a multiplicity of photometric methods in a wide variety of matrices. Three methods that are applicable to silicate rocks and minerals are given in detail below. These are based respectively upon the yellow colour given by vanadium with tungstate in phosphoric acid solution, the purple colour of the complex formed with N-benzoyl-o-tolylhydroxylamine in 6 N hydrochloric acid solution, and the brown product formed by vanadate oxidation of diaminobenzidine.

PHOSPHOTUNGSTATE METHOD

Vanadium forms a yellow-coloured complex with alkali tungstate in solutions containing phosphoric acid. The yellow solution has a maximum optical density in the ultraviolet region of the spectrum, but measurements can conveniently be made at 400 nm within the range of even the simplest spectrophotometer. The Beer–Lambert Law is valid over the concentration range normally encountered in silicate rocks (Fig. 101). The reaction is not sufficiently sensitive for direct application to silicate rocks and it is necessary to use a procedure for concentrating the vanadium. In the method described below, adapted from Bennett and Pickup,[9] a solvent extraction stage is incorporated removing vanadium from dilute acetic acid solution as its complex with 8-hydroxyquinoline.

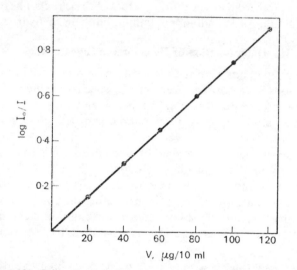

FIG. 101. Calibration graph for vanadium using a phosphotungstate method (2-cm cells, 400 nm).

Method

Reagents: *8-Hydroxyquinoline solution*, dissolve 1 g of the reagent in 100 ml of 2 N acetic acid.
Chloroform.
Sulphuric acid, 6 N.
Phosphoric acid, 5 N.
Sodium tungstate solution, dissolve 8·25 g of the dihydrate in 50 ml of water.

Standard vanadium stock solution, dissolve 0·2296 g of dried ammonium metavanadate in water and dilute to 500 ml with water. This solution contains 200 μg V/ml.

Standard vanadium working solution, dilute the stock solution with water to give working standard containing 10μg V/ml.

Procedure. Accurately weigh approximately 0·5 g of the finely ground rock material into a platinum crucible and add 3 g of anhydrous sodium carbonate and 0·1 g of potassium nitrate (note 1). Fuse the mixture over a Bunsen burner for 30 minutes, or longer if refractory minerals are present, and allow to cool. Extract the melt with hot water, filter using a medium or close-textured paper, and wash the residue well with hot 2 per cent (w/v) sodium carbonate solution. Discard the residue and combine the filtrate and washings.

Transfer this solution, or a suitable aliquot of it containing not more than 60 μg vanadium, to a 100-ml separating funnel, add 2 drops of methyl orange indicator solution and titrate with 6 N sulphuric acid until the equivalence point is reached. Swirl the solution to remove as much as possible of the liberated carbon dioxide, and add 1 ml of 8-hydroxyquinoline solution and 3 ml of chloroform. Shake the solution for 1 minute to extract the dark-coloured vanadium complex, allow the phases to separate and remove the chloroform layer. Rinse the funnel with a little chloroform. Add a further 0·5 ml of 8-hydroxyquinoline solution and 3 ml of chloroform and again extract by shaking for 1 minute. If the extract shows an appreciable dark colour, repeat the extraction for a third time. Discard the aqueous solution.

Collect all the chloroform extracts in a small platinum crucible, add 0·1 g of sodium carbonate and allow the chloroform to evaporate. Burn off the organic matter and fuse the residue to convert all vanadium to sodium vanadate. Dissolve the melt by warming with 2–3 ml of water, transfer to a 10-ml volumetric flask, add 1 ml of 6 N sulphuric acid, 1 ml of 5 N phosphoric acid and 0·5 ml of sodium tungstate solution. Heat the solution just to boiling, cool and dilute to the mark with water. Measure the optical density of the solution relative to water in 2-cm cells using the spectrophotometer set at a wavelength of 400 nm. Measure also the optical density of a reagent blank solution, prepared in the same way as the sample solution but omitting the rock powder.

Calibration. Transfer aliquots of 1–6 ml of the standard vanadium working solution containing 10–60 μg V, to separate 10-ml volumetric flasks and add to each sulphuric acid, phosphoric acid and sodium tungstate solution as described for the sample solution above. Plot the relation of optical density to vanadium concentration (Fig. 101).

Note: 1. The exclusion of potassium nitrate leads to low recoveries of vanadium, particularly from rocks rich in ferrous iron. The prolonged heating of rock samples with a flux containing excessive potassium nitrate or other oxidising agent should, however, be avoided as this will result in an appreciable attack of the platinum crucible.

METHOD USING *N*-BENZOYL-*o*-TOLYLHYDROXYLAMINE

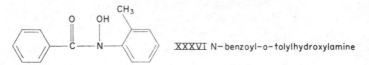

XXXVI N-benzoyl-o-tolylhydroxylamine

A number of hydroxylamine derivatives have been suggested as reagents for vanadium, but very few combine a sufficient sensitivity with the necessary specificity to enable vanadium to be determined in silicate rocks without a prior separation from interfering elements. One reagent that has been used for this is *N*-benzoyl-*o*-tolylhydroxylamine (XXXVI) which reacts with vanadium in strongly acid solution to give a purple-coloured complex,[10] readily soluble in organic solvents. The molar absorptivity is 5250, and the Beer–Lambert Law is valid in the range 0–60 μg vanadium in 10 ml of extract (Fig. 102). The maximum optical density occurs at a wavelength of 510 nm (Fig. 103). The reaction occurs only with vanadium(V), and somewhat lower calibration curves

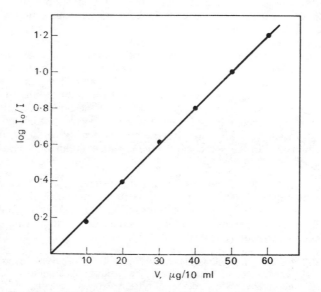

Fig. 102. Calibration graph for vanadium using *N*-benzoyl-*o*-tolyl-hydroxylamine (2-cm cells, 510 nm).

are obtained unless the aliquots of the standard vanadium solution are reoxidised prior to colour formation. For this oxidation an aqueous potassium permanganate solution is used, added drop by drop until an excess is present. Any chlorine liberated on adding hydrochloric acid is removed with sulphamic acid.

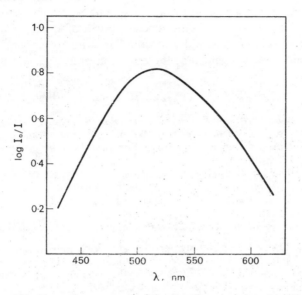

FIG. 103. Absorption spectrum of the vanadium complex with N-benzoyl-o-tolylhydroxylamine (2-cm cells, 40 μg V/10 ml).

N-benzoyl-o-tolylhydroxylamine has been used for the determination of vanadium in silicate rocks and minerals by Jeffery and Kerr,[11] whose procedure is given in detail below. The most serious interference is from titanium which gives an intense yellow colour with the reagent. This colour can be completely suppressed by adding a small quantity of sodium fluoride.

This reagent also gives yellow colours with gold and platinum. Whilst the small amounts of gold and platinum that occur in silicate rocks are insufficient to interfere with the determination of vanadium, any platinum introduced from the platinum apparatus may completely obscure the violet colour from rocks containing only a few ppm of vanadium. For this reason prolonged fusion in platinum should be avoided and the alternative decomposition in PTFE and silica used— particularly for samples low in vanadium.

The complex formed between vanadium and *N*-benzoyl-*o*-tolylhydroxylamine is soluble in a number of organic solvents. Solutions in carbon tetrachloride, chloroform, isobutylmethyl ketone and toluene all give similar calibration curves. Chloroform should be avoided because of possible interference from the presence of traces of ethanol which give rise to a yellow-coloured complex of vanadium with the reagent. Carbon tetrachloride is recommended as the most convenient solvent to use.

Method

Reagents: *N-benzoyl-o-tolylhydroxylamine solution,* dissolve 0·02 g of the recrystallised reagent in 100 ml of carbon tetrachloride. (The preparation of this reagent is described by Majundar and Das.[10])

Potassium permanganate solution, approximately 0·02 M.

Sulphamic acid solution, approximately 0·05 M.

Sodium fluoride solution, saturated aqueous, store in a polythene bottle.

Standard vanadium stock solution, prepare as described above.

Procedure. Decompose a 100-mg portion of the finely powdered silicate rock or mineral (note 1) by evaporation in platinum or PTFE apparatus with sulphuric, nitric and hydrofluoric acids in the usual way, removing excess sulphuric acid by heating on a hot plate. Fuse the dry residue in platinum or preferably in silica with potassium pyrosulphate, and extract the fused melt with 10 ml of water containing 2 drops of 20 N sulphuric acid. Transfer the solution to a separating funnel and add potassium permanganate solution drop by drop until an excess is present, giving a pink colour that persists for 5 minutes.

The volume of the solution at this stage should be about 20 ml. Add 2 ml of sulphamic acid solution, 2 ml of sodium fluoride solution and 20 ml of concentrated hydrochloric acid. Now add by pipette 10 ml of the *N*-benzoyl-*o*-tolylhydroxylamine reagent solution, stopper and shake for 30 seconds. Allow the two layers to separate and filter the lower organic layer through a small wad of cotton wool into a 2-cm spectrophotometer cell. Measure the optical density relative to carbon tetrachloride at a wavelength of 510 nm. Determine also the optical density of a reagent blank solution similarly prepared but omitting the sample material.

Calibration. Transfer aliquots of 1–6 ml of the standard vanadium solution containing 10–60 μg vanadium to a series of separating funnels and dilute each with water to a volume of 20 ml. Add potassium permanganate solution drop by drop until a slight excess is present, then add sulphamic acid solution, sodium fluoride solution and hydrochloric acid, and continue as described above for the sample solution. Plot the relation of optical density to vanadium concentration (Fig. 103).

Note: 1. This sample weight is sufficient for rock samples containing 50–500 ppm V. For rocks containing less vanadium than this, use a larger sample weight, for rocks containing more, dilute the rock solution to volume and use an aliquot for the determination.

DIAMINOBENZIDINE METHOD

Cheng[12] has described a procedure for the determination of vanadium based upon the ability of vanadate solutions to oxidise diaminobenzidine. The product solutions are reddish-brown in colour, exhibiting absorption maxima at 340, 380 and 470 nm (Fig. 104). The oxidation is carried out in the presence of a fixed amount of phosphoric acid, when the Beer–Lambert Law is followed up to at least 10 ppm V. Although the coloured product is fairly stable, unnecessary delay in measuring the solutions should be avoided to minimise air-oxidation of the reagent.

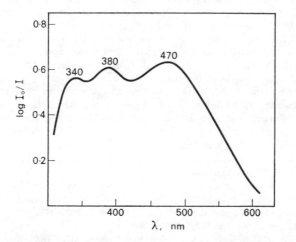

FIG. 104. Absorption spectrum of oxidized benzidine.

Chromium(VI) and other oxidising agents interfere, and the vanadate oxidation of the reagent is partially inhibited by the presence of sulphuric and hydrochloric acids.

Chan and Riley[13] have used this reaction as the basis of a method for the analysis of silicate and other materials. A cation exchange procedure is used to separate vanadium from large quantities of iron, and serves also to concentrate the vanadium in a small volume.

Method

Reagents: *Hydrogen peroxide*, 0·3 per cent solution, dilute 10 ml of 30 per cent w/v solution to 1 litre.

Perchloric acid, 0·05 N.

Diaminobenzidine solution, dissolve 0·1 g of the reagent in 25 ml of water as required.

Ion-exchange resin, Zeo Karb 225, in the form of a short column 8-cm length, 0·8-cm diameter, washed several times with warm 3 N hydrochloric acid and then thoroughly with water. Regenerate after use by washing with 300 ml of 3 N hydrochloric acid and 200 ml of water.

Procedure. Accurately weigh approximately 0·5 g of the finely powdered silicate rock material into a platinum crucible and decompose by evaporation almost to dryness with 4 ml of concentrated perchloric acid and 15 ml of concentrated hydrofluoric acid. Add 2 ml of concentrated perchloric acid and repeat the evaporation, this time to complete dryness. Add a further 1 ml of concentrated perchloric acid and 10 ml of water and warm the crucible until all soluble material has dissolved. Rinse the solution into a beaker and dilute with water to about 100 ml.

Transfer the solution to the top of the ion-exchange column, and allow to pass through at a rate of about 5–10 ml per minute. Rinse the resin with 50 ml of 0·1 N hydrochloric acid and discard both the eluate and the washings. Elute the vanadium from the column with 40 ml of 0·3 per cent hydrogen peroxide solution. Transfer the eluate to a platinum basin, add 2–3 drops of concentrated perchloric acid and evaporate to strong fumes of perchloric acid.

Using 0·05 N perchloric acid transfer the solution to a 10-ml volumetric flask, add 1 ml of concentrated phosphoric acid, mix well and allow to cool. Add 1 ml of diaminobenzidine solution, and dilute to volume with water. Mix the solution, allow to stand for 30 minutes, then measure the optical density in 4-cm cells with the spectrophotometer set at a wavelength of 470 nm. Measure also the optical density of a reagent blank solution prepared in the same way as the sample solution but omitting the rock material.

Calibration. Transfer aliquots of a standard vanadium solution containing up to 10 μg vanadium to separate 10-ml volumetric flasks, add 1 ml of concentrated phosphoric acid and 1 ml of diaminobenzidine solution to each, dilute to volume and measure the optical densities after 30 minutes as described above for the sample solution. Plot the relation of optical density to vanadium concentration.

References

1. BERTRAND D., *Bull. Amer. Nat. Hist. Museum* (1950) **94**, No. 7.
2. PONDAL P. I., *An. Soc. Espanola Fis. Quim.* (1930) **28**, 438.
3. TAYLOR S. R., *Phys. Chem. Earth* (1965) **6**, 133.
4. TUREKIAN K. K. and WEDEPOHL K. H., *Bull. Geol. Soc. Amer.* (1961) **72**, 175.
5. WAGER L. R. and MITCHELL R. L., *Rept. Int. Geol. Congr. XVIII*, 1948, Pt. II, 140.
6. AHRENS L. H., *Geochim. Cosmochim. Acta* (1954) **5**, 52.
7. GOLDSCHMIDT V. M., *Geochemistry*, Oxford, 1954, p. 489.
8. YOSHIMURA J. and MURAKAMI Y., *Bull. Chem. Soc. Japan* (1962) **35**, 1001.
9. BENNETT W. H. and PICKUP R., *Colon. Geol. Min. Res.* (1952) **3**, 171.
10. MAJUMDAR A. K. and DAS G., *Anal. Chim. Acta* (1964) **31**, 147.
11. JEFFERY P. G. and KERR G. O., *Analyst* (1967) **92**, 763.
12. CHENG K. L., *Talanta* (1961) **8**, 658.
13. CHAN K. M. and RILEY J. P., *Anal. Chim. Acta* (1966) **34**, 337.

CHAPTER 49

ZINC

Occurrence

The occurrence of zinc in silicate rocks was examined by Sandell and Goldich,[1] who found that the amounts of zinc present in rocks of the plutonic and volcanic series could be related to the sum of the magnesium and ferrous iron, and also to the manganese content. Rader[2] has also shown that the zinc content of basaltic rocks can be related to the total iron content.

In general, basic rocks contain more zinc than acidic rocks as shown in Table 47, based upon work by Wedepohl.[3] This has been confirmed by Vinogradov[4] and by Rosman,[5] although a somewhat higher abundance of about 60 ppm was indicated by both these authors for acidic rocks and lower average values of 30 and 39 ppm respectively given for ultrabasic rocks. Shales contain similar amounts of zinc to igneous silicate rocks, but sandstones contain appreciably less.

TABLE 47. THE ZINC CONTENT
OF SOME SILICATE ROCKS

Rock type	Zn, ppm
Gabbro	79
Diabase	137
Basalt	121
Syenite, trachyte, etc.	26
Diorite, andesite, etc.	40
Granite*	34
Grandiorite, etc.	46
Rhyolite, liparite	60

* A series of twelve European granites gave an average of 20 ppm Zn, fourteen granites from U.S.A. gave 58 ppm Zn, average of all results 41 ppm Zn.

Zinc appears to be concentrated in the ferromagnesian minerals—examples are known of amphiboles and pyroxenes containing 600 ppm and more of zinc, probably substituting for iron and magnesium. Zinc silicate minerals are known from igneous rocks, but these are very rare. The silicates willemite and hemimorphite are secondary minerals, formed from the sulphide, sphalerite ZnS (known also as blende, zinc blende, and black jack). Sphalerite is the commonest of zinc minerals, often occurring with galena PbS and pyrite FeS_2 in both silicate and carbonate rocks. It is the main ore of zinc, although there is a small production of the carbonate, now known as smithsonite.

The Determination of Zinc in Silicate Rocks

Dithizone extraction procedures have been used[6] for the determination of zinc in silicate rocks, but Carmichael and McDonald[7] have shown that such procedures are subject to interference from other metals, particularly copper, cobalt and nickel. This interference leads to high results—in some analyses twice the accepted values were obtained. This conclusion is in agreement with the work of Greenland,[8] who particularly implicated nickel as the cause of high zinc values when a dithizone extraction was followed by photometric determination with dithizone.

A number of spectrophotometric procedures for small amounts of zinc were investigated by Margerum and Santacana,[9] who preferred a single colour dithizone procedure with bis(2-hydroxyethyl)dithiocarbamate as a masking agent. The use of zincon was not recommended when certain impurities were present. When zinc was isolated by the use of a column of anion exchange resin, either dithizone or zincon could be used. Such procedures have been described by Rader et al.[2] and Huffman et al.[10]

ION-EXCHANGE—ZINCON METHOD

Zincon is the trivial name for the red organic reagent 5-(2-carboxyphenyl)-1-(2-hydroxy-5-sulphophenyl)-3-phenylformazan (XXXVII) which forms a blue complex with zinc in alkaline solution.[11, 12] The maximum absorption occurs at a wavelength of 620 nm, at which the reagent has only very slight absorption. The Beer–Lambert Law is

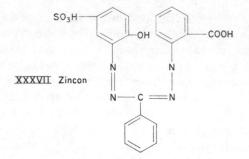

XXXVII Zincon

followed in the range 0–120 μg zinc in 50 ml. The reagent solution is stable for at least a week, but the zinc complex for only a few hours. Some slight fading is observed after about 24 hours.

Many ions interfere with the determination of zinc, including aluminium, beryllium, bismuth, cadmium, cobalt, chromium(III), copper, iron (III), manganese, mercury, molybdenum(VI), nickel and titanium, which all react with the reagent, and must be separated before the zinc complex is formed. Anion-exchange separation[11, 13] can conveniently be used for this.

The procedure given below is essentially that of Huffman et al., [10] designed to determine zinc in rocks containing as much as 20 per cent iron. The rock material is decomposed by evaporation with hydrofluoric, nitric, perchloric and sulphuric acids, and the residue dissolved in dilute hydrochloric acid. On passing the hydrochloric acid solution through a column of anion exchange resin, zinc, lead, molybdenum, iron, uranium and cadmium are absorbed on the resin, and the remaining constituents eluted. Iron is removed by washing the column with 1·2 M hydrochloric acid, and the zinc collected by elution with 0·01 M hydrochloric acid. Zinc is precipitated from this eluate with diethyldithiocarbamate at a pH of 8·5, and the precipitate extracted into chloroform. Dilute hydrochloric acid is used to strip the zinc from the organic solution, and the determination is then completed photometrically with zincon.

Method

Apparatus. *Ion-exchange column*, set up a small ion-exchange column 8-cm in length and 0·8-cm in diameter, of a strongly basic ion exchange resin, such as Dowex 1 × 8, and wash thoroughly with 1·2 M hydrochloric acid, followed by 0·01 M hydrochloric acid and then water. Leave the column 1·2 M in hydrochloric acid ready for use.

Reagents: *Ammonium citrate buffer solution,* dissolve 25 g of citric acid in 300 ml of water, bring pH of the solution to 8·5 by adding dilute aqueous ammonia, and dilute to 500 ml with water.

Sodium borate buffer solution, dissolve 14·4 g of sodium hydroxide. 24·74 g of boric acid and 29·82 g of potassium chloride in about 950 ml of water, cool to room temperature and dilute to 1 litre. The pH should be 10·2.

Zincon solution, transfer 0·130 g of the solid zincon reagent to a 100-ml volumetric flask, add 2 ml of 1 M sodium hydroxide solution and dilute to volume with water after the zincon is in solution. The solution is stable for about 10 days.

Standard zinc stock solution, ignite zinc sulphate heptahydrate to constant weight in an electric muffle furnace set at a temperature of 450°. Allow to cool, and weigh 0·2469 g of the anhydrous salt into a 200-ml volumetric flask and dilute to volume with water. This solution contains 500 μg zinc per ml.

Standard zinc working solution, dilute 5 ml of the stock solution to 500 ml with 1·2 M hydrochloric acid. This solution contains 5 μg zinc per ml.

Procedure. Weigh approximately 0·5 g of the finely ground sample material into a small platinum dish or crucible, moisten with water and add 10 ml of concentrated nitric acid and 10 ml of hydrofluoric acid. Cover the vessel, and set it aside for several hours, preferably overnight.

Add to the dish 5 ml of perchloric acid and 10 ml of 20 N sulphuric acid. Transfer the dish to a hot plate and evaporate to a volume of about 3 ml. Cool the dish, wash down the sides with a little water, add a further 5 ml of perchloric acid and again evaporate, this time just to dryness—but avoiding baking the residue. Add 10 ml of concentrated hydrochloric acid and 25 ml of water to the residue, rinse the solution into a beaker and digest on a steam bath for 30 minutes. Cool the solution and transfer to a 100-ml volumetric flask.

If any insoluble residue remains, collect it on a small filter, wash with a little water, dry, ignite and fuse with the minimum quantity of sodium carbonate. Dissolve the melt in the minimum amount of hydrochloric acid and combine with the solution in the 100-ml volumetric flask before dilution to volume.

Transfer the entire 100 ml of solution (or a suitable portion of it containing from 20 to 100 μg zinc and diluted to a volume of 100 ml with 1·2 M hydrochloric acid) to the resin column. Regulate the flow rate of the solution through the column to about 1 ml per minute by adjusting the stopcock at the bottom of the column. When the flow ceases, discard the solution that has passed through, wash the column with 50 ml of $1 \cdot 2$ M hydrochloric acid and discard the wash solution. Place a clean 150-ml beaker under the column and elute the zinc by passing 45 ml of $0 \cdot 01$ M hydrochloric acid through the resin also at a flow rate of about 1 ml per minute.

Add 3 drops of phenolphthalein indicator solution to the beaker containing the zinc solution, and adjust the pH to $8 \cdot 5 \pm 0 \cdot 5$ by adding dilute ammonia solution dropwise until the pink indicator colour just but only just forms. Quantitatively transfer the solution to a 125 ml separating funnel, add 2 ml of the diethyldithiocarbamate solution, stopper the funnel and shake the solution. Add 10 ml of chloroform to the funnel and shake to extract the zinc–carbamate complex, draining the chloroform solution into a clean separating funnel. Rinse the stem of the extraction funnel with about 2 ml of chloroform and add this chloroform to the organic extract. Repeat the extraction and rinsing operations once more using 5 and 2 ml volumes of chloroform respectively, and then discard the aqueous solution.

Add 10 ml of water to the chloroform solution and wash by shaking the funnel for about 30 seconds. Drain the chloroform into a clean 125-ml separating funnel, rinse the stem with about 2 ml of chloroform and add the chloroform wash liquor to the main portion of chloroform. Add 10 ml of the $0 \cdot 16$ M hydrochloric acid to the combined chloroform solution and strip the zinc from the organic layer by shaking for at least 1 minute. Remove and discard the lower chloroform layer. Wash the aqueous phase with 10 ml of chloroform by shaking for 30 seconds and again drain, remove and discard the organic layer.

Filter the aqueous solution through a $5 \cdot 5$-cm filter paper previously washed with $0 \cdot 16$ M hydrochloric acid, and collect the filtrate in a 50-ml volumetric flask. Wash the separating funnel twice with about 3 ml of water, passing the washings through the paper into the flask. Add 10 ml of the sodium borate buffer solution to the flask and mix with the solution. A pH of $9 \pm 0 \cdot 5$ should be obtained. Add 3 ml of the zincon reagent solution to the flask to give a brownish mixed colour, with a transition to blue as the zinc concentration increases. Dilute the solution to 50 ml with water. The intensity of the colour complex reaches a maximum very quickly and is stable for a few hours.

Measure the optical density of this solution in 1-cm cells, against the reagent blank as the reference solution, using the spectrophotometer set at a wavelength of 620 nm.

Calibration. Transfer aliquots of 0–25 ml of the standard solution containing 0–125 μg Zn to separate 150-ml beakers and add $1 \cdot 2$ M hydrochloric acid to dilute each solution to about 100 ml in volume. Transfer each solution in turn to a resin column, elute and process the zinc fraction as described in the method above. Measure the optical densities as before, and plot these values against zinc concentration to obtain a standard working curve.

Determination of Zinc by Atomic Absorption Spectroscopy

The ease and simplicity and freedom from interference which characterises the determination of zinc by atomic absorption spectroscopy is in direct contrast to the lengthy, difficult determination by spectrophotometry, for which a careful separation from interfering elements is required. The limit of detection by the procedure outlined below is about 0·5 ppm zinc, which is adequate for most silicate rocks. The calibration curve for zinc is slightly convex towards the concentration axis.[14] A procedure for determining zinc and copper in the same solution has been given by Belt[15] and the technique has been used by Burrell[16] for determining zinc in amphibolites.

Procedure. Accurately weigh approximately 0·5 g of the finely powdered silicate rock material into a small platinum dish or crucible, and evaporate to dryness with 1 ml of concentrated perchloric acid and 5 ml of concentrated hydrofluoric acid. Moisten the dry residue with a further 1 ml of perchloric acid and again evaporate to dryness. Allow to cool and add 0·3 ml of perchloric acid and rinse the residue into a small beaker. Warm until dissolution is complete, then cool, and dilute to 25 ml with water in a volumetric flask.

Using an atomic absorption spectrophotometer fitted with a zinc lamp and operating according to the manufacturers instructions, measure the absorption at a wavelength of 213·86 nm. Use a 10 ppm zinc standard solution to check the calibration of the instrument.

References

1. SANDELL E. B. and GOLDICH S. S., *J. Geol.* (1943) **51**, 99 and 167.
2. RADER L. F., SWADLEY W. C., LIPP H. H. and HUFFMAN C. Jr., *U.S. Geol. Surv. Prof. Paper* 400-B, p. B437, 1960.
3. WEDEPOHL K. H., *Geochim. Cosmochim. Acta* (1953) **3**, 93.
4. VINOGRADOV A. P., *Geochemistry* (1962) No. 7, 641.
5. ROSMAN K. J. R., *Geochim. Cosmochim. Acta* (1972) **36**, 801.
6. SANDELL E. B., *Ind. Eng. Chem., Anal. Ed.* (1937) **9**, 464.
7. CARMICHAEL I. and McDONALD A., *Geochim. Cosmochim. Acta* (1961) **22**, 87.
8. GREENLAND L., *Geochim. Cosmochim. Acta* (1963) **27**, 269.
9. MARGERUM D. W. and SANTACANA F., *Analyt. Chem.* (1960) **32**, 1151.
10. HUFFMAN C. Jr., LIPP H. H. and RADER L. F., *Geochim. Cosmochim. Acta* (1963) **27**, 209.
11. YOE J. H. and RUSH R. M., *Anal. Chim. Acta* (1952) **6**, 526.
12. YOE J. H. and RUSH R. M., *Analyt. Chem.* (1954) **26**, 1345.
13. JACKSON R. K. and BROWN J. G., *Proc. Amer. Soc. Hort. Sci.* (1956) **68**, 1.
14. ERDEY L., SVEHLA G. and KOLTAI L., *Talanta* (1963) **10**, 531.
15. BELT C. B. Jr., *Econ. Geol,* (1964) **59**, 240.
16. BURRELL D. C., *Norsk. Geol. Tidsskr.* (1965) **45**, 21.

CHAPTER 50

ZIRCONIUM AND HAFNIUM

Occurrence

Basic rocks contain very little zirconium, and these small amounts are concentrated largely in the pyroxenes and the accessory ore minerals. In the course of magmatic differentiation both zirconium and hafnium are concentrated in the residual liquid fraction, giving rise to successive crystalline rocks with increasing amounts of both elements.[1] In the later rocks these elements appear very largely in the mineral zircon. As a group of the alkali-rich rocks contain more zirconium than the calc-alkali rocks.[2] No significant difference in zirconium content has been observed between plutonic rocks and their volcanic equivalents.[2] Typical values for a number of rock types are given in Table 48, which has been compiled from data by Dagenhardt,[2] Turekian and Wedepohl,[3] Chao and Fleischer[4] and other workers.

TABLE 48. THE ZIRCONIUM AND HAFNIUM CONTENT OF SOME ROCK TYPES

Rock type	Zr, ppm	Hf, ppm
Ultrabasic	50	0·6
Basalt	140	2
Syenite	500	11
Granodiorite	140	2
Granite	175	4
Shale	160	3
Sandstone	260	5
Limestone	25	0·5

The chief source of zirconium is the mineral zircon, $ZrSiO_4$. There are a small number of other zirconium minerals including silicates eudialyte and catapleite and the oxide baddleyite, ZrO_2.

500

Hafnium does not form separate minerals. The ratio of zirconium to hafnium in rocks and minerals varies somewhat, ranging from about 30:1 in granite pegmatites to about 80:1 in ultrabasic rocks. Late-stage pegmatites often contain zircons with abnormally high hafnium content.

Gravimetric Determination in Silicate Rocks

One of the oldest procedures still in use for the determination of zirconium (+ hafnium) is based upon precipitation as phosphate from dilute sulphuric acid solution. This determination can readily be combined with those of a number of other minor constituents of silicate rocks, such as chromium, vanadium, sulphur and chlorine in an initial alkaline filtrate, and the rare earths and barium with the zirconium in the residue. This method is given in textbooks of rock analysis, but the published procedures do not stress the difficulties in making accurate determinations of these small amounts of zirconium.[5]

PHOSPHATE METHOD

The method outlined by Bennett and Pickup[5] is used as the basis for the procedure described in detail below. A 5-g sample weight is taken for the analysis, fused with sodium carbonate and potassium nitrate, and the melt extracted with water. The insoluble residue is separated, dissolved in dilute sulphuric acid and the zirconium precipitated as phosphate after separation of barium as barium sulphate.

Method

Reagents: *Sodium carbonate wash solution*, dissolve 10 g of the anhydrous reagent in 500 ml of water.

Sulphurous acid solution, pass a slow stream of sulphur dioxide gas from a siphon or cylinder through 50 ml of water for about 15 minutes. Prepare freshly as required.

Hydrogen peroxide solution, 30 per cent.

Diammonium hydrogen phosphate solution, dissolve 5 g of the reagent in 50 ml of water.

Ammonium nitrate wash solution, dissolve 25 g of the reagent in 500 ml of water.

Procedure. Accurately weigh approximately 5 g of the finely powdered silicate rock material into a platinum dish and mix thoroughly with 25 g of anhydrous sodium carbonate and 1 g of potassium nitrate. Cover the dish with a platinum lid and transfer to a cold electric muffle furnace. Gradually raise the temperature of the furnace to about 1000°, allowing time for the carbon dioxide to be evolved at the sintering stage. Maintain the temperature of the furnace at 1000° for about 30 minutes before removing the platinum dish. Allow to cool. Digest the melt with hot water to which a little ethanol has been added, until all soluble material has passed into solution, and all green colour due to the presence of manganate has disappeared. Remove the platinum dish, rinse with water and set it aside. Digest the solution and residue on a steam bath for 30 minutes, breaking up the residue with the flattened end of a glass rod.

Collect the residue on a medium-textured filter paper and wash it well with hot sodium carbonate wash solution. Combine the filtrate and washings and set aside for the determination of chromium, vanadium, sulphur and chlorine.

Rinse the residue back into the original beaker with water, add 150 ml of water and with stirring 20 N sulphuric acid, drop by drop until the residue has largely dissolved. Add 6 ml of 20 N sulphuric acid in excess and sufficient sulphurous acid to discharge the brown colour due to manganese. Dissolve any residue remaining on the surface of the platinum dish by washing with a few ml of hot, dilute sulphuric acid, add the washings to the solution and dilute to a volume of about 250 ml with water. Filter the solution through the paper previously used and wash with very dilute sulphuric acid. Reserve the paper and residue for the determination of barium. Allow the combined filtrate and washings to stand overnight, and examine for any further trace of precipitated barium sulphate. If present it should be recovered and added to the main barium residue for purification.   Dilute the filtrate to a volume of about 400 ml and add 100 ml of 20 N sulphuric acid, 5 ml of hydrogen peroxide solution (note 1) and 10 ml of the ammonium phosphate solution. Heat to about 45°, and keep at this temperature for 2 to 3 hours. Zirconium phosphate is normally precipitated as a gelatinous but flocculent precipitate. Occasionally, particularly at low zirconium concentrations, the precipitate is very finely divided. In these instances, the solution should be allowed to stand overnight before filtration.

Collect the precipitated zirconium phosphate on a close-textured filter paper and wash three or four times with small quantities of ammonium nitrate wash solution. Transfer the paper to a small platinum crucible, dry and burn off the paper but avoid strong ignition of the residue. This consists largely of zirconium and hafnium phosphates, but may be contaminated with silica, and small amounts of niobium, tantalum, titanium and thorium. It can be purified by the following procedure. Moisten the residue with water, add 2 ml of concentrated hydrofluoric acid and a few drops of 20 N sulphuric acid. Evaporate to dryness on a hot plate to remove silica in the usual way. Now add a small quantity of anhydrous sodium carbonate, fuse and extract the melt with water. Collect the residue on a small filter, wash with a little hot water and discard the filtrate and washings. Transfer the paper to a small platinum crucible, dry, burn off the paper and fuse the residue with a little potassium pyrosulphate. Cool, dissolve the melt in a little 4 N sulphuric acid, add a little hydrogen peroxide solution and repeat the precipitation with ammonium phosphate solution.

Allow the beaker to stand for 2 hours at a temperature of about 50°. Allow to cool, collect the precipitate on a small close-textured filter paper, wash 3 or 4 times with small amounts of ammonium nitrate wash solution, and transfer to a small weighed platinum crucible. Dry, ignite and weigh the residue as zirconium (+ hafnium) pyrophosphate (Zr,Hf)P$_2$O$_7$ (note 2).

Notes: 1. Hydrogen peroxide is added to form a yellow-coloured complex with titanium, preventing the precipitation of titanium phosphate. If the yellow colour is discharged during the digestion stage, a further small amount of hydrogen peroxide should be added.

2. The ignited phosphates may still be contaminated with small amounts of thorium, niobium and tantalum, but this is not likely to give rise to serious error in the analysis of silicate rocks.

OTHER GRAVIMETRIC METHODS

Although frequently used for the gravimetric determination of zirconium, precipitation with mandelic or *p*-bromomandelic acid, does not appear to have been widely applied to silicate rock analyses. Tserkovnitskaya and Borovaya[6] describe an extraction of the zirconium complex of mandelic acid for the determination in low-grade ores, but the determination was completed photometrically with arsenazo I.

Precipitation with phenylarsonic acid, followed by re-precipitation with *p*-(*p*-dimethylaminophenylazo)benzenearsonic acid was described by Tuzova and Nemodruk[7] for the analysis of silicates. After ignition of the complex to give the dioxides, the determination of zirconium and hafnium was completed by X-ray spectrography.

Photometric Methods for the Determination of Zirconium and Hafnium

Although a large number of reagents have been suggested for the photometric determination of zirconium, none of the reactions is completely specific and few are even selective. In their application to silicate rocks it is usually necessary to make a separation from interfering elements. Babko and Vasilenko [8] studied a total of eighteen reagents for zirconium and considered xylenol orange and methylthymol blue to be the best. Arsenazo I[6] and III[9] have been used to determine zirconium in rocks and minerals, and more recently quinalizarin sulphonic acid has been proposed[10] for this purpose.

The procedure described below is based upon the use of xylenol orange (XXXVIII). This reagent, which is yellow in colour, forms a red-coloured complex with zirconium, that has a maximum absorption at a

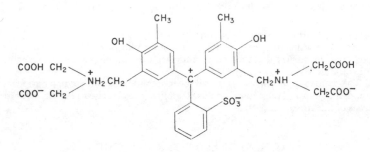

XXXVIII   Xylenol orange

wavelength of 535 nm. In a dilute acid medium the only other elements to form complexes with the reagent are hafnium, bismuth, tin, molybdenum and iron(III).[11] In the determination of zirconium (+ hafnium), ferric iron is reduced with ascorbic acid. None of the remaining elements are likely to be present in amounts sufficient to interfere. The separation of zirconium from other elements, other than from silicon, is therefore not required with this reagent.

Method

Reagents: *Ascorbic acid solution*, dissolve 2·5 g of the reagent in 50 ml of water. Prepare freshly as required.

Xylenol orange solution, dissolve 50 mg of the reagent in water and dilute to 100 ml.

Procedure. Accurately weigh approximately 0·5 g of the finely powdered silicate rock into a platinum dish and add 5 ml of 20 N sulphuric acid, 1 ml of concentrated nitric acid and 10 ml of concentrated hydrofluoric acid. Transfer the dish to a hot plate and evaporate to fumes of sulphuric acid. Allow to cool, dilute with a few ml of water, add 5 ml of concentrated hydrofluoric acid and evaporate to fumes. Allow to cool, dilute with water and again evaporate to fumes. Repeat the evaporation once more, this time to dryness (note 1).

Allow to cool, moisten the residue with 5 ml of water, add 10 ml of 6 N hydrochloric acid, warm to dissolve most of the residue and then rinse into a 150-ml beaker containing 50 ml of water. Warm the solution on a hot plate until all soluble material has dissolved. If any residue remains undecomposed, collect on a small filter, wash with a little water, dry and ignite in a small platinum crucible. Fuse with a little sodium carbonate, extract with water, filter, dissolve the residue in a little hydrochloric acid and add to the main rock solution.

Transfer the solution to a 100-ml volumetric flask, add a further 10 ml of 6 N hydrochloric acid and dilute to volume with water. Mix well and transfer an aliquot of the solution containing not more than 40 μg of zirconium to a 50-ml volumetric flask. Add sufficient hydrochloric acid to bring the final concentration to 0·8 N, followed by 5 ml of the ascorbic acid solution to reduce ferric iron and 2 ml of the xylenol orange solution. Dilute to volume with water, mix well and measure the optical density using the spectrophotometer set at a wavelength of 525 nm, against a reagent blank solution prepared in the same way as the sample solution but omitting the sample material.

Calibration. For the calibration graph, use aliquots of a standard zirconium solution containing 0–40 μg zirconium.

Note: 1. A major part of the zirconium may be present in the form of the mineral zircon, which is decomposed by acid only with some difficulty.

The Determination of Hafnium

As shown in Table 47, hafnium is present in silicate rocks to a smaller extent than zirconium, although by no means as small as the amounts of some other elements that are readily determined by photometric methods. Chemical methods, including spectrophotometry, have not been successfully applied to the determination of hafnium in silicate rocks. For this purpose emission spectrography,[10] X-ray spectrography[7] and neutron activation analysis[12-14] are commonly employed.

References

1. GOLDSCHMIDT V. M., *Geochemistry*, Oxford University Press, London, p. 423, 1958.
2. DEGENHARDT H., *Geochim. Cosmochim. Acta* (1957) **11**, 279.
3. TUREKIAN K. K. and WEDEPOHL W. H., *Bull. Geol. Soc. Amer.* (1961) **72**, 175.
4. CHAO E. C. T. and FLEISCHER M., *Rept. Int. Geol. Congr. XXI Session* **1**, p. 106, 1960.
5. BENNETT W. H. and PICKUP R., *Colon. Geol. Min. Res.* (1952) **3**, 171.

6. TSERKOVNITSKAYA I. A. and BOROVAYA N. S., *Vestn. Leningr. Univ.* (1962) No. 16, *Ser. Fiz. i. Khim.* (3), 148.
7. TUZOVA A. M. and NEMODRUK A. M., *Zhur. Anal. Khim.* (1958) 13, 674.
8. BABKO A. K. and VASILENKO V. T., *Zavod. Lab.* (1961) 27, 640.
9. GORYUSHINA V. G. and ROMANOV E. V., *Zavod. Lab.* (1960) 26, 415.
10. CULKIN F. and RILEY J. P., *Anal. Chim. Acta* (1965) 32, 197.
11. CHENG K. L., *Talanta* (1959) 2, 61.
12. SETSER J. L. and EHMANN W. D., *Geochim. Cosmochim. Acta* (1964) 28, 769.
13. BUTLER J. R. and THOMPSON A. J., *Geochim. Cosmochim. Acta* (1965) 29, 167.
14. MORRIS D. F. C. and SLATER D. N., *Geochim. Cosmochim. Acta* (1963) 27, 285.

NOTE ADDED IN PROOF

THE method described below, due to Murphy et al.,[1] involving titration of ferric iron in solution with ferrocene (dicyclopentadienyliron) is based upon the reaction

$$Fe(C_5H_5)_2 + Fe^{3+} = Fe^{2+} + Fe(C_5H_5)_2^+.$$

In the presence of thiocyanate ions, the end point is indicated by the disappearance of the red colour of ferric thiocyanate and its replacement by the blue colour of the ferricenium ion.

Method

Reagents: *Ammonium thiocyanate solution*, dissolve 5 g in 100 ml water.

Lissapol NDB solution, dissolve 5 ml of Lissapol NDB detergent liquid (available from Hopkin & Williams Ltd) in 95 ml of 95 per cent ethanol.

Ferrocene solution, dissolve 0·5830 g of ferrocene (dicyclopentadienyliron) in 2-methoxyethanol and dilute with this solvent to 500 ml. Store in a glass stoppered bottle.

Boric acid solution, saturated aqueous.

Potassium permanganate solution, 0·1 per cent w/v aqueous.

Standard ferric iron solution, dissolve 0·4911 g of ferrous ammonium sulphate in about 25 ml of water containing 10 ml of 20 N sulphuric acid and add a small excess of bromine water. Evaporate the solution to fumes of sulphuric acid, cool, cautiously add about 100 ml of water. Heat until a clear solution is obtained, then transfer to a 500 ml volumetric flask, cool to room temperature and dilute to volume with water. This solution contains 0·2 mg Fe_2O_3 per ml. Prepare also a ferric iron spike solution by diluting one volume of this standard ferric iron solution with four volumes of water.

If required, the iron content of the ferrous ammonium sulphate can be checked by dissolving 1 g in water, oxidizing with bromine water and precipitating ferric hydroxide with ammonia. The precipitate should be filtered off, ignited at 1000° and weighed as Fe_2O_3.

Procedure. To a clear plastic beaker of about 250 ml capacity, add 100 ml of saturated boric acid and to a similar 100 ml beaker add 50 ml of the same solution. Set these aside in readiness for the next stage.

To 0·5 g of the finely powdered rock material in a large platinum crucible ("Pratt crucible", see p. 279) add a few ml of water and 10 ml of 20 N sulphuric acid. Add more water to about half fill the crucible and heat to boiling over a small flame protected from draughts. After a few seconds boiling add 10 ml of concentrated hydrofluoric acid without interrupting the heating, replace the lid immediately and note the time when boiling recommences. After 10 minutes boiling remove the flame and quickly rinse the lid into the crucible with boric acid solution from the 100 ml beaker. Add further boric acid solution almost to fill the crucible and without delay tip the contents into the 100 ml of boric acid solution in the 250 ml beaker. Rinse the crucible into the beaker with the boric acid solution remaining in the 100 ml beaker and then transfer the whole contents to a 200 ml volumetric flask using boiled, air-free water for the rinsing and dilution to the mark.

Transfer a 20 ml aliquot of this solution to a 150 ml glass beaker and dilute to about 75 ml with boiled water. Stir the solution with a magnetic stirrer and, after adding 10 ml of ammonium thiocyanate solution and 2 ml of the Lissapol NDB solution, titrate with the ferrocene solution. The end point is indicated in good white light by the appearance of a blue colour without any trace of red ferric thiocyanate colour. Add 1 ml of the ferric iron spike solution and again titrate to the same end point, noting the total volume of titrant added. Carry out the above titration procedure using 1 ml of the ferric iron spike solution to which has been added 1 ml of 20 N sulphuric acid before diluting to 75 ml, and note the volume of titrant required.

To standardize the ferrocene solution, transfer 10 ml of the standard ferric iron solution by pipette to a 150 ml beaker and add potassium permanganate solution drop at a time until there is a small visible excess, then dilute to 75 ml and proceed as described above, including the addition of the 1 ml of the ferric iron spike solution.

1. MURPHY J. M., READ J. I. and SERGEANT, G. A., *Analyst* (1974), **99**, 273.

INDEX OF ROCK AND MINERAL SPECIES

509

AUTHOR INDEX

SUBJECT INDEX

Entries in italic type refer to complete chapters.

523